Random Eigenvalue Problems

NORTH HOLLAND SERIES IN

Probability and Applied Mathematics

A.T. Bharucha-Reid, *Editor*

AN INTRODUCTION TO STOCHASTIC PROCESSES
D. Kannan

STOCHASTIC METHODS IN QUANTUM MECHANICS
S. P. Gudder

APPROXIMATE SOLUTION OF RANDOM EQUATIONS
A. T. Bharucha-Reid (ed.)

PROBABILITY DISTRIBUTIONS ON LINEAR SPACES
N. N. Vakhania

PROBABILISTIC METRIC SPACES
B. Schweizer and A. Sklar

RANDOM EIGENVALUE PROBLEMS
J. vom Scheidt and W. Purkert

Random Eigenvalue Problems

JÜRGEN vom SCHEIDT
Ingenieurhochschule Zwickau
WALTER PURKERT
Karl-Marx-Universität Leipzig

NORTH HOLLAND
New York · Amsterdam · Oxford

Elsevier Science Publishing Co., Inc.
52 Vanderbilt Avenue, New York, New York 10017

Distributed outside the United States and Canada, except in Socialist countries by:
Elsevier Science Publishers B.V.
P.O. Box 211, 1000 AE, Amsterdam, The Netherlands

Distributed in Socialist countries by:
Akademie-Verlag, DDR-1086 Berlin, Postfach 1233

Library of Congress Cataloging in Publication Data

Vom Scheidt, Jürgen.
Random eigenvalue problems.
(North Holland series in probability and applied mathematics)
Includes bibliographical references and indexes.
1. Stochastic matrices. 2. Eigenvalues. I. Purkert, Walter. II. Title. III. Series.
QA188.V65 1983 512.9′434 82-18225
ISBN 0-444-00769-5

Printed in GDR

Preface

Since the times of Euler eigenvalue problems have been very important in physics and the technical sciences. At present stochastic influences are more and more considered. This leads to eigenvalue problems for random operators.

In the present book we deal with eigenvalue problems for random matrices and random ordinary differential operators using expansion methods. We have been inspired by problems of physics, especially by the mixed-crystal theory, as well as by the pioneer work of W. E. Boyce and his coworkers in the field of random eigenvalue problems.

In Chapter 1 the expectations, variances and correlation of eigenvalues and eigenfunctions are calculated. Chapter 2 contains results about the limit distributions for eigenvalue problems involving weakly correlated processes. This book is in particular written for readers interested in random eigenvalue problems from a practical point of view. Thus numerical examples are given. But we believe that the limit theorems given in Chapter 2 are also of theoretical interest. Limit theorems of this kind are also applicable to boundary- and initial-value problems.

The writers thank Professors H. Beckert (Leipzig), H. Bunke, O. Bunke (Berlin), J. Kerstan (Jena), G. Laßner, A. Uhlmann, K. Unger and E. Zeidler (Leipzig) for their continuous interest and inspiring discussions.

We also wish to thank Prof. A. T. Bharucha-Reid (Atlanta) for inviting us to write this book for the North-Holland Series in Probability and Applied Mathematics.

We thank the Akademie-Verlag Berlin, especially Dr. Höppner for the agreeable cooperation, and last but not least, the Ingenieurhochschule Zwickau for generously supporting this project.

Zwickau, May 1982

J. vom Scheidt
W. Purkert

Contents

Introduction

In recent years the application of mathematical methods in the natural and technological sciences is characterized among others by the fact that the mathematical models based on the most simplified assumptions are being replaced by models based on more realistic assumptions, which in turn are more complicated. This fact leads to intensify the investigations of nonlinear problems. These considerations also result in the stochastic nature of many quantities being taken into account. The development in stochastic analysis during the last 10 to 15 years proves the importance of the stochastic nature of many quantities.

Basic notions and methods of the functional analysis can be often transferred very simply to the stochastic case. However, characteristic difficulties concerning the essential stochastic questions are liable to occur, e.g. the distribution law of the solution of a stochastic equation (cf. BHARUCHA-REID [3]). Attemps to overcome these difficulties have been only successful in certain very special cases; hence we are obliged to pursue simpler goals, for instance the calculation of the first two moments.

These general remarks are also true for problems of spectral theory. Notions such as resolvent set and spectrum, and spectral theorems (such as the disjoint decomposition of the spectrum into the continuous spectrum, point spectrum and residual spectrum) can be transferred to the case of a stochastic operator. The spectral theory of stochastic operators is in its early stages of development, hence only a few general results are known (cf. BHARUCHA-REID [3]). This book is not aimed at describing or further developing the spectral theory of random operators. We confine ourselves to random operators having a discrete spectrum. Operators with a discrete spectrum play an important role in applications in physics and technology. Many quantities involved in physical and technological eigenvalue problems are of a stochastic nature. Because of the practical importance of random operators with a discrete spectrums we have restricted our attentions to the development of procedures for calculating quantities of practical importance. We suppose, however, that the theory of the asymptotic distribution laws for eigenvalues and eigenfunctions given in Chapter 2 are also of theoretical interest.

For some classes of random matrices WIGNER [1] has given asymptotic statements for the spectrum suggested by investigations concerning the statistical theory of energy levels, and initiated a field of research which has turned out to be fruitful. The results of this field are not dealt with in the present book. For the sake of information we give a very incomplete short review on the asymptotic distribution of the eigenvalues of random matrices.

Let $(\alpha_n A_n)_{n=1,2,\ldots}$ be a sequence of matrices where the matrix A_{n+1} is constituted from the matrix A_n by addition of a new row and a new column and $(\alpha_n)_{n=1,2,\ldots}$ denotes a suitable sequence of real numbers. Let the elements of A_n be real-valued random variables where $a_{ik}(\omega) = a_{ki}(\omega)$ almost surely for all i, k. The real eigenvalues of $\alpha_n A_n$ are denoted by $\lambda_i^{(n)}(\omega)$, and we assume $\lambda_1^{(n)} \leqq \lambda_2^{(n)} \leqq \ldots \leqq \lambda_n^{(n)}$. The function

$$N_n(\lambda, \omega) \doteq \frac{1}{n} \sum_{\lambda_i^{(n)}(\omega) < \lambda} 1$$

is called the *empirical distribution function* of the eigenvalues of the matrix $\alpha_n A_n$. This function is a random step function which has jumps of the level $k_i(\omega)/n$ in the points $\lambda_i^{(n)}(\omega)$ where $k_i(\omega)$ denotes the multiplicity of $\lambda_i^{(n)}(\omega)$.

The first result, the now classical semi-circle law, was obtained by WIGNER [1]: Let the elements $a_{ij}(\omega)$ of the matrix A_n be independent random variables for $1 \leqq i \leqq j \leqq n$ and symmetric, $a_{ij}(\omega) = a_{ji}(\omega)$ almost surely. Furthermore, let $a_{ij}(\omega)$ have the distribution function $F_{ij}(x)$ where

(a) $F_{ij}(-x + 0) = 1 - F_{ij}(x)$,

(b) $\int_{-\infty}^{\infty} x^2 \, \mathrm{d}F_{ij}(x) = \sigma^2 > 0$,

(c) $\int_{-\infty}^{\infty} x^{2l} \, \mathrm{d}F_{ij}(x) \leqq B_l < \infty \quad$ for $\quad l = 1, 2, \ldots$

Then the empirical distribution function $N_n(\lambda, \omega)$ of the matrix system $(\alpha_n A_n)$ with $\alpha_n = 1/2\sigma\sqrt{n}$ converges in probability to the deterministic function

$$N(\lambda) = \int_{-\infty}^{\lambda} n(t) \, \mathrm{d}t \tag{0.1}$$

as $n \to \infty$ where

$$n(t) = \begin{cases} \frac{2}{\pi}\sqrt{1 - t^2} & \text{for} \quad |t| \leqq 1 , \\ 0 & \text{for} \quad |t| > 1 , \end{cases}$$

$$\lim_{n\to\infty} N_n(\lambda, \omega) = N(\lambda) \quad \text{in probability} .$$

This statement was proved by Wigner with the help of the method of moments.

ARNOLD [1] has shown the almost sure convergence of $N_n(\lambda, \omega)$ to $N(\lambda)$; and has reduced the conditions to weaker ones. The conditions of Arnold determine the exact limits of the validity of the semi-circle law.

Pastur and his co-worker (BENDERSKIJ, PASTUR [1]; PASTUR [1, 2]) have essentially generalized the results of Wigner. They introduced the definition of a self-averaging system of matrices. A system of matrices (A_n) is called weak (strong) self-averaging if for the empirical distribution function $N_n(\lambda, \omega)$ the following hold:

$$\lim_{n\to\infty} N_n(\lambda, \omega) = N(\lambda) \quad \text{in probability}$$

$$\left(\lim_{n\to\infty} N_n(\lambda, \omega) = N(\lambda) \quad \text{a.s.}\right),$$

where the limit function $N(\lambda)$ denotes a nonrandom function. An important result of Pastur is the following:

Let the matrices

$$A_n = A_n^{(0)} + \frac{1}{\sqrt{n}} V_n$$

be symmetric random matrices of the order n and $A_n^{(0)}$ be a diagonal matrix where the diagonal elements h_i, $i = 1, 2, \ldots, n$, are real, independent random variables which have the distribution function $F(x)$. Let V_n denote a real symmetric random matrix possessing independent elements $v_{ij}(\omega)$ for $1 \leqq i \leqq j \leqq n$. Suppose the conditions

$$\langle v_{ij} \rangle = 0, \quad \langle v_{ij}^2 \rangle = \sigma^2 > 0,$$

$$\lim_{n\to\infty} \frac{1}{n} \sum_{i=1}^{n} \int\limits_{|x|>\tau\sqrt{n}} x^2 \,\mathrm{d}F_{in}(x) = 0 \quad \text{for} \quad \tau > 0$$

are satisfied where

$$F_{ij}(x) = \mathsf{P}(v_{ij} < x).$$

Then it follows that:

(a) The system (A_n) is self-averaging with the limit distribution $N(\lambda)$. The convergence in probability of $N_n(\lambda, \omega)$ to $N(\lambda)$ can be shown for all $\lambda \in \mathbb{R}$ for which $N(\lambda)$ is continuous and $N(\pm\infty) = F(\pm\infty)$.

(b) The Stieltjes transform $m(z)$ of the function $N(\lambda)$ is given by the solution of the equation

$$m(z) = \int\limits_{-\infty}^{\infty} \frac{\mathrm{d}F(t)}{t - z - \sigma^2 m(z)},$$

which is analytic for $\operatorname{Im} z \neq 0$ and satisfies $\operatorname{Im} m(z) > 0$ for $\operatorname{Im} z > 0$. The solution of this equation under the conditions mentioned exists and is unique.

Pastur [3] has also proved the self-averaging property for the empirical distribution function of the Schrödinger operator with a random potential in a cube of volume V. Further results can be found in the papers of Benderskij, Pastur [1] and Pastur [2].

Expositions of results for random matrices which are connected with the semi-circle law are given in the books of Metha [1] and Girko [1]. In addition, the book of Girko contains a series of results concerning the joint distribution function of the eigenvalues of random matrices if special conditions for the distribution of the elements of these matrices are assumed.

In Chapter 1 we deal with the calculation of the statistical characteristics (expectations, variances, correlation quantities) associated with the eigenvalues and eigenelements. First, we show that the eigenvalues are random variables and the eigenelements random elements in a suitable Hilbert space. The measurability of the roots for random polynomials is proved by HAMMERSLEY [1], from which the measurability of the eigenvalues of random matrices followed. In Section 1.2 the measurability for the components of the eigenvectors of random matrices is shown. In this section we also prove that the eigenvalues of random operators which belong to a very general class of random operators with a discrete spectrum are measurable. The random matrices are contained in this class of operators. In this case it is possible to derive the measurability of the eigenelements if the simplicity of the eigenvalues is supposed.

A general classification of possible methods for the determination of moments of the solution of a random problem leads to two categories, depending upon whether the statistical or nonstatistical part of the analysis is carried out first. One type consists of first expressing the solution in terms of the coefficients in the problem, without regard to whether these coefficients are random or determinate. Having obtained such a relation the statistical properties are then calculated. This type of procedure has been termed by KELLER [2] to be „honest". This method is applied if the solution can be expressed exactly or at least approximatively by a representation which involves the coefficients of the problem.

A second method capitalizes upon the fact that the moments of the random solution are sought and not the solution itself. In order to do this, we average the given problem. For further calculations statistical assumptions (for example, the independence of certain quantities) must be introduced which are usually not justified. Such methods are characterized by Keller as „dishonest". Dishonest methods have been used for a long time in statistical mechanics. There for instance the exact equations are too numerous to handle, and are replaced by other equations involving various averaged quantities. Dishonest methods are applied in the study of turbulence (KRAICHNAN [1]), as well as in the propagation of waves in random media (KELLER [1, 2]; KORAL, KELLER [1]). RICHARDSON [1] has used the dishonest methods in the study of initial value problems of special ordinary differential equations and HAINES [1, 2] in the study of an example of an eigenvalue problem.

In Chapter 1 we deal with perturbation methods which belong to the class of honest methods. Before summing up the results obtained we will discribe the most important results for the calculation of moments of eigenvalues and eigenelements of random operators contained in literature.

In the following we give a short review of the papers of BOYCE [1, 2, 3, 4, 5], BOYCE, GOODWIN [1], GOODWIN, BOYCE [1], and HAINES [1, 2] which deal with the random eigenvalue problem

$$M(\omega)\, u = \lambda N(\omega)\, u\,, \qquad U_i[u] = 0\,, \quad i = 1\,,\, 2,\, \ldots\,,\, 2m\,, \tag{0.2}$$

where $M(\omega)$ and $N(\omega)$ are random differential operators of the orders $2m$ and $2m'$, respectively ($m > m' \geqq 0$), extended over the interval $0 \leqq x \leqq 1$. We assume that M and N can be written as

$$M(\omega)\, u = \sum_{k=0}^{m} (-1)^k \,[f_k(x, \omega)\, u^{(k)}]^{(k)} \,,$$

$$N(\omega)\, u = \sum_{k=0}^{m'} (-1)^k \,[g_k(x, \omega)\, u^{(k)}]^{(k)}$$

where $f_m(x, \omega) \neq 0$ a.s., $\langle f_m(x, \omega)\rangle \neq 0$. The boundary conditions $U_i[u] = 0$ consist of $2m$ linearly independent equations of the form

$$U_i[u] \doteq \sum_{k=0}^{2m-1} \left(\alpha_{ik} u^{(k)}(0) + \beta_{ik} u^{(k)}(1)\right) = 0 \,, \qquad i = 1, 2, \dots, 2m \,.$$

Furthermore, the eigenvalue problem (0.2) is supposed to be selfadjoint and positive definite.

Both types of methods applied to eigenvalue problems can be classified in the following form:

(1) Honest methods
(1.1) Perturbation methods
(1.2) Variational methods
(1.3) Asymptotic methods
(1.4) Integral equation methods

(2) Dishonest methods
(2.1) Iteration methods
(2.2) Hierarchy methods.

Perturbation methods

Boyce [4] considers the eigenvalue problem

$$\left(M_0 + M_1(\omega)\right) u = \lambda\left(N_0 + N_1(\omega)\right) u \tag{0.3}$$

with the deterministic boundary conditions

$$U_i[u] = 0, \qquad i = 1, 2, \dots, 2m \,,$$

where $\langle M_1\rangle = \langle N_1\rangle = 0$. It is often convenient to seek the eigenvalues $\lambda_g(\omega)$ and the eigenfunctions $u_g(x, \omega)$ in the form of a series of homogeneous terms in the coefficients of the perturbation operators M_1 and N_1, that is, solutions of the form

$$\lambda_g(\omega) = \sum_{k=0}^{\infty} \lambda_{gk}(\omega) \,, \qquad u_g(x, \omega) = \sum_{k=0}^{\infty} u_{gk}(x, \omega)$$

are sought. It follows that

$$\lambda_g(\omega) = \mu_g + b_{gg}(\omega) + O_2$$

where

$$b_{gg}(\omega) = \langle\!\langle M_1(\omega)\, w_g, w_g\rangle\!\rangle - \mu_g \langle\!\langle N_1(\omega)\, w_g, w_g\rangle\!\rangle;$$

and μ_g denote the eigenvalues of the unperturbed (or averaged) eigenvalue problem

$$M_0 w = \mu N_0 w, \qquad U_i[w] = 0 \,, \qquad i = 1, 2, \dots, 2m \,,$$

and $w_g(x)$ the eigenfunctions belonging to μ_g normalized by $\langle\!\langle N_0 w_g, w_g \rangle\!\rangle = 1$. Hence, the mean of $\lambda_g(\omega)$ can be calculated by

$$\langle \lambda_g \rangle = \mu_g + O_2 \tag{0.4}$$

and the variance by

$$\operatorname{var} \lambda_g = \langle b_{gg}^2 \rangle + O_3 . \tag{0.5}$$

The symbol O_k denotes the terms of the k-th and higher order which were not written. This expansion up to terms of the first order leads to the trivial statement

$$\langle \lambda_g(\omega) \rangle \approx \mu_g ,$$

but this relation does not give information concerning the averaging problem (cf. below and Section 1.3.1).

By means of the correlation functions of the coefficients of M_1 and N_1 the variance of the eigenvalue $\lambda_g(\omega)$ can be determined. For example, let M_1 and N_1 have the form

$$M_1 u = \sum_{k=0}^{m} (-1)^k [\tilde{f}_k(x, \omega)\, u^{(k)}]^{(k)} , \qquad N_1 = 0$$

where $\langle \tilde{f}_k(x, \omega) \rangle = 0$ for $k = 0, 1, \dots, m$ and $\tilde{f}_j, \tilde{f}_k$ are uncorrelated for $k \neq j$. Then Boyce obtains

$$\operatorname{var} \lambda_g = \sum_{k=0}^{m} \int_0^1 \int_0^1 \tilde{K}_{kk}(x, y) \left(w_g^{(k)}(x)\, w_g^{(k)}(y)\right)^2 \mathrm{d}x\, \mathrm{d}y + O_3 .$$

The function $\tilde{K}_{kk}(x, y)$ denotes the correlation function of $\tilde{f}_k(x, \omega)$.

Before these general considerations Boyce [1, 4] and Boyce, Goodwin [1] considered some special problems which have applications in mechanics:

$$-u'' = \lambda(1 + m(x, \omega))\, u , \qquad u(0) = u(1) = 0 , \tag{0.6}$$

$$[(1 + a(x, \omega))\, u']' = -\lambda(1 + a(x, \omega))\, u , \qquad u(0) = u(1) = 0 , \tag{0.7}$$

$$\begin{gathered} [(1 + r(x, \omega))^4\, u'']'' = \lambda(1 + r(x, \omega))^2\, u , \\ u(0) = u(1) = u''(0) - \delta_1 u'(0) = u''(1) - \delta_2 u'(1) = 0 , \end{gathered} \tag{0.8}$$

$$u'''' = \lambda u , \qquad u(0) = u''(0) = u''(1) + \varepsilon(\omega)\, u'(1) = u(1) = 0 . \tag{0.9}$$

The following results can be given:

for (0.6): $\langle g\pi \lambda_g^{-1/2} \rangle \approx 1$,

$$\langle (g\pi \lambda_g^{-1/2})^2 \rangle \approx 1 + \tfrac{1}{4} \int_0^1 \int_0^1 \langle m(x, \omega)\, m(y, \omega) \rangle \sin^2 (g\pi x) \sin^2 (g\pi y)\, \mathrm{d}x\, \mathrm{d}y ;$$

for (0.7): $\left\langle \frac{1}{g\pi} \lambda_g^{1/2} \right\rangle \approx 1$,

$$\left\langle \left(\frac{1}{g\pi} \lambda_g^{1/2} \right)^2 \right\rangle \approx 1 + \int_0^1 \int_0^1 \langle a(x, \omega)\, a(y, \omega) \rangle \cos (2g\pi x) \cos (2g\pi y)\, \mathrm{d}x\, \mathrm{d}y ;$$

for (0.8): for the case $\delta_1 = \delta_2 = 0$:

$$\langle g\pi\lambda_g^{-1/4}\rangle \approx 1\,,$$

$$\operatorname{var}(g\pi\lambda_g^{-1/4}) \approx \int_0^1\int_0^1 \langle r(x,\omega)\, r(y,\omega)\rangle \sin^2(g\pi x)\sin^2(g\pi y)\,dx\,dy;$$

and

for (0.9): $$\langle\lambda_g^{-1}\rangle \approx \frac{1}{g^4\pi^4}\left\{1 - \frac{2}{g^2\pi^2}\varepsilon + \frac{1}{g^4\pi^4}\left(\frac{7}{2} + g\pi\tanh(g\pi)\right)\varepsilon^2\right\}.$$

The expressions for the cases (0.6) and (0.7) are not quite correct, since they were expanded up to terms of the second order, but one term of the second order does not appear.

Variational methods

The eigenvalues of the problem (0.2) can be calculated by the minimum principle

$$\lambda_g(\omega) = \inf_{u\in\mathfrak{U}_{g-1}} \langle\!\langle Mu, u\rangle\!\rangle\,,$$

where

$$\mathfrak{U}_{g-1} = \{u \in \mathbf{C}^{2m}[0,1]:\ U_i[u] = 0,\ i = 1, 2, \ldots, 2m;\ \langle\!\langle Nu, u\rangle\!\rangle = 1,$$
$$\langle\!\langle Nu, u_s\rangle\!\rangle = 0 \quad \text{for all eigenfunctions } u_s(x,\omega) \text{ belonging to}$$
$$\lambda_s(\omega),\ s = 1, 2, \ldots, g-1\}$$

and $\langle\!\langle\cdot\,,\cdot\rangle\!\rangle$ denotes the scalar product on $\mathbf{L}_2(0, 1)$.

From the definition of $\mathfrak{U}_g$ it can be seen that this set is dependent on ω for $g > 1$, and if also N is supposed to be deterministic. Hence, it is only possible to make statements about the first eigenvalue. For example, for a deterministic operator N the inequalities

$$\langle\lambda_1\rangle \leqq \mu_1\,,$$
$$\langle\lambda_1^2\rangle \leqq \mu_1^2 + \sum_{r,s=0}^{m}\int_0^1\int_0^1 \langle\tilde f_r(x,\omega)\,\tilde f_s(y,\omega)\rangle \left(w_1^{(r)}(x)\, w_1^{(s)}(y)\right)^2 dx\,dy$$

can be derived where

$$Mu = M_0u + M_1u$$

with

$$M_0u = \sum_{k=0}^{m} (-1)^k\,[\langle f_k\rangle\, u^{(k)}]^{(k)}$$

and

$$M_1u = \sum_{k=0}^{m'} (-1)^k\,[\tilde f_k u^{(k)}]^{(k)}\,,$$

$\tilde f_k(x,\omega) \doteq f_k(x,\omega) - \langle f_k(x)\rangle$. μ_1 denotes the first eigenvalue and $w_1(x)$ the eigenfunction belonging to μ_1 of the averaged problem

$$M_0w = \mu Nw\,, \quad U_i[w] = 0\,, \qquad i = 1, 2, \ldots, 2m\,.$$

BOYCE [1, 4] has investigated special eigenvalues problem with variational principles. For eigenvalue problems with random boundary conditions variational principles are not suitable because the variational domain is a random set.

Asymptotic methods

The eigenvalue problem

$$\sum_{k=0}^{m} (-1)^k [f_k(x, \omega)\, u^{(k)}]^{(k)} = \lambda g_0(x, \omega)\, u \tag{0.10}$$

can be transformed by a transformation of x and u,

$$v = (g_0^{2m-1} f_m)^{1/4m}\, u\,, \qquad \xi = \frac{1}{K} \int_0^x \left(\frac{g_0}{f_m}\right)^{1/2m} \mathrm{d}t$$

with $K = \int_0^1 (g_0/f_m)^{1/2m}\, \mathrm{d}t$ to the problem

$$\sum_{k=0}^{m} P_{2m-2k}(\xi)\, v^{(2m-2k)}(\xi) = K^{2m} \lambda v(\xi)$$

with $P_{2m}(\xi) \equiv 1$. If for the differential equation (0.10) boundary conditions of the form

$$U_i[u] = \sum_{j=0}^{2m-1} \alpha_{ij} u^{(j)}(0) = 0 \qquad \text{for} \quad i = 1, 2, \ldots, s\,,$$

$$U_i[u] = \sum_{j=0}^{2m-1} \beta_{ij} u^{(j)}(1) = 0 \qquad \text{for} \quad i = s+1\,, s+2, \ldots, 2m$$

are given then

$$\frac{1}{\lambda_g} (g\pi)^{2m} \sim K^{2m} = \left(\int_0^1 \left(\frac{g_0}{f_m}\right)^{1/2m} \mathrm{d}t \right)^{2m}.$$

Put

$$g_0(x, \omega) = 1 + \tilde{g}_0(x, \omega)\,, \qquad f_m(x, \omega) = 1 + \tilde{f}_m(x, \omega)$$

where $\langle \tilde{g}_0 \rangle = \langle \tilde{f}_m \rangle = 0$, $|\tilde{g}_0|$, $|\tilde{f}_m|$ sufficiently small and $\tilde{g}_0$, $\tilde{f}_m$ are uncorrelated. Then, BOYCE [4] has shown that

$$\operatorname{var}\left(\frac{1}{\lambda_g} (g\pi)^{2m}\right) \sim \int_0^1 \int_0^1 \{ \langle \tilde{g}_0(x, \omega)\, \tilde{g}_0(y, \omega) \rangle + \langle \tilde{f}_m(x, \omega)\, \tilde{f}_m(y, \omega) \rangle \}\, \mathrm{d}x\, \mathrm{d}y$$

up to terms of the second order. BOYCE [1, 4] has investigated the eigenvalue problem (0.6) by the asymptotic method. The result is

$$\operatorname{var} \lambda_g^{-1} \sim \frac{\sigma^2}{(g\pi)^4} \int_0^1 \int_0^1 \varrho(x - y)\, \mathrm{d}x\, \mathrm{d}y\,,$$

where $m(x, \omega)$ is supposed to be a wide-sense stationary process having zero-mean and the correlation function $\sigma^2\varrho(x - y)$.

The perturbation method applied to the problem (0.6) leads to

$$\operatorname{var} \lambda_g^{-1} = \frac{4\sigma^2}{(g\pi)^4} \int_0^1\int_0^1 \varrho(x - y) \sin^2 (g\pi x) \sin^2 (g\pi y) \, dx \, dy + O_3 .$$

This result yields the same result as the asymptotic method if the relation

$$\lim_{g\to\infty} \int_0^1\int_0^1 \varrho(x - y) \sin^2 (g\pi x) \sin^2 (g\pi y) \, dx \, dy = \tfrac{1}{4} \int_0^1\int_0^1 \varrho(x - y) \, dx \, dy$$

is used.

Integral equation methods

The eigenvalue problem

$$Mu = \lambda(1 + g_0(x, \omega)) \, u \, , \quad U_i[u] = 0 \, , \quad i = 1, 2, \dots, 2m \tag{0.11}$$

has been investigated by BOYCE [4] and GOODWIN, BOYCE [1] where M is a deterministic operator of the order $2m$ and the boundary conditions are also deterministic. Using the Green's function $G(x, y)$ corresponding to the operator M and the given boundary conditions $U_i[u] = 0$, $i = 1, 2, \dots, 2m$, then

$$\sum_{g=1}^{\infty} \langle \lambda_g^{-1} \rangle = \int_0^1 G(x, x) \, dx \, ,$$

$$\sum_{g=1}^{\infty} \langle \lambda_g^{-2} \rangle = \int_0^1\int_0^1 G^2(x, y) \left(1 + \langle g_0(x, \omega) \, g_0(y, \omega) \rangle\right) dx \, dy \, .$$

By means of the assumption $|g_0(x, \omega)| \leqq \delta < 1$ almost surely estimates for $\langle \lambda_g^{-k} \rangle$ can be obtained. From

$$G(x, y) = \sum_{r=1}^{\infty} \frac{w_r(x) \, w_r(y)}{\mu_r} \approx \frac{w_1(x) \, w_1(y)}{\mu_1}$$

the approximate relation

$$\operatorname{var} \lambda_1^{-1} \approx \mu_1^{-2} \int_0^1\int_0^1 \langle g_0(x, \omega) \, g_0(y, \omega) \rangle \, w_1^2(x) \, w_1^2(y) \, dx \, dy$$

can be derived. This relation can also be obtained by means of the perturbation expansions. μ_g, $w_g(x)$ are the eigenvalues and the normalized eigenfunctions of the averaged problem

$$Mw = \mu w \, , \quad U_i[w] = 0 \, , \quad i = 1, 2, \dots, 2m \, ,$$

associated with the random eigenvalue problem (0.11).

Statistical characteristics of the eigenvalues of the eigenvalue problem (0.9) are determined in (BOYCE, GOODWIN [1]) by integral equation methods. The disadvantage of this method for the calculation of $\langle \lambda_1^{-1} \rangle$ is that the distribution function of $\varepsilon(\omega)$ must be known, and knowledge of the first moments of $\varepsilon(\omega)$ is not sufficient.

Iteration methods

BOYCE [3] has applied an iteration method to the problem (0.11). The eigenvalue problem (0.11) is equivalent to the integral equation

$$\lambda^{-1}u(x) = \int_0^1 G(x, y) \left(1 + g_0(y, \omega)\right) u(y)\, dy .$$

$G(x, y)$ denotes the Green's function defined above. This equation is averaged

$$\langle \lambda^{-1}u(x)\rangle = \int_0^1 G(x, y) \langle u(y)\rangle\, dy + \int_0^1 G(x, y) \langle g_0(y)\, u(y)\rangle\, dy \tag{0.12}$$

and it is supposed that

$$\langle \lambda^{-1}u(x)\rangle = \langle \lambda^{-1}\rangle \langle u(x)\rangle ,$$

$$\langle g_0(x)\, u(x)\rangle = \langle g_0(x)\rangle \langle u(x)\rangle ,$$

so that the averaged integral equation can be solved. These and further analogous assumptions can be explained by the concept of „local independence" (cf. BOYCE [3, 4]). With the help of these assumptions, from (0.12) it follows that

$$\langle u_g(x)\rangle = w_g(x) , \qquad \langle \lambda_g^{-1}\rangle = \mu_g$$

where $(\!(w_g, w_h)\!) = \delta_{gh}$. If the integral equation is iterated then this leads to

$$\lambda^{-2}u(x) = \int_0^1\!\!\int_0^1 G(x, y)\, G(y, z)\, [1 + g_0(y, \omega)]\, [1 + g_0(z, \omega)]\, u(z)\, dy\, dz .$$

Averaging this equation one obtains

$$\langle \lambda^{-2}u(x)\rangle = \int_0^1\!\!\int_0^1 G(x, y)\, G(y, z)\, \{\langle u(z)\rangle + \langle g_0(y)\, u(z)\rangle + \langle g_0(z)\, u(z)\rangle + \langle g_0(z)\, g_0(y)\, u(z)\rangle\}\, dy\, dz .$$

On the basis of similar assumptions given above this averaged equation is reduced to

$$\langle \lambda^{-2}\rangle \langle u(x)\rangle = \int_0^1\!\!\int_0^1 G(x, y)\, G(y, z)\, \{1 + \langle g_0(z)\, g_0(y)\rangle\} \langle u(z)\rangle\, dy\, dz .$$

Replacing $\langle u(x)\rangle$ by $w_g(x)$ and forming the scalar product of the resulting equation with $w_G(x)$ leads to

$$\langle \lambda_g^{-2}\rangle \approx \mu_g^{-2} + \mu_g^{-1} \int_0^1\!\!\int_0^1 G(y, z)\, w_g(y)\, w_g(z)\, \langle g_0(y)\, g_0(z)\rangle\, dy\, dz .$$

From this equation it follows that

$$\operatorname{var} \lambda_g^{-1} \approx \mu_g^{-1} \int_0^1\!\!\int_0^1 G(y, z)\, w_g(y)\, w_g(z)\, \langle g_0(y)\, g_0(z)\rangle\, dy\, dz .$$

This result can be generalized to the eigenvalue problem

$$Mu = \lambda N(\omega)\, u , \qquad U_i[u] = 0 , \qquad i = 1, 2, \dots , 2m ,$$

where

$$N(\omega)\, u \doteq \sum_{k=0}^{m'} (-1)^k \, [g_k(x, \omega)\, u^{(k)}]^{(k)} ,$$

$m' \leqq m - 1$. The order of the deterministic operator M is $2m$.

Hierarchy methods

The iteration method is based on some assumptions for the correlation relations in order to solve the averaged integral equations. The hierarchy method takes into consideration further equations so that all occurring quantities can be calculated. Applications of this method to eigenvalue problems have been made by HAINES [1, 2].

Consider the eigenvalue problem (0.11) where every boundary condition only refers to $x = 0$ or to $x = 1$. On averaging (0.11) it follows that

$$\begin{aligned} &M\langle u(x)\rangle = \langle \lambda\big(1 + g_0(x, \omega)\big)\, u(x)\rangle , \\ &U_i[\langle u\rangle] = 0 , \qquad i = 1, 2, \ldots, 2m . \end{aligned} \tag{0.13}$$

This problem contains the unknown quantities $\langle u\rangle$, $\langle \lambda u\rangle$, and $\langle \lambda g_0 u\rangle$. To calculate these quantities further equations must be added. The problem (0.11) is multiplied by $\lambda\big(1 + g_0(y)\big)$ and averaged so that

$$\begin{aligned} &M\langle \lambda\big(1 + g_0(y)\big)\, u(x)\rangle = \langle \lambda^2\big(1 + g_0(x)\big)\, \big(1 + g_0(y)\big)\, u(x)\rangle , \\ &U_i[\langle \lambda\big(1 + g_0(y)\big)\, u\rangle] = 0 , \qquad i = 1, 2, \ldots, 2m . \end{aligned} \tag{0.14}$$

In this equation new unknown quantities are contained, so that additional equations must be added. To terminate this hierarchy of equations, and to solve the system, additional assumptions have to be made. Stopping at the first equation and assuming the relation

$$\langle \lambda(1 + g_0)\, u\rangle = \langle \lambda\rangle\, \langle u\rangle ,$$

the solution of the obtained equation is

$$\langle u_g(x)\rangle = w_g(x) , \qquad \langle \lambda_g\rangle = \mu_g ;$$

i.e. the eigenvalues and the eigenfunctions of the averaged problem

$$Mw = \mu w , \qquad U_i[w] = 0 , \qquad i = 1, 2, \ldots, 2m ,$$

result. Using the assumption

$$\langle \lambda^2\big(1 + g_0(x)\big)\, \big(1 + g_0(y)\big)\, u(x)\rangle = \langle \lambda^2\rangle\, [1 + \langle g_0(x)\, g_0(y)\rangle]\, \langle u(x)\rangle$$

the system of equations (0.13), (0.14) changes to

$$\begin{aligned} &\langle u(x)\rangle = \int_0^1 G(x, y)\, \langle \lambda\big(1 + g_0(y)\big)\, u(y)\rangle\, dy , \\ &\langle \lambda\big(1 + g_0(y)\big)\, u(x)\rangle = \langle \lambda^2\rangle \int_0^1 G(x, z)\, [1 + \langle g_0(x)\, g_0(z)\rangle]\, \langle u(z)\rangle\, dz . \end{aligned}$$

The second equation for $x = y$ is substituted in the first equation and one obtains

$$\langle u(x)\rangle = \langle \lambda^2\rangle \int_0^1\int_0^1 G(x, y)\, G(y, z)\, [1 + \langle g_0(y)\, g_0(z)\rangle]\, \langle u(z)\rangle\, dz\, dy\,.$$

This result is similar to the result obtained by the iteration method.

If we compare these methods we notice that the perturbation method has priority. Variational methods and integral equation methods lead only to statements for the first eigenvalue. Moreover, in order to apply the integral equation methods very strong conditions for the calculation of the mean of the eigenvalues are required. Both of these methods can only be used on a limited scale.

In the literature little attention has been paid so far to random eigenfunctions. Van der Linde [1] has investigated the variance of the eigenfunctions of the eigenvalue problem (0.11) where $\langle g_0(x, \omega)\rangle \equiv 0$ is assumed, and the sample functions of $g_0(x, \omega)$ should be sufficiently small a.s., $|g_0(x, y)| \leqq \eta$. Furthermore, let the eigenvalue problem be self-adjoint and positive definite. Then the variance of the eigenfunction $u_g(x, \omega)$ can be calculated by

$$\operatorname{var} u_g(x) = \mu_g^2 \sum_{\substack{k, r=1\\ k, r\neq g}}^{\infty} \frac{\langle b_{kg} b_{rg}\rangle}{(\mu_k - \mu_g)\,(\mu_r - \mu_g)}\, w_k(x)\, w_r(x) + O_3$$

where $b_{ik} \doteq (\!(g_0 w_i, w_k)\!)$. These results are applied to the special case

$$-u'' = \lambda(1 + g_0)\, u\,, \qquad u(0) = u(1) = 0$$

and $g_0(x, \omega)$ is supposed to be a white noise process.

Random boundary conditions have been considered in the application of eigenvalue problems. But in literature only some very special examples of eigenvalue problems with random boundary conditions have been treated (cf. Boyce [3, 4] and Goodwin, Boyce [1]).

Section 1.3.1 is concerned with randomly perturbed symmetric matrices with simple eigenvalues of the unperturbed matrix. The case of a random matrix $A_0 + B(\omega)$ is treated where A_0 denotes a diagonal matrix (the general case of a symmetric matrix A_0 can be reduced easily to this case by a deterministic transformation). The matrix of the perturbations $B(\omega) = \big(b_{ij}(\omega)\big)_{1\leqq i, j\leqq n}$ is supposed to fulfil the condition $|B(\omega)| < \varepsilon$ where ε is sufficiently small. Then the expansions for the eigenvalues $\lambda_g(\omega)$ and the normalized eigenvectors ${}^g y(\omega)$

$$\begin{aligned} \lambda_g(\omega) &= \mu_g - \sum_{k=1}^{\infty} \lambda_{gk}(\omega)\,, \\ {}^g y(\omega) &= \sum_{k=0}^{\infty} {}^g y_k(\omega) \end{aligned} \tag{0.15}$$

exist. The values μ_g denote the eigenvalues of A_0 and ${}^g y_0 = (0, \ldots, 0, \overset{g}{1}, 0, \ldots, 0)^{\mathsf T}$ the normalized eigenvectors belonging to μ_g. $\lambda_{gk}(\omega)$, ${}^g y_k(\omega)$ are the terms of $\lambda_g(\omega)$ and ${}^g y(\omega)$, respectively, which are homogeneous of the k-th order in the perturbations $b_{ij}(\omega)$, $k = 1, 2, \ldots$ These honmogeneous terms λ_{gk}, ${}^g y_k$ are given explicitly up to the fourth order, that is, for $k = 1, 2, 3, 4$. From this the means and the correlation relations of $\lambda_g(\omega)$ and ${}^g y(\omega)$ can be calculated, the means up to the fourth order and the correlation relations up to the fifth order. Subsequently, the general random eigenvalue problem

$$(A_0 + B(\omega))\, x = \lambda(C_0 + D(\omega))\, x \tag{0.16}$$

is considered where the eigenvectors are normalized by

$$\langle\!\langle (C_0 + D(\omega))\, x, x \rangle\!\rangle = 1 .$$

It is assumed that the investigated eigenvalue of the unperturbed (deterministic) problem

$$A_0 x = \mu C_0 x$$

is simple. Let the matrices A_0 and C_0 be symmetric and the norms of the random matrices of the perturbations $B(\omega)$ and $D(\omega)$ be sufficiently small. Expansions of the form (0.15) lead to a system of equations which can be solved successively. The mean of the eigenvalue $\lambda_g(\omega)$ is calculated up to terms of the second order in the random perturbations $b_{ij}(\omega)$ and $d_{ij}(\omega)$.

Section 1.3.2 considers the case of a randomly perturbed symmetric matrix with multiple eigenvalues of the unperturbed matrix. For this case the eigenvalues are only considered and the expansions are deduced from the characteristic equation. The main result of this section consists in the formulation of a convergence condition and a convergence proof for the perturbation expansions. This general convergence theorem makes valid the considerations of Section 1.3.1.

In Section 1.3.3 the expansions if the preceding sections are applied to concrete cases. First, a symmetric matrix $A_0 + B(\omega)$ is considered where A_0 is supposed to have simple eigenvalues. The perturbations $b_{ij}(\omega)$ are assumed to be statistically independent except for a symmetry condition so that b_{ij} and b_{rs} are independent if $(i, j) \neq (r, s)$ and $(i, j) \neq (s, r)$ and bounded almost surely. Furthermore, let all the b_{ij} have the same distribution function $F(x)$. This will be the case if all the elements of the matrix A_0 are measured with small independent errors. The means of the eigenvalues $\lambda_g(\omega)$ of the matrix $A_0 + B(\omega)$ are given up to terms of the fourth order with dependence on the quantities

$$\sigma^2 \doteq \int_{-\infty}^{\infty} x^2 \,\mathrm{d}F(x)\,, \qquad \sigma_4 \doteq \int_{-\infty}^{\infty} x^4 \,\mathrm{d}F(x), \quad \text{and} \quad r \doteq \sigma_4 - 3\sigma^4 .$$

In the same way, the correlation relations of the eigenvalues can be represented by these quantities σ^2, σ_4 and r. A numerical example uses uniformly distributed random variables $b_{ij}(\omega)$ on $[-d, d]$.

If the above case of independent random variables $b_{ij}(\omega)$ is under consideration but these $b_{ij}(\omega)$ are normally distributed, then the expansions (0.15) can be taken formally at first where the λ_{gk} are determined as functions of the $b_{ij}(\omega)$ by the formulas given in Section 1.3.1. However, using normally distributed perturbations the convergence condition of the perturbation expansions are not fulfilled with positive probability since the perturbations $b_{ij}(\omega)$ are greater than each given number with positive probability. Section 1.3.3.2 shows that the expressions derived for independent bounded $b_{ij}(\omega)$ in Section 1.3.3.1 can be used as approximations for $\langle\lambda_g(\omega)\rangle$ also in the case of normally distributed $b_{ij}(\omega)$. The additional occurring error is estimated.

The expansions for multiple eigenvalues of the unperturbed matrix are applied in Section 1.3.3.3 to calculate the means for eigenvalues with multiplicity two. Numerical examples are given for various distribution laws of the b_{ij}.

Section 1.4.1 is concerned with random eigenvalue problems of the form

$$\begin{aligned} &\big(M_0 + M_1(\omega)\big)\, u = \lambda\big(r(x) + N_1(\omega)\big)\, u\,, \qquad 0 \leqq x \leqq 1\,, \\ &U_i[u] \doteq U_{i0}[u] + U_{i1}(\omega)\,[u] = 0\,, \qquad i = 1, 2, \dots, 2m\,. \end{aligned} \tag{0.17}$$

Let M_0 be a differential operator of the order $2m$,

$$M_0 u \doteq \sum_{k=0}^{m} (-1)^k \left[f_k(x)\, u^{(k)}\right]^{(k)}$$

and $M_1(\omega)$, $N_1(\omega)$ be random differential operators of the order p or q, respectively, with $p, q \leqq 2m$. In the boundary conditions

$$U_{i0}[u] \doteq \sum_{k=0}^{2m-1} \big(a_{ik} u^{(k)}(0) + b_{ik} u^{(k)}(1)\big)$$

denotes the deterministic part and

$$U_{i1}(\omega)\,[u] \doteq \sum_{k=0}^{2m-1} \big(\alpha_{ik}(\omega)\, u^{(k)}(0) + \beta_{ik}(\omega)\, u^{(k)}(1)\big)$$

the random part. The eigenvalue problem (0.17) contains as special cases all the eigenvalue problems investigated in papers of Boyce [1, 3, 4]; Boyce, Goodwin [1]; Goodwin, Boyce [1]; Haines [1, 2] and van der Linde [1]. Furthermore, it is assumed that the unperturbed eigenvalue problem associated with the random problem (0.17)

$$M_0 w = \mu r w\,, \qquad U_{i0}[u] = 0\,, \qquad i = 1, 2, \dots, 2m\,,$$

is self-adjoint and positive definite. The eigenvalues of the unperturbed problem are denoted by μ_l and its normalized eigenfunctions by $w_l(x)$. Assume that the eigenvalue μ_g belonging to the eigenvalue $\lambda_g(\omega)$ which is to be investigated is simple. Let such conditions for the random operators $M_1(\omega)$, $N_1(\omega)$ and for the coefficients $\alpha_{ik}(\omega)$, $\beta_{ik}(\omega)$ of the boundary conditions be fulfilled so that the expansions

$$\lambda_g(\omega) = \sum_{k=0}^{\infty} \lambda_{gk}(\omega)\,, \qquad u_g(x, \omega) = \sum_{k=0}^{\infty} u_{gk}(x, \omega) \tag{0.18}$$

exist. The terms $\lambda_{gk}(\omega)$ and $u_{gk}(x, \omega)$ are given for $k = 1, 2$ with the random character of the boundary conditions being taken into account. By means of these expansions the expectations $\langle\lambda_g(\omega)\rangle$ and $\langle u_g(x, \omega)\rangle$ can be calculated up to terms of the second order. These investigations lead to nontrivial statements for the so-called „averaging problem“ which requires the calculation of the differences $\langle\lambda_g(\omega)\rangle - \mu_g$ and $\langle u_g(x, \omega)\rangle - w_g(x)$ since up to terms of the second order the equations

$$\langle \lambda_g(\omega) \rangle - \mu_g = \langle \lambda_{g2}(\omega) \rangle ,$$
$$\langle u_g(x, \omega) \rangle - w_g(x) = \langle u_{g2}(x, \omega) \rangle$$

are obtained.

The correlation relations, particularly, the variance of $\lambda_g(\omega)$ and $u_g(x, \omega)$ can be derived from the given perturbation expansions up to terms of the third order in the perturbations.

In Section 1.4.2 self-adjoint eigenvalue problems with deterministic boundary conditions are considered. For this special case approximations of order higher than the second are given by the aid of the method of Ritz; and convergence estimates are investigated. These considerations refer to self-adjoint, positive definite, random eigenvalue problems of the form

$$M(\omega)\, u = \lambda u , \qquad U_{i0}[u] = 0 , \qquad i = 1, 2, \dots , 2m , \tag{0.19}$$

where $0 \leqq x \leqq 1$ and

$$M(\omega)\, u \doteq \sum_{k=0}^{m} (-1)^k \, [f_k(x, \omega)\, u^{(k)}]^{(k)} .$$

This problem is written as

$$\langle M \rangle\, u + M_1(\omega)\, u = \lambda u , \qquad U_{i0}[u] = 0 , \qquad i = 1, 2, \dots , 2m ,$$

and $M_1(\omega) \doteq M(\omega) - \langle M \rangle$ is interpreted as a perturbation operator. The eigenvalues $\lambda_l(\omega)$ and the eigenfunctions $u_l(x, \omega)$ of (0.19) are assumed to have expansions as in (0.18). If the eigenvalue μ_g of the averaged problem

$$\langle M \rangle\, w = \mu w , \qquad U_{i0}[w] = 0 , \qquad i = 1, 2, \dots , 2m , \tag{0.20}$$

belonging to the eigenvalue $\lambda_g(\omega)$ is simple and the random perturbation operator $M_1(\omega)$ fulfils the condition

$$||M_1 u|| \leqq a(\omega)\, ||u|| + b(\omega)\, ||\langle M \rangle\, u||$$

for all $u \in \mathrm{D}(\langle M \rangle)$ where $a(\omega)$, $b(\omega)$ are nonnegative constants almost surely then the expansions of $\lambda_g(\omega)$ and $u_g(x, \omega)$ converge almost surely on suitable domains depending on $a(\omega)$, $b(\omega)$ and $\min \{\mu_g - \mu_{g-1}, \mu_{g+1} - \mu_g\}$. The expansions of $\lambda_g(\omega)$ and $u_g(x, \omega)$ are written in dependence on $b_{ij}(\omega) = (\!(M_1(\omega)\, w_i, w_j)\!)$ up to terms of the fourth order. The functions $w_i(x)$ are the normalized eigenfunctions of the problem (0.20). These expansions follow from the results derived for matrices by means of the method of Ritz. For example, the first terms for the eigenvalue are

$$\begin{aligned} \lambda_{g1} &= b_{gg} , \\ \lambda_{g2} &= - \sum_{\substack{i=1 \\ i \neq g}}^{\infty} \frac{b_{ig}^2}{\mu_i - \mu_g} , \\ \lambda_{g3} &= \sum_{\substack{i, j=1 \\ i, j \neq g}}^{\infty} \frac{b_{gi} b_{ij} b_{jg}}{(\mu_i - \mu_g)\,(\mu_j - \mu_g)} - b_{gg} \sum_{\substack{i=1 \\ i \neq g}}^{\infty} \frac{b_{ig}^2}{(\mu_i - \mu_g)^2} . \end{aligned} \tag{0.21}$$

From the expansions of $\lambda_g(\omega)$ and $u_g(x, \omega)$ the means and the correlation relations can again be calculated approximately.

Section 1.4.3 considers the eigenvalue problem

$$\begin{aligned} -u'' + a(x, \omega)\, u &= \lambda(1 + b(x, \omega))\, u\,, \\ u(0) = u(1) + \eta(\omega)\, u'(1) &= 0\,. \end{aligned} \tag{0.22}$$

This example is characteristic for eigenvalue problems of the form (0.17) and contains interesting applications. For instance, the one-dimensional Schrödinger equation with a random potential follows for $b(x, \omega) \equiv \eta(\omega) = 0$. Assume the relations $\langle a(x, \omega)\rangle \equiv \langle b(x, \omega)\rangle \equiv \langle \eta(\omega)\rangle = 0$ so that the averaged or unperturbed problem

$$-w'' = \mu w\,, \qquad w(0) = w(1) = 0$$

has the simple eigenvalues $\mu_l = (\pi l)^2$ and the eigenfunctions $w_l(x) = \sqrt{2}\sin(l\pi x)$ belonging to μ_l. The general expansions of Section 1.4.1 can be used for the calculation of the means and the correlation relations of the eigenvalues and eigenfunctions. These contain infinite series given in (0.21) as examples. For the eigenfunctions $w_g(x)$ of the averaged problem these series are calculated explicitly. Thus, expressions for $\langle \lambda_g(\omega)\rangle$, $\langle u_g(x, \omega)\rangle$, $\operatorname{var} \lambda_g$, $\operatorname{var} u_g(x)$ are obtained up to terms of the second order in the perturbations in which the correlation functions of $a(x, \omega)$ and $b(x, \omega)$, the cross-correlation function of $\big(a(x, \omega), b(x, \omega)\big)$ and the correlation quantities $\langle a(x)\,\eta\rangle$, $\langle b(x)\,\eta\rangle$ and $\langle \eta^2\rangle$ are involved. If the wide-sense stationarity of the vector process $\big(a(x, \omega), b(x, \omega)\big)$ is also assumed then, for instance, $\operatorname{var} \lambda_g$ is given by

$$\begin{aligned} \operatorname{var} \lambda_g = \int_0^1 \tilde{R}(u) \left[(1 - u)\,(2 + \cos(2g\pi u)) + \frac{3}{2g\pi} \sin(2g\pi u)\right] du \\ - 4\mu_g \int_0^1 \langle \eta \tilde{a}(u)\rangle\,\big(1 - \cos(2g\pi u)\big)\, du + 4\mu_g^2\, \langle \eta\rangle^2 \end{aligned}$$

where $\tilde{a}(x, \omega)$ is defined by $\tilde{a} \doteq a - \mu_g b$ and $\tilde{R}(u)$ denotes the correlation function of $\tilde{a}(x, \omega)$. Furthermore, in Section 1.4.3.1 the asymptotic behaviour of the eigenvalues of the problem (0.22) is investigated after the behaviour of the means and the variances of the eigenvalues as $g \to \infty$. For instance, it follows that

$$\lim_{g\to\infty} \frac{1}{\mu_g^2} \operatorname{var} \lambda_g = \int_0^1 R_b(u)\,(2 - 2u)\, du + 4 \int_0^1 \langle \eta b(u)\rangle\, du + 4\langle \eta^2\rangle$$

where $R_b(u)$ is the correlation function of $b(x, \omega)$. This result can be found for the special case $\eta = 0$ in the papers of Boyce [1, 4].

The computations of $\langle \lambda_g\rangle - \mu_g$, $\langle u_g\rangle - w_g$ (averaging problem), $\operatorname{var} \lambda_g$, $\operatorname{var} u_g$ for the eigenvalue problem (0.22) are made in Section 1.4.3.2 in each case up to terms of the second order in the random perturbations. Numerical calculations for various correlation assumptions are given and plotted. In particular, for the one-dimensional Schrödinger equation a probabilistic interpretation of the ,,bowing-effect", well-known in mixed crystal theory (cf. SCHULZE, UNGER [1]), is treated.

Section 1.4.3.3 is concerned with the eigenvalue problem (0.22) for the case that $a(x, \omega)$, $b(x, \omega)$ and $\eta(\omega)$ are uncorrelated and $a(x, \omega)$, $b(x, \omega)$ are assumed to be white noise processes, that is, $a(x, \omega)$, $b(x, \omega)$ are generalized wide-sense stationary processes where $\langle a(x, \omega)\rangle = \langle b(x, \omega)\rangle = 0$ and the ,,correlation functions" have the form

$$\langle a(x, \omega)\, a(y, \omega)\rangle = \tfrac{1}{2}\, S_1 \delta(x - y)\,, \qquad \langle b(x, \omega)\, b(y, \omega)\rangle = \tfrac{1}{2}\, S_2 \delta(x - y)\,. \tag{0.23}$$

The realizations of such processes are not continuous so that the existence and uniqueness theorems of classical analysis can only be applied with difficulty to the eigenvalue problem (0.22) with such processes as coefficients. On the other hand, reasonable results which can easily be calculated are obtained if the correlation functions (0.23) are substituted into the expressions for $\langle \lambda_{g2}\rangle$, $\langle u_{g2}\rangle$ and var u_g derived in Section 1.4.3.1. With the aid of the weakly correlated process to be introduced in Chapter 2 the expressions found this way undergo a mathematically exact interpretation. Furthermore, for the considered special case the critical points of var $u_g(x, \omega)$ are determined where results are given which contain the results derived by VAN DER LINDE [1], as a special case.

In Section 1.4.4 the general expansions of Section 1.4.1 are applied to transverse vibrations of elastic beams for which the cross-sectional area and the moment of inertia of the cross-section are random processes. Numerical results are given for the eigenvalues in the case of an elastically supported beam.

The basic concept for the considerations of Chapter 2 is the concept of a weakly correlated random field. The definition of such a field can be illustrated as follows: In physics and technological sciences random fields are often found which have the property that the influence of the field does not reach far, i.e. the values of the field at two points do not correlate when the distance of these points exceeds a certain quantity $\varepsilon > 0$. This number ε is called the correlation length of the random field. In applications ε is always assumed to be sufficiently small. Hence, weakly correlated fields can also be characterized as fields without ,,distant effect" or as fields of ,,noise-natured character". Examples for such fields are the force which acts on a Brownian particle, the fluctuation of the force which is produced by a turbulent media (for example, on a plate or a beam), the fluctuation of a technological surface, the fluctuation of parameters within technological materials, and many others.

The correlation function of a weakly correlated field $f(x, \omega)$, $x \in \mathbb{R}^n$, can be written as

$$\langle f(x, \omega)\, f(y, \omega)\rangle = \begin{cases} R(x, y) & \text{for} \quad |x - y| \leqq \varepsilon\,, \\ 0 & \text{for} \quad |x - y| > \varepsilon\,. \end{cases}$$

This property is not sufficient for a definition. In order to develop an adequate technique for the proof of asymptotic distribution laws of quantities in which weakly correlated fields are involved, we have given an exact mathematical definition of the notion of a „weakly correlated field" and adequate existence theorems have been proved. In the literature weakly correlated processes were only used in a more or less heuristic way.

The classical paper of ORNSTEIN, UHLENBECK [1] which deals with Brownian motion must be mentioned first. Ornstein and Uhlenbeck investigated the differential equation

$$m \frac{\mathrm{d}u}{\mathrm{d}t} + fu = F(t) \tag{0.24}$$

for the velocity of a Brownian particle where $F(t)$ denotes the random force which is caused by the impacts of molecules of the surrounding media. This force $F(t)$ is assumed to be weakly correlated (in the present terminology) since Ornstein and Uhlenbeck write: „There will be correlation between the values of $F(t)$ at different times t_1 and t_2 only when $|t_1 - t_2|$ is very small." According to that, it is supposed that the correlation function $\varphi(x)$ „is a function with a very small maximum at $x = 0$". For the higher moments of $F(t)$ they assume a decomposition into groups of lower moments according to the model of the „cluster condition" in modern physics. The exact definition of a weakly correlated field in Chapter 2 of this book is based on this idea.

The main result of Ornstein und Uhlenbeck consists in the fact that the solution $u(x)$ of (0.24) for each fixed x has an asymptotic normal distribution. The method of moments is used for the proof. Another paper which deals with similar problems was written by VAN LEAR, UHLENBECK [1].

With reference to Ornstein and Uhlenbeck, BOYCE [2] has shown that the solution $u(x)$ of a boundary value problem for a deterministic Sturm-Liouville operator for each fixed x converges to a normally distributed random variable as $\varepsilon \downarrow 0$ if the inhomogeneous term of the differential equation is assumed to be a weakly correlated process with the correlation lenght ε.

The essential results of Chapter 2 may be summed up briefly. First, the definition of a weakly correlated field is given and some existence theorems are proved. For instance, for each given $\varepsilon > 0$ a weakly correlated field exists having the correlation length ε. Furthermore, weakly correlated fields exist possessing sufficiently smooth realizations or realizations which are bounded almost surely. Finally, weakly correlated connected vector fields are also introduced.

All further results are based upon the limit theorems for linear functionals of weakly correlated fields proved in Section 2.1.2. For a given weakly correlated field $f_\varepsilon(x, \omega)$ defined on $\mathcal{D} \times \Omega \subset \mathbb{R}^m \times \Omega$ it is assumed that the limit value

$$\lim_{\varepsilon \downarrow 0} \frac{1}{\varepsilon^m} \int_{\mathcal{K}_\varepsilon(0)} \langle f_\varepsilon(x) f_\varepsilon(x + y) \rangle \, \mathrm{d}y \doteq a(x) < \infty \tag{0.25}$$

exists where $\mathcal{K}_\varepsilon(0) \doteq \{y \in \mathcal{D}: |y| \leqq \varepsilon\}$. $a(x)$ is called the intensity of the field $f_\varepsilon(x, \omega)$. Then the random processes

$$r_{i\varepsilon}(x, \omega) \doteq \frac{1}{\sqrt{\varepsilon^m}} \int_{\mathcal{D}} F_i(x, y) f_\varepsilon(y, \omega)\, dy\,, \qquad i = 1, 2, \dots, t\,, \tag{0.26}$$

for $x \in \mathcal{G} \subset R^l$, where $F_i(x, y) \in \mathbf{L}_2(\mathcal{D})_y$ relative to y, satisfies

$$\lim_{\varepsilon \downarrow 0} \big(r_{1\varepsilon}(x, \omega), \dots, r_{t\varepsilon}(x, \omega)\big) = \big(\xi_1(x, \omega), \dots, \xi_t(x, \omega)\big) \tag{0.27}$$

in distribution. The vector process $\big(\xi_1(x, \omega), \dots, \xi_t(x, \omega)\big)$ is a Gaussian vector process having zero mean and the correlation relations

$$\langle \xi_i(x, \omega)\, \xi_j(y, \omega)\rangle = \int_{\mathcal{D}} F_i(x, z)\, F_j(y, z)\, a(z)\, dz\,, \qquad i, j = 1, 2, \dots, t\,. \tag{0.28}$$

By means of the method of moments the proof is given, where the convergence in distribution then follows from the convergence of moments. The limit theorem given here is the prototype of a series of limit theorems which can be obtained from this by generalizations or specializations.

To apply the limit theorems for weakly correlated fields to eigenvalue problems the generalization to functions of linear functionals of the type (0.26) is necessary. In simplest case of such a generalized limit theorem has the following form: Let $f_\varepsilon(x, \omega)$ be a weakly correlated process on $\mathcal{D} \subset R^1$ having continuous realizations, and intensity $a(x)$; and let $\bar{r}_\varepsilon(\omega)$ be a random variable defined by

$$\bar{r}_\varepsilon(\omega) \doteq \int_{\mathcal{D}} F(y)\, f_\varepsilon(y, \omega)\, dy$$

where $F \in \mathbf{L}_2(\mathcal{D})$. Assume that $\bar{r}_\varepsilon(\omega) \in \mathcal{J}$ a.s. Furthermore, let $d(y)$ be a real function on $\mathcal{J}$ satisfying the following conditions:

(1) d can be represented by

$$d(y) = d_0 + d_1 y + y^2 g(y)$$

where $g(y)$ is bounded on $\mathcal{K}_\eta(0) = \{y: |y| \leqq \eta\}$, where η is an arbitrarily small positive number.

(2) All moments of $d_\varepsilon(\omega) \doteq d\big(\bar{r}_\varepsilon(\omega)\big)$ exist.

Then, the random variable $\big(d_\varepsilon(\omega) - d_0\big)/\sqrt{\varepsilon}$ converges in distribution as $\varepsilon \downarrow 0$ to a normally distributed random variable $\xi(\omega)$ with

$$\langle \xi(\omega)\rangle = 0\,,$$
$$\langle \xi^2(\omega)\rangle = \sigma^2 = d_1^2 \int_{\mathcal{D}} F^2(y)\, a(y)\, dy\,.$$

In Section 2.1.3 an essentially more general theorem of this type for weakly correlated connected vector fields is given.

Section 2.2 is concerned with limit theorems for the eigenvalues and eigenvectors of random matrices whose elements are linear functionals of weakly correlated processes or weakly correlated fields, respectively. Such matrices are obtained if the method of Ritz is applied to an eigenvalue problem for differential operators with weakly correlated coefficients. These results are of importance from two points of view: (1) Many eigenvalue problems for differential operators are only accessible by means of approximation methods, particularly in practical applications. The method of Ritz has been found to be an efficient method. (2) It is possible to prove limit theorems for eigenvalues and eigenfunctions of differential operators having weakly correlated coefficients (see below), but these theorems have always been proved by the aid of perturbation theory. The application of the perturbation theory assumes that the random coefficients of the operators are sufficiently small. Such conditions are not necessary in the theorems for matrices since these can be proved without perturbation theory. Hence, limit theorems for the eigenvalues and eigenfunctions of random differential operators without conditions concerning the smallness of the weakly correlated coefficients can be applied approximately if the results of Section 2.2 are used.

In Section 2.2.1 some results concerning the discrete spectrum of differential operators and the method of Ritz are given, and applied to some concrete examples of ordinary differential operators which are investigated from the probabilistic point of view in the following sections.

The central result of Section 2.2.2 is the following: Let $(f_{1\varepsilon}(x, \omega), \dots, f_{l\varepsilon}(x, \omega))$ be a weakly correlated connected vector field on $\mathcal{D} \subset R^m$ having continuous sample functions and intensities

$$a_{uv}(x) \doteq \lim_{\varepsilon \downarrow 0} \frac{1}{\varepsilon^m} \int_{\mathcal{D}} \langle f_{u\varepsilon}(x) f_{v\varepsilon}(x + y) \rangle \, dy ,$$

$u, v = 1, 2, \dots, l$. Then, the random eigenvalue problem

$$(A_0 + B(\omega)) U = \Lambda(C_0 + D(\omega)) U \tag{0.29}$$

is considered. A_0 and C_0 are deterministic symmetric n by n matrices and $B(\omega)$ and $D(\omega)$ random matrices whose elements are written in the form

$$b_{ij}(\omega) = \sum_{u=1}^{l} \sum_{v=1}^{t} \beta_{ij,uv} \bar{r}^{\varepsilon}_{uv}(\omega) , \qquad i, j = 1, 2, \dots, n ,$$

$$d_{ij}(\omega) = \sum_{u=1}^{l} \sum_{v=1}^{t} \delta_{ij,uv} \bar{r}^{\varepsilon}_{uv}(\omega) , \qquad i, j = 1, 2, \dots, n ,$$

where $\beta_{ij,uv}$ and $\delta_{ij,uv}$ denote constants, and

$$\bar{r}^{\varepsilon}_{uv}(\omega) = \int_{\mathcal{D}} F_{uv}(x) f_{u\varepsilon}(x, \omega) \, dx \quad \text{with} \quad F_{uv} \in \mathbf{L}_2(\mathcal{D}) .$$

Let the eigenvalues of the problem (0.29) be

$$\Lambda_{1\varepsilon}(\omega) \leqq \Lambda_{2\varepsilon}(\omega) \leqq \dots \leqq \Lambda_{n\varepsilon}(\omega) ,$$

and let $U_{1\varepsilon}, U_{2\varepsilon}, \dots, U_{n\varepsilon}$ be the eigenvectors which are orthonormalized by $\langle\!\langle (C_0 + D(\omega)) U_{i\varepsilon}, U_{j\varepsilon} \rangle\!\rangle = \delta_{ij}$. The matrix $C_0 + D(\omega)$ is assumed to be positive definite. The averaged problem $A_0 U_0 = \Lambda_0 C_0 U_0$ belonging to the eigenvalue problem (0.29) has the eigenvalues $\Lambda_{10} \leqq \Lambda_{20} \leqq \dots \leqq \Lambda_{n0}$ and the eigenvectors

$U_{10}, U_{20}, \dots, U_{n0}$, where $U_{i0} = (u_{ij0})^{\mathrm{T}}_{1 \leqq j \leqq n}$ and $\langle\!\langle C_0 U_{i0}, U_{j0} \rangle\!\rangle = \delta_{ij}$. Let the eigenvalues

$$\Lambda_{i_1 0}, \Lambda_{i_2 0}, \dots, \Lambda_{i_s 0}, \qquad \{i_1, i_2, \dots, i_s\} \subset \{1, 2, \dots, n\},$$

be simple. Then, convergence in distribution

$$\lim_{\varepsilon \downarrow 0} \frac{1}{\sqrt{\varepsilon^m}} (\Lambda_{i_1 \varepsilon} - \Lambda_{i_1 0}, \Lambda_{i_2 \varepsilon} - \Lambda_{i_2 0}, \dots, \Lambda_{i_s \varepsilon} - \Lambda_{i_s 0}) = (\xi_{i_1}, \xi_{i_2}, \dots, \xi_{i_s})$$

is proved. The random vector $(\xi_{i_1}, \xi_{i_2}, \dots, \xi_{i_s})$ is a Gaussian random vector having the first two moments

$$\langle \xi_g \rangle = 0,$$
$$\langle \xi_g \xi_h \rangle = \sum_{u,p=1}^{l} \sum_{v,q=1}^{t} d_{gguv} d_{hhpq} \int_{\mathcal{D}} F_{uv}(x) F_{pq}(x) a_{up}(x) \, \mathrm{d}x,$$

$g, h \in \{i_1, \dots, i_s\}$, where

$$d_{ghuv} = \sum_{i,j=1}^{n} (\beta_{ij,uv} - \Lambda_{h0} \delta_{ij,uv}) u_{gi0} u_{hj0}.$$

Similar results are given for the eigenvectors of the eigenvalue problem (0.29).

Section 2.2.3 is concerned with the comparision of the asymptotic distribution laws with simulation results. First, a formula is given for the simulation of weakly correlated processes with an arbitrary correlation length ε. Then the simple eigenvalue problem $(A_0 + B(\omega)) U = \Lambda U$ is considered where

$$A_0 = \begin{pmatrix} 2 & -2 \\ -2 & -1 \end{pmatrix}, \qquad B(\omega) = \begin{pmatrix} b_{11}(\omega) & 0 \\ 0 & b_{22}(\omega) \end{pmatrix}$$

and

$$b_{11}(\omega) = \int_{\alpha}^{\beta} f_\varepsilon(x, \omega) \, \mathrm{d}x, \qquad b_{22}(\omega) = \int_{\alpha}^{\beta} x f_\varepsilon(x, \omega) \, \mathrm{d}x.$$

The process $f_\varepsilon(x, \omega)$ is a weakly correlated process simulated by the above mentioned formula. For example, we see that the calculation of the distribution laws of the random eigenvalues by means of 3000 realizations gives a very good correspondence with the asymptotic distribution laws derived in Section 2.2.2 for larger correlation lengths ε. A series of numerical results for different numbers of realizations and different correlation lengths are obtained, and are compared with the asymptotic distribution laws.

In Section 2.2.4 the limit theorems of Section 2.2.2 are applied to the eigenvalue problem

$$\begin{aligned} &(fu'')'' = \lambda(-g_1 u'' + gu), \qquad 0 \leqq x \leqq 1, \\ &u(0) = u(1) = u''(0) = u''(1) = 0. \end{aligned} \tag{0.30}$$

It is assumed that the random processes $f(x, \omega)$ and $g(x, \omega)$ constitute a weakly correlated connected vector process (f, g). The method of Ritz leads to an eigenvalue problem for matrices of the form considered in Section 2.2.2. From problem (0.30) the problem of the bending vibrations of bars and the buckling problem for a bar can be derived by specialization. For both problems the asymptotic distribution laws for the eigenvalues and the eigenfunctions are calculated corresponding to the limit theorems of Section 2.2.2. The obtained results are analyzed numerically for different conditions on the respective processes. The statements concerning the probability of fatigue of a bar with random parameters loaded by a given force are of particular interest in applications to technological mechanics.

Asymptotic distribution laws are proved for the eigenvalues and eigenfunctions of random differential operators in Section 2.3. The random eigenvalue problem

$$M(\omega)\,u = \lambda N(\omega)\,u\,, \qquad U_i[u] = 0\,, \qquad i = 1, 2, \dots, 2m\,, \qquad a \leqq x \leqq b\,, \tag{0.31}$$

is considered, where M and N are random differential operators of the form

$$M(\omega)\,u \doteq \sum_{i=0}^{m} (-1)^i\,[f_i(x, \omega)\,u^{(i)}]^{(i)}\,,$$

$$N(\omega)\,u \doteq \sum_{i=0}^{m'} (-1)^i\,[g_i(x, \omega)\,u^{(i)}]^{(i)}$$

with $m \geqq m'$. Let the processes $f_i(x, \omega)$, $i = 0, 1, \dots, m$, and $g_i(x, \omega)$, $i = 0, 1, \dots, m'$, be sufficiently smooth. The conditions for a discrete spectrum are assumed. The eigenvalues of the problem (0.31) are denoted by $\lambda_1(\omega) \leqq \lambda_2(\omega) \leqq \dots$ where $\lim_{g\to\infty} \lambda_g(\omega) = \infty$ a.s. and the eigenfunctions by $u_g(x, \omega)$. Put

$$f_i(x, \omega) \doteq f_{0i}(x) + f_{1i}(x, \omega)\,, \qquad i = 0, 1, \dots, m\,,$$

$$f_{0i}(x) \doteq \langle f_i(x, \omega)\rangle\,, \qquad f_{1i}(x, \omega) \doteq f_i(x, \omega) - f_{0i}(x)$$

and similarly for $g_i(x, \omega)$, $i = 0, 1, \dots, m'$. The operators M_k, N_k are defined by

$$M_k u \doteq \sum_{i=0}^{m} (-1)^i\,[f_{ki} u^{(i)}]^{(i)} \quad \text{for} \quad k = 0, 1\,,$$

$$N_k u \doteq \sum_{i=0}^{m'} (-1)^i\,[g_{ki} u^{(i)}]^{(i)} \quad \text{for} \quad k = 0, 1\,.$$

Hence, the eigenvalue problem (0.31) can be written in the form

$$(M_0 + M_1(\omega))\,u = \lambda(N_0 + N_1(\omega))\,u\,, \qquad U_i[u] = 0\,, \quad i = 1, 2, \dots, 2m\,. \tag{0.32}$$

The averaged problem associated with (0.32)

$$M_0 w = \mu N_0 w\,, \qquad U_i[w] = 0\,, \qquad i = 1, 2, \dots, 2m\,,$$

shall have eigenvalues $\mu_1 \leqq \mu_2 \leqq \dots$ and eigenfunctions $w_g(x)$.

Now, the operators $M_1(\omega)$ and $N_1(\omega)$ of (0.32) are considered as „small" perturbations, and for the solutions of the eigenvalue problem (0.32) expansions of the form (0.18) are given. For the terms λ_{gk} and u_{gk} representation formulas are derived which are used for the proof of the following limit theorem:

Let the random coefficients f_{1i}, $i = 0, 1, \dots, m$, g_{1i}, $i = 0, 1, \dots, m'$, of the operators M_1, N_1 in the eigenvalue problem (0.32) constitute a weakly correlated connected vector process having the intensities

$$a_{ij}^{11}(x) = \lim_{\varepsilon \downarrow 0} \frac{1}{\varepsilon} \int_{-\varepsilon}^{\varepsilon} \langle f_{1i}(x) f_{1j}(x + y) \rangle \, dy ,$$

$$a_{ij}^{12}(x) = \lim_{\varepsilon \downarrow 0} \frac{1}{\varepsilon} \int_{-\varepsilon}^{\varepsilon} \langle f_{1i}(x) g_{1j}(x + y) \rangle \, dy ,$$

$$a_{ij}^{22}(x) = \lim_{\varepsilon \downarrow 0} \frac{1}{\varepsilon} \int_{-\varepsilon}^{\varepsilon} \langle g_{1i}(x) g_{1j}(x + y) \rangle \, dy .$$

Assume that the conditions for the application of perturbation theory are fulfilled, and the eigenvalues $\mu_{g_1}, \mu_{g_2}, \dots, \mu_{g_s}$ are simple. Then, convergence in distribution

$$\lim_{\varepsilon \downarrow 0} \frac{1}{\sqrt{\varepsilon}} \left(\lambda_{g_1}(\omega) - \mu_{g_1}, \lambda_{g_2}(\omega) - \mu_{g_2}, \dots, \lambda_{g_s}(\omega) - \mu_{g_s}\right)$$
$$= \left(\xi_{g_1}(\omega), \xi_{g_2}(\omega), \dots, \xi_{g_s}(\omega)\right)$$

is obtained, where $(\xi_{g_1}, \dots, \xi_{g_s})$ denotes a zero mean Gaussian random vector with the correlation relations

$$\langle \xi_g \xi_h \rangle = \sum_{j,i=0}^{m-1} \int_a^b b_{ij}^{gh}(y) \left(w_g^{(i)}(y)\, w_h^{(j)}(y)\right)^2 dy ,$$

$g, h \in \{g_1, \dots, g_s\}$. The quantities b_{ij}^{gh} are determined by the intensities $a_{ij}^{\cdot\cdot}$.

Analogous asymptotic distribution laws can be shown for the eigenfunctions. In this case the generalized Green's functions of the operators $M_0 - \mu_g N_0$, $M_0 - \mu_h N_0$ (with the boundary conditions $U_i[.] = 0$, $i = 1, 2, \dots, 2m$) are contained in the moments of the limit distribution. Limit distributions are also written for random vectors whose components are eigenvalues $\lambda_g(\omega)$ or eigenfunctions $u_h(x, \omega)$, respectively.

In Section 2.3.3 the random eigenvalue problem

$$-cu'' + f_{01}(x, \omega)\, u = \lambda\left(1 + g_{01}(x, \omega)\right) u , \qquad 0 \leqq x \leqq l ,$$
$$u(0) = u(l) = 0$$

is investigated as an example of the application of the limit theorem given above. The eigenvalues μ_g and the eigenfunctions $w_g(x)$ of the averaged problem

$$-cw'' = \mu w , \qquad w(0) = w(l) = 0$$

can be given explicitly in this case. Under corresponding conditions the asymptotic normal distribution follows for the distributions of the eigenvalues and eigenfunctions from the general limit theorem. The asymptotic density function of the eigenvalues and the asymptotic variances of the eigenfunctions are plotted for some special cases. Different stability questions are discussed.

The theory of the weakly correlated fields and the corresponding limit theorems can also be applied to initial value problems, boundary value problems and boundary-initial value problems (cf. PURKERT, VOM SCHEIDT [6, 10, 11]; VOM SCHEIDT [1, 2, 4]). These applications are of especial interest in mechanics when the random behaviour of a material has to be investigated. A monograph by vom Scheidt on this subject is now in preparation.

The formulas, lemmas and theorems are numbered separately in all chapters. All random variables are considered on a fixed probability space $\{\Omega, \mathfrak{A}, \mathsf{P}\}$, and characterized by the notation ω. For instance, a random process is denoted by $f(x, \omega)$, $x \in \mathbb{R}^1$. The expectation is written by peaked brackets, that is, for example, $\langle \lambda(\omega) \rangle = \langle \lambda(.) \rangle = \langle \lambda \rangle$ denotes the mean of the random variable $\lambda(\omega)$. The norm of vectors and matrices is denoted by bold-faced vertical lines $|.|$, and the norm on function spaces by the usual notation.

1. Statistical moments

1.1. General notations

The fundamental studies on a theory of random equations and random operators are contained in the classical papers of Frechét [1, 2], Mourier [1], Hanš [1, 2, 3, 4] and Spaček [1], as well as in the books of Bharucha-Reid [1, 2, 3, 4] and Skorohod [1]. In this section we will give some necessary definitions, concepts and theorems. Let $\{\Omega, \mathfrak{A}, \mathsf{P}\}$ be a complete probability space and $\mathcal{X}$ a separable Banach space with norm $||.||$, and let $\mathfrak{B}$ denote the σ-algebra of all Borel subsets of $\mathcal{X}$.

Definition 1.1. A single-value mapping $x\colon \Omega \to \mathcal{X}$ is said to be a *random variable with values in* $\mathcal{X}$ if the inverse image under the mapping x of every Borel set $\mathcal{B} \in \mathfrak{B}$ belongs to $\mathfrak{A}$; that is, $x^{-1}(\mathcal{B}) = \{\omega\colon x(\omega) \in \mathcal{B}\} \in \mathfrak{A}$ for all $\mathcal{B} \in \mathfrak{B}$.

Because of the separability of the space $\mathcal{X}$ the following statements are equivalent:

(1) A mapping $x\colon \Omega \to \mathcal{X}$ is a random variable if and only if the functions $x^*\big(x(\omega)\big)$ are real-valued random variables for each $x^* \in \mathcal{X}^*$. $\mathcal{X}^*$ denotes the adjoint space of $\mathcal{X}$.

(2) A mapping $x\colon \Omega \to \mathcal{X}$ is a random variable if and only if there exists a sequence $\big(x_n(\omega)\big)$ of simple random variables which converges to $x(\omega)$ almost surely, that is, there exists a set A_0 with $\mathsf{P}(A_0) = 0$ such that

$$\lim_{n\to\infty} ||x_n(\omega) - x(\omega)|| = 0 \quad \text{for every} \quad \omega \in \Omega \setminus A_0 .$$

A random variable $y\colon \Omega \to \mathcal{X}$ is said to be *simple* if it is constant on each of a finite number of disjoint sets $A_i \in \mathfrak{A}$ and equal to 0 on $\Omega \setminus (\bigcup_{i=1}^{n} A_i)$.

Since $\mathcal{X}$ is assumed to be separable, the norm of a random variable is a non-negative real-valued random variable.

Definition 1.2. The *$\mathcal{X}$-valued random variables* $x(\omega)$ and $y(\omega)$ are said to be *equivalent* if $\mathsf{P}\big(\omega\colon x(\omega) \neq y(\omega)\big) = 0$.

The definition of the important normal or Gaussian random variable which plays a central role in probability theory is reduced to the Gaussian distribution of real-valued random variables.

Definition 1.3. A random variable $x(\omega)$ with values in a Banach space $\mathcal{X}$ is said to be a *Gaussian* or *normal random variable* if $x^*(x(\omega))$ is a scalar-valued Gaussian random variable for every $x^* \in \mathcal{X}^*$.

Definition 1.4. Let $x(\omega)$ be a random variable. The *expectation* or the *mean* of $x(\omega)$, denoted by $\langle x(\omega)\rangle$, is defined as the Bochner integral of $x(\omega)$ over Ω (if this integral exists), that is

$$\langle x(\omega)\rangle \doteq {}^{(\mathrm{B})}\int_{\Omega} x(\omega)\, \mathsf{P}(\mathrm{d}\omega)\,. \tag{1.1}$$

If the Bochner integral of $x(\omega)$ over Ω does not exist then $x(\omega)$ does not have an expectation. A necessary and sufficient condition that $\langle x(\omega)\rangle$ exists is that

$$\langle \|x(\omega)\|\rangle \doteq {}^{(\mathrm{L})}\int_{\Omega} \|x(\omega)\|\, \mathsf{P}(\mathrm{d}\omega) < \infty\,.$$

If $\langle x_1(\omega)\rangle$ and $\langle x_2(\omega)\rangle$ exist and a_1, a_2 are constants, then $\langle a_1x_1(\omega) + a_2x_2(\omega)\rangle$ exists, and

$$\langle a_1x_1(\omega) + a_2x_2(\omega)\rangle = a_1\langle x_1(\omega)\rangle + a_2\langle x_2(\omega)\rangle\,.$$

Furthermore, if $\langle x(\omega)\rangle$ exists, then

$$\|\langle x(\omega)\rangle\| \leqq \langle \|x(\omega)\|\rangle\,.$$

Now we give the generalization of the dominated convergence theorem for $\mathcal{X}$-valued random variables.

Theorem 1.1. *Let $(x_n(\omega))$ be a sequence of $\mathcal{X}$-valued random variables which converges almost surely to some limit random variable $x(\omega)$; that is*

$$\lim_{n\to\infty} \|x_n(\omega) - x(\omega)\| = 0 \quad \textit{almost surely.}$$

If $\langle x_n(\omega)\rangle$ exists for $n = 1, 2, \ldots$, and there is a fixed nonnegative random variable $y(\omega)$ such that $\|x_n(\omega)\| \leqq y(\omega)$ for all n and ω and $\langle y(\omega)\rangle < \infty$, then $\langle x(\omega)\rangle$ exists and

$$\lim_{n\to\infty} \langle x_n(\omega)\rangle = \langle x(\omega)\rangle\,.$$

The following theorem is often applied.

Theorem 1.2. *If $x(\omega)$ is a Hilbert space-valued random variable and $\langle \|x(\omega)\|\rangle$ exists, then, for all $y \in \mathcal{H}$,*

$$\langle (\!(x(\omega), y)\!)\rangle = (\!(\langle x(\omega)\rangle, y)\!)\,. \tag{1.2}$$

The scalar product of the Hilbert space $\mathcal{H}$ is denoted by $(\!(.\,,.)\!)$. To introduce random operators we consider an additional separable Banach space $\mathcal{Y}$, and let $\mathfrak{C}$ denote the σ-algebra of all Borel subsets of $\mathcal{Y}$.

Definition 1.5. A single-valued mapping $T(\omega)\colon \Omega \times \mathcal{X}_1 \to \mathcal{Y}$ is said to be a *random operator* with the domain $\mathcal{X}_1 = \mathrm{D}(T)$ if $T(\omega)\,x$ is a $\mathcal{Y}$-valued random variable for every $x \in \mathcal{X}_1$.

Definition 1.6. A *random operator* $T(\omega)$ with the domain $\mathrm{D}(T)$ is said to be *linear* if $\mathrm{D}(T)$ is a linear set and

$$T(\omega)\,[a_1x_1 + a_2x_2] = a_1T(\omega)\,x_1 + a_2T(\omega)\,x_2$$

almost surely for all $x_1, x_2 \in \mathrm{D}(T)$, where a_1, a_2 are scalars. A *random operator* $T(\omega)$ is said to be *bounded* if $\mathrm{D}(T) = \mathcal{X}$ and there exists an a.s. nonnegative real-valued random variable $M(\omega)$ such that for all $x \in \mathcal{X}$

$$||T(\omega)\,x|| \leqq M(\omega)\,||x|| \quad \text{a.s.}$$

In a manner similar to Definition 1.6 we can define the random analogues of an *inverse operator*, an *adjoint operator*, a *compact operator*, and in this case of a Hilbert space, a *symmetric operator*, a *self-adjoint operator*, and a *positive definite operator*. These definitions follow from the usual deterministic theory since the defining relations are demanded to be true only for almost all $\omega \in \Omega$. We can show that the inverse operator $T^{-1}(\omega)$ and also the adjoint operator $T^*(\omega)$ defined in this way are random operators if $T(\omega)$ is a random operator.

Now we give the definition of the averaged operator.

Definition 1.7. Let $T(\omega)$ be a random operator with the domain $\mathrm{D}(T)$. The operator $\langle T(\omega)\rangle$ defined by

$$\langle T(\omega)\rangle\, x \doteq {}^{(\mathrm{B})}\!\int_{\Omega} T(\omega)\, x\mathsf{P}(\mathrm{d}\omega) \tag{1.3}$$

for $x \in \mathrm{D}(\langle T\rangle) \doteq \{x \in \mathrm{D}(T) : \langle T(\omega)\, x\rangle \text{ exists}\}$ is said to be the *averaged operator* associated with $T(\omega)$.

Theorem 1.3. *Let $\mathcal{X}$ be a Hilbert space with the scalar product $(\!(\cdot\,, \cdot)\!)$. Then*

$$\langle(\!(T(\omega)\, x, y)\!)\rangle = (\!(\langle T(\omega)\rangle\, x, y)\!) \tag{1.4}$$

for all $x \in \mathrm{D}(\langle T\rangle)$ and $y \in \mathcal{X}$.

Theorem 1.4. *Let $\mathcal{X}$ be a Hilbert space with the scalar product $(\!(\cdot\,, \cdot)\!)$ and $T(\omega)$ a random operator where $\mathrm{D}(T) = \mathrm{D}(\langle T\rangle)$. If $T(\omega)$ is a self-adjoint random operator then $\langle T(\omega)\rangle$ is also. If $T(\omega)$ is a positive definite operator then $\langle T(\omega)\rangle$ is also.*

Example 1.1. Let $\mathcal{X} = \mathbb{R}^n$, $\mathcal{Y} = \mathbb{R}^n$. In this case a random linear operator $T(\omega)$ has a unique representation as an n by n random matrix, $T(\omega) = \big(a_{ij}(\omega)\big)_{1\leqq i,j\leqq n}$. If $a_{ij}(\omega)$, $i, j = 1, 2, \ldots, n$, are real-valued random variables then $T(\omega)$ denotes a random operator with $\mathrm{D}(T) = \mathbb{R}^n$. For the averaged operator $\langle T(\omega)\rangle$ we have

$$\langle T(\omega)\rangle = \big(\langle a_{ij}(\omega)\rangle\big)_{1\leqq i,j\leqq n}\,, \qquad \mathrm{D}(\langle T\rangle) = \mathbb{R}^n$$

if $\langle a_{ij}(\omega)\rangle < \infty$ for $i, j = 1, 2, \ldots, n$.

Example 1.2. Let $a_i(x, \omega)$, $i = 0, 1, \ldots, n$, be measurable stochastic processes with

$$\int_a^b \langle a_i^2(x, \omega)\rangle\, \mathrm{d}x < \infty \quad \text{for} \quad i = 0, 1, \ldots, n\,.$$

Then the differential operator

$$T(\omega) \doteq \sum_{i=0}^{n} a_i(x, \omega) \frac{\mathrm{d}^i}{\mathrm{d}x^i} \tag{1.5}$$

is a random operator as a mapping of $\Omega \times \mathbf{L}_2(a, b)$ to $\mathbf{L}_2(a, b)$ where we have $\mathrm{D}(T) = \mathbf{C}^n[a, b]$. The averaged operator is given by

$$\langle T(\omega)\rangle = \sum_{i=0}^{n} \langle a_i(x, \omega)\rangle \frac{\mathrm{d}^i}{\mathrm{d}x^i}.$$

Example 1.3. Let $K(x, y, \omega)$ be a random field where

$$\int_a^b \int_a^b K^2(x, y, \omega)\, \mathrm{d}x\, \mathrm{d}y \leqq M(\omega) < \infty \quad \text{a.s.}$$

and

$$\int_a^b \int_a^b \langle K^2(x, y, \omega)\rangle\, \mathrm{d}x\, \mathrm{d}y < \infty .$$

The operator $T(\omega)$ defined by

$$T(\omega)\, u(x) \doteq \int_a^b K(x, y, \omega)\, u(y)\, \mathrm{d}y$$

is a random operator with $\mathrm{D}(T) = \mathbf{L}_2(a, b)$. We have the averaged operator

$$\langle T(\omega)\rangle\, u(x) = \int_a^b \langle K(x, y, \omega)\rangle\, u(y)\, \mathrm{d}y$$

and its domain $\mathrm{D}(\langle T\rangle) = \mathbf{L}_2(a, b)$.

Let $T(\omega)$ be a linear operator on a Banach space $\mathscr{X}$ to itself, and let $\mathbb{C}$ denote the space of complex numbers.

Definition 1.8. The set of values of $(\omega, \lambda) \in \Omega \times \mathbb{C}$ for which the range of $(\lambda I - T(\omega))$ is dense in $\mathscr{X}$, $(\lambda I - T(\omega))^{-1}$ exists and is continuous forms the *resolvent set* $\varrho(T(\omega))$ of the operator $T(\omega)$. The complement of $\varrho(T(\omega))$ is called the *spectrum* $\sigma(T(\omega)) = \Omega \times \mathbb{C} \setminus \varrho(T(\omega))$ of the operator $T(\omega)$.

We state the following theorem.

Theorem 1.5. *Let $T(\omega)$ be a $\mathfrak{L}(\mathscr{X})$-valued random variable, where $\mathfrak{L}(\mathscr{X})$ denotes the collection of all bounded linear operators on $\mathscr{X}$ into $\mathscr{X}$. Then for every $\lambda \in \mathbb{C}$ the set $\{\omega : (\omega, \lambda) \in \varrho(T(\omega))\}$ is measurable, that is*

$$\{\omega : (\omega, \lambda) \in \varrho(T(\omega))\} \in \mathfrak{A} .$$

As deterministic spectral theory shows, the spectrum of a random operator $T(\omega)$ can be divided into mutually exclusive components:

$$\sigma(T(\omega)) = \mathsf{C}(T(\omega)) \cup \mathsf{R}(T(\omega)) \cup \mathsf{P}(T(\omega)) .$$

The *continuous spectrum* $\mathsf{C}(T(\omega))$ is formed by all values of $(\omega, \lambda) \in \Omega \times \mathbb{C}$ for which range $(\lambda I - T(\omega))$ is dense in $\mathscr{X}$, $(\lambda I - T(\omega))^{-1}$ exists but is not continuous. The values of $(\omega, \lambda) \in \Omega \times \mathbb{C}$ for which the range $(\lambda I - T(\omega))$ is not dense in $\mathscr{X}$ and $(\lambda I - T(\omega))^{-1}$ exists form the *residual spectrum* $\mathsf{R}(T(\omega))$. The values (ω, λ) of $\Omega \times \mathbb{C}$ for which $(\lambda I - T(\omega))^{-1}$ does not exist form the *point spectrum* $\mathsf{P}(T(\omega))$ of $T(\omega)$.

The spectral theory of random operators is in its early stages of development, hence only a few general results are known. An important result in this theory was proved by Ryll-Nardzewski [1]. Results of the spectral theory of random operators are given in the book of Bharucha-Reid [3]. In this book we will deal with random operators which only have a point spectrum $\mathsf{P}(T(\omega)) = \sigma(T(\omega))$.

1.2. Measurability of eigenvalues and eigenelements

First we state the following theorem which is important for this section.

Theorem 1.6. *Let $T(\omega)$ be a compact, self-adjoint, random operator on a separable Hilbert space $\mathscr{H}$. Then two sequences of eigenvalues*

$$\lambda_1^- \leqq \lambda_2^- \leqq \dots \leqq 0\,,$$

$$\lambda_1^+ \geqq \lambda_2^+ \geqq \dots \geqq 0$$

exist almost surely, and these eigenvalues are random variables. In particular, the eigenvalues of a symmetric matrix ordered according to increasing magnitude are random variables.

Proof. A compact, self-adjoint random operator $T(\omega)$ possesses a pure point spectrum for almost all $\omega \in \Omega$. We also say in this case $T(\omega)$ has a discrete spectrum. Then the eigenvalues can be determined by the Courant's principle:

$$\lambda_g^-(\omega) = \sup_{\tilde{\mathscr{J}}_g} \inf_{\tilde{\mathscr{S}}_g(V)} ⟪T(\omega)\, u, u⟫\,, \tag{1.6}$$

$$\lambda_g^+(\omega) = \inf_{\tilde{\mathscr{J}}_g} \sup_{\tilde{\mathscr{S}}_g(V)} ⟪T(\omega)\, u, u⟫\,, \tag{1.7}$$

where

$$\tilde{\mathscr{J}}_g = \{V = (v_1, \dots, v_{g-1}) \colon v_i \in \mathscr{H},\ \{v_1, \dots, v_{g-1}\} \text{ linear independent}\}$$

$$\tilde{\mathscr{S}}_g(V) = \{u \in \mathscr{H} \colon \|u\| = 1\,,\ ⟪u, v_i⟫ = 0 \text{ for } i = 1, 2, \dots, g-1\}\,.$$

From the definition of a random operator it follows that the value $⟪T(\omega)\, u, u⟫$ denotes a random variable. The measurability of $\lambda_g^-(\omega)$ and $\lambda_g^+(\omega)$ is then obtained from well-known theorems of the measure theory if we can prove that the infimum and the supremum in (1.6), (1.7) can be taken over a countable number of ω independent sets. The following Lemma 1.1 establishes this. ◀

Lemma 1.1. *Let $T(\omega)$ be a compact, self-adjoint operator. The sets $\mathscr{H}_i$ denote arbitrarily countable sets dense in $\mathscr{H}$, $\mathscr{J}_g$ is defined by*

$$\mathscr{J}_g = \{V = (v_1, \dots, v_{g-1}) \colon v_i \in \mathscr{H}_i,\ \{v_1, \dots, v_{g-1}\} \textit{ linear independent}\},$$

and $\mathcal{S}_g(V)$ *is a countable set being dense in* $\tilde{\mathcal{S}}_g(V)$. *Then*

$$\lambda_g^-(\omega) = \sup_{\mathcal{J}_g} \inf_{\mathcal{S}_g(V)} \langle\!\langle T(\omega)\, u, u \rangle\!\rangle\,, \tag{1.8}$$

$$\lambda_g^+(\omega) = \inf_{\mathcal{J}_g} \sup_{\mathcal{S}_g(V)} \langle\!\langle T(\omega)\, u, u \rangle\!\rangle\,. \tag{1.9}$$

Proof. It is sufficient to prove (1.8) since the proof of (1.9) is analogous. At first, we give the proof for $g = 2$. The case $g = 1$ is contained in this proof Then the equation

$$\sup_{\tilde{\mathcal{J}}_2} \inf_{\tilde{\mathcal{S}}_2(v)} \langle\!\langle T(\omega)\, u, u \rangle\!\rangle = \sup_{\mathcal{J}_2} \inf_{\mathcal{S}_2(v)} \langle\!\langle T(\omega)\, u, u \rangle\!\rangle \tag{1.10}$$

has to be established where

$\mathcal{J}_2$ denotes a countable set which is dense in $\mathcal{H}$,

$\tilde{\mathcal{S}}_2(v) = \{u \in \mathcal{H} \colon ||u|| = 1\,,\ \langle\!\langle u, v \rangle\!\rangle = 0\}$ and

$\mathcal{S}_2(v)$ is a countable set which is dense in $\tilde{\mathcal{S}}_2(v)$.

The sets $\mathcal{S}_2(v)$ and $\mathcal{J}_2$ are independent of $T(\omega)$ and we can proceed with fixed sets $\mathcal{J}_2$ and $\mathcal{S}_2(v)$ for every realization $T(\omega_0)$ of $T(\omega)$. Therefore, we do not write the argument ω of T. To prove (1.10) we use the following Lemma 1.2.

Lemma 1.2. *For every* v, $v_0 \in \mathcal{H}$, $||v|| = ||v_0|| = 1$, *with* $||v - v_0|| < \varepsilon$ *and* $u \in \mathcal{H}$, $||u|| = 1$, *with* $\langle\!\langle u, v \rangle\!\rangle = 0$ *an element* u_0 *from* $\mathcal{H}$ *exists having the properties:*

$$\langle\!\langle u_0, v_0 \rangle\!\rangle = 0\,, \qquad ||u_0|| = 1$$

and $||u - u_0|| < \tilde{\varepsilon}(\varepsilon)$ *where* $\lim_{\varepsilon \downarrow 0} \tilde{\varepsilon}(\varepsilon) = 0$.

Proof. We put

$$\tilde{u}_0 = u - \langle\!\langle u, v_0 \rangle\!\rangle\, v_0 \quad \text{and} \quad u_0 = \frac{\tilde{u}_0}{||\tilde{u}_0||}$$

then

$$\langle\!\langle u_0, v_0 \rangle\!\rangle = 0 \quad \text{and} \quad ||u_0|| = 1\,.$$

Furthermore, we have

$$||\tilde{u}_0 - u|| = |\langle\!\langle u, v_0 \rangle\!\rangle|\, ||v_0|| = |\langle\!\langle u, v_0 - v \rangle\!\rangle| < \varepsilon\,,$$

$$||\tilde{u}_0||^2 = ||u||^2 - \langle\!\langle u, v_0 \rangle\!\rangle^2 = 1 - \langle\!\langle u, v_0 - v \rangle\!\rangle^2;$$

and from this

$$\sqrt{1 - \varepsilon^2} \leqq ||\tilde{u}_0|| \leqq 1\,.$$

The above inequality gives

$$||u_0 - u|| = \frac{1}{||\tilde{u}_0||}\, ||\tilde{u}_0 - ||\tilde{u}_0||\, u|| \leqq \frac{1}{||\tilde{u}_0||}\, (||\tilde{u}_0 - u|| + 1 - ||\tilde{u}_0||)$$

$$\leqq \frac{1 + \varepsilon}{\sqrt{1 - \varepsilon^2}} - 1 \doteq \tilde{\varepsilon}(\varepsilon)\,,$$

and Lemma 1.2 is proved. ◀

To establish (1.10) we now prove the relation

$$\sup_{\tilde{\mathcal{J}}_2} \inf_{\tilde{\mathcal{S}}_2(v)} \langle\!\langle Tu, u \rangle\!\rangle = \sup_{\mathcal{J}_2} \inf_{\tilde{\mathcal{S}}_2(v)} \langle\!\langle Tu, u \rangle\!\rangle . \tag{1.11}$$

We assume that

$$a \doteq \sup_{\mathcal{J}_2} \inf_{\tilde{\mathcal{S}}_2(v)} \langle\!\langle Tu, u \rangle\!\rangle < \sup_{\tilde{\mathcal{J}}_2} \inf_{\tilde{\mathcal{S}}_2(v)} \langle\!\langle Tu, u \rangle\!\rangle \doteq a + b ,$$

where $b > 0$. Then from a property of the supremum the inequality

$$\inf_{\tilde{\mathcal{S}}_2(v)} \langle\!\langle Tu, u \rangle\!\rangle \leqq a < a + b$$

for all $v \in \mathcal{J}_2$ follows; and there exists a $v_0 \in \tilde{\mathcal{J}}_2$ such that

$$\inf_{\tilde{\mathcal{S}}_2(v)} \langle\!\langle Tu, u \rangle\!\rangle \leqq a < a + \frac{b}{2} \leqq \inf_{\tilde{\mathcal{S}}_2(v_0)} \langle\!\langle Tu, u \rangle\!\rangle . \tag{1.12}$$

The set $\mathcal{J}_2$ is dense in $\tilde{\mathcal{J}}_2$ so that an element $v_1 \in \mathcal{J}_2$ exists where $||v_1 - v_0|| < \varepsilon$. Because of a property of the infimum we can find an element $u_1 \in \tilde{\mathcal{S}}_2(v_1)$ such that

$$\inf_{\tilde{\mathcal{S}}_2(v_1)} \langle\!\langle Tu, u \rangle\!\rangle \leqq \langle\!\langle Tu_1, u_1 \rangle\!\rangle < a + \frac{b}{4} .$$

Using Lemma 1.2, an element $u_0 \in \tilde{\mathcal{S}}_2(v_0)$ for $u_1 \in \tilde{\mathcal{S}}_2(v_1)$ with $||u_0 - u_1|| < \tilde{\varepsilon}(\varepsilon)$ can be found. We get the inequality

$$\begin{aligned} |\langle\!\langle Tu_1, u_1 \rangle\!\rangle - \langle\!\langle Tu_0, u_0 \rangle\!\rangle| &\leqq |\langle\!\langle T(u_1 - u_0), u_1 \rangle\!\rangle + \langle\!\langle Tu_0, u_1 - u_0 \rangle\!\rangle| \\ &\leqq 2||T||\, \tilde{\varepsilon}(\varepsilon) < \frac{b}{4} \end{aligned}$$

for small ε; and therefore

$$\langle\!\langle Tu_0, u_0 \rangle\!\rangle \leqq \langle\!\langle Tu_1, u_1 \rangle\!\rangle + [\langle\!\langle Tu_0, u_0 \rangle\!\rangle - \langle\!\langle Tu_1, u_1 \rangle\!\rangle] < a + \frac{b}{2} .$$

This inequality is a contradiction to the inequality (1.12), and (1.11) is proved.

To finish the proof of (1.10) we show that

$$A \doteq \inf_{\tilde{\mathcal{S}}_2(v)} \langle\!\langle Tu, u \rangle\!\rangle = \inf_{\mathcal{S}_2(v)} \langle\!\langle Tu, u \rangle\!\rangle \doteq B .$$

Let $A < B \leqq \langle\!\langle Tu, u \rangle\!\rangle$ hold for all $u \in \mathcal{S}_2(v)$. Then an element $u_0 \in \tilde{\mathcal{S}}_2(v)$ exists such that

$$A < \langle\!\langle Tu_0, u_0 \rangle\!\rangle < B . \tag{1.13}$$

Since $\mathcal{S}_2(v)$ is dense in $\tilde{\mathcal{S}}_2(v)$, we have a sequence (u_n), $u_n \in \mathcal{S}_2(v)$, with $\lim\limits_{n\to\infty} u_n = u^0$ and

$$B \leqq \lim_{n\to\infty} \langle\!\langle Tu_n, u_n \rangle\!\rangle = \langle\!\langle Tu_0, u_0 \rangle\!\rangle .$$

This contradiction to inequality (1.13) proves (1.10).

To complete the proof of Lemma 1.1 we use a lemma analogous to Lemma 1.2 for $g > 2$.

Lemma 1.3. *For* $V_0 = (v_{01}, \dots, v_{0,g-1})$, $V = (v_1, \dots, v_{g-1})$ *with* $v_{0i}, v_i \in \mathcal{H}$, $\|v_{0i}\| = \|v_i\| = 1$, $\|v_{0i} - v_i\| < \varepsilon$ *for* $i = 1, 2, \dots, g-1$ *and* $\{v_{01}, \dots, v_{0,g-1}\}$, $\{v_1, \dots, v_{g-1}\}$ *linear independent, respectively, and* $u \in \mathcal{H}$ *with* $\|u\| = 1$, $⟪u, v_i⟫ = 0$ *for* $i = 1, 2, \dots, g-1$, *an element* $u_0 \in \mathcal{H}$ *exists which has the properties:*

$$⟪u_0, v_{0i}⟫ = 0 \quad \text{for} \quad i = 1, 2, \dots, g-1\,, \qquad \|u_0\| = 1\,,$$

and

$$\|u_0 - u\| < \tilde{\varepsilon}(\varepsilon)\,, \quad \text{where} \quad \lim_{\varepsilon \downarrow 0} \tilde{\varepsilon}(\varepsilon) = 0\,.$$

Proof. We put

$$\tilde{u}_0 = u - \sum_{i=1}^{g-1} r_i v_{0i}\,, \qquad u_0 = \frac{\tilde{u}_0}{\|\tilde{u}_0\|}$$

where $\{r_i\}$ denotes the solution of the system of equations

$$\sum_{i=1}^{g-1} r_i ⟪v_{0i}, v_{0l}⟫ = ⟪u, v_{0l}⟫\,, \qquad l = 1, 2, \dots, g-1\,.$$

This solution $\{r_i\}$ is unique since, from the independence of the $\{v_{0i}\}$ the relation $\det(⟪v_{0i}, v_{0l}⟫) \neq 0$ is obtained. That the element u_0 satisfies the conditions $⟪u_0, v_{0i}⟫ = 0$ for $i = 1, 2, \dots, g-1$ and $\|u_0\| = 1$ is evident. Furthermore, we have

$$\|\tilde{u}_0 - u\| = \Big\|\sum_{i=1}^{g-1} r_i v_{0i}\Big\| \leqq \sum_{i=1}^{g-1} |r_i| \leqq c(v_{01}, \dots, v_{0,g-1})\,\varepsilon$$

because

$$|r_i| = \left|\frac{\begin{vmatrix} \dots ⟪u, v_{01}⟫ \dots \\ \vdots \\ \dots ⟪u, v_{0,g-1}⟫ \dots \end{vmatrix}}{\det(⟪v_{0r}, v_{0s}⟫)}\right| = \left|\frac{\begin{vmatrix} \dots ⟪u, v_{01} - v_1⟫ \dots \\ \vdots \\ \dots ⟪u, v_{0,g-1} - v_{g-1}⟫ \dots \end{vmatrix}}{\det(⟪v_{0r}, v_{0s}⟫)}\right| \leqq c_i(v_{01}, \dots, v_{0,g-1})\,\varepsilon\,,$$

and

$$\|\tilde{u}_0\|^2 = 1 - 2\sum_{i=1}^{g-1} r_i ⟪v_{0i} - v_i, u⟫ + \sum_{i,j=1}^{g-1} r_i r_j ⟪v_{0i}, v_{0j}⟫$$

yields $\big|\, \|\tilde{u}_0\| - 1 \,\big| \leqq c_0 \varepsilon$.

Now, the norm $\|u - u_0\|$ can be estimated as in the proof of Lemma 1.2 if we note that for small ε the norm of $\tilde{u}_0$ is greater than $1/2$. Lemma 1.3 is now proved. ◀

By means of Lemma 1.3 the relation

$$\sup_{\tilde{\mathcal{J}}_g} \inf_{\tilde{\mathcal{S}}_g(V)} ⟪Tu, u⟫ = \sup_{\mathcal{J}_g} \inf_{\mathcal{S}_g(V)} ⟪Tu, u⟫$$

can be shown as in the case of $g = 2$. Hence, Lemma 1.1 and also Theorem 1.1 are proved. ◀

We have already noted that from Theorem 1.1 the measurability of the eigenvalues of a symmetric random matrix

$$A(\omega) = \big(a_{ik}(\omega)\big)_{1 \leqq i,\, k \leqq n}$$

follows if we order the eigenvalues according to increasing magnitude, $\lambda_1(\omega) \leqq \lambda_2(\omega) \leqq \dots \leqq \lambda_n(\omega)$. We now show that the components of the eigenvectors of such matrices are measurable. The eigenvectors $u = (u_1, \dots, u_n)^{\mathsf{T}}$ are calculated from the system of linear equations

$$\big(A(\omega) - \lambda(\omega)\, I\big)\, u = 0$$

and they are functions of ω.

Theorem 1.7. *The components of the eigenvectors of a symmetric random matrix are random variables if the parameters appearing in the vectors are assumed to be measurable.*

Proof. Since the measurability of the eigenvalues $\lambda_g(\omega)$ has been proved the matrix $\big(b_{ij}(\omega)\big) \doteq \big(a_{ij}(\omega) - \lambda_g(\omega)\, \delta_{ij}\big)$ denotes a random matrix for every g. Therefore, we now consider (in general) a random matrix $B(\omega) = \big(b_{ij}(\omega)\big)_{1 \leqq i,\, j \leqq n}$ which is not zero almost surely. We will establish the measurability of the vector $u(\omega) = \big(u_1(\omega), \dots, u_n(\omega)\big)^{\mathsf{T}}$ where $u(\omega)$ denotes the solution of the system of linear equations

$$\sum_{j=1}^{n} b_{ij}(\omega)\, u_j = 0\,, \qquad i = 1, 2, \dots, n\,.$$

The set A_k is defined as follows:

$$A_k \doteq \{\omega\colon \operatorname{rank} B(\omega) = k\}\,, \quad \text{for} \quad k = 1, 2, \dots, n\,.$$

Since the rank of a matrix is determined by rational operations, the sets A_k are measurable. We have

$$\mathsf{P}\big(\Omega \setminus \bigcup_{k=1}^{n} A_k\big) = 0$$

because the matrix $B(\omega)$ only vanishes on a set with probability zero. For a fixed k the set A_k will be decomposed into sets $C^k_{i_1 \dots i_k}$. The set $C^k_{i_1 \dots i_k}$ is formed by all $\omega \in A_k$ for which the Gaussian elimination applied to $B(\omega)\, u = 0$, with the i_1-th, i_2-th, ..., i_k-th row as the successive pivot rows, can be realized. Thus

$$A_k = \bigcup C^k_{i_1 \dots i_k}$$

where the union is taken over all subsets $\{i_1, \dots, i_k\}$ from $\{1, \dots, n\}$ with the condition $i_1 < i_2 < \dots < i_k$. The sets $C^k_{i_1 \dots i_k}$ are defined by inequalities involving random variables; and therefore these sets are measurable. Now, those sets $C^k_{i_1 \dots i_k}$ are omitted for which $\mathsf{P}(C^k_{i_1 \dots i_k}) = 0$ and the remaining sets are denoted by $C^k_1, \dots, C^k_{m_k}$. For all $\omega \in C^k_i$ we obtain a system of linear equations

$$\begin{aligned}
\hat{u}_1 + \hat{b}_{12}\hat{u}_2 + \dots + \hat{b}_{1k}\hat{u}_k + \dots + \hat{b}_{1n}\hat{u}_n &= 0\,,\\
\hat{u}_2 + \dots + \hat{b}_{2k}\hat{u}_k + \dots + \hat{b}_{2n}\hat{u}_n &= 0\,,\\
&\ \vdots\\
\hat{u}_k + \dots + \hat{b}_{kn}\hat{u}_n &= 0\,.
\end{aligned}$$

The vector $(\hat{u}_1, \dots, \hat{u}_n)^\mathsf{T}$ is a permutation of $(u_1, \dots, u_n)^\mathsf{T}$ and the $\hat{b}_{ij}$ denote random variables since they are computed by rational operations from the random variables b_{ij}. We have

$$\mathsf{P}(A_k \setminus \bigcup_{i=1}^{m_k} C_i^k) = 0 .$$

If the values $\hat{u}_{k+1}, \dots, \hat{u}_n$ are chosen as measurable functions of ω, then the vector $(u_1, \dots, u_n)^\mathsf{T}$ is measurable on C_i^k, that is

$$\{\omega : u_1 < x_1, \dots, u_n < x_n\} \cap C_i^k \in \mathfrak{A} .$$

Thus

$$\begin{aligned}\{\omega : u_1 < x_1, \dots, u_n < x_n\} &= \bigcup_{k=1}^{n} [\{\omega : u_1 < x_1, \dots, u_n < x_n\} \cap A_k] \\ &= \bigcup_{k=1}^{n} [\bigcup_{i=1}^{m_k} [\{\omega : u_1 < x_1, \dots, u_n < x_n\} \cap C_i^k]] \\ &\in \mathfrak{A} .\end{aligned}$$

The measurability is preserved if the set of the eigenvectors is transformed to a set of orthogonal eigenvectors. ◀

The proof is also valid for an nonhomogeneous system of linear equations if the nonhomogeneous terms are random variables. For systems of linear equations $A(\omega)\, x = b(\omega)$, with $\mathsf{P}(\det A(\omega) \neq 0) = 1$, the measurability of the solution was proved by Nake [1] with the aid of Cramer's rule.

In the following we consider the class of symmetric, positive definite operators on a separable Hilbert space $\mathcal{H}$. Let T belong to this class. The completion of the domain $\mathrm{D}(T)$ relative to the metric which is defined by $(\!(x, y)\!)_T \doteq (\!(Tx, y)\!)$ is called the *energetic space* $\mathcal{H}_T$ of the operator T. The norm $\|x\|_T \doteq \sqrt{(\!(x, x)\!)_T}$ is said to be the *energetic norm*. An element $x \in \mathcal{H}_T$ is called a *generalized eigenelement* relative to the eigenvalue λ of the symmetric, positive definite operator T if for all $y \in \mathcal{H}_T$ the relation $(\!(x, y)\!)_T = \lambda (\!(x, y)\!)$ is satisfied. These generalized eigenvalues and eigenelements are the usual eigenvalues and eigenelements of the *Friedrich's extension* $\tilde{T}$ of T. The element x denotes a usual eigenelement of the operator T associated with the eigenvalue λ if $x \in \mathrm{D}(T)$, i.e. we have $Tx = \lambda x$.

A symmetric, positive definite operator T on $\mathcal{H}$ ($\mathcal{H}$ is infinite-dimensional) is said to have a *generalized discrete spectrum*, if

(1) the operator T has an infinite sequence of generalized eigenvalues which have no finite points of accumulation,

$$0 < \lambda_1 \leqq \lambda_2 \leqq \dots \leqq \lambda_n \leqq \dots , \qquad \text{and}$$

(2) the sequence (x_g) of the generalized eigenelements is dense in $\mathcal{H}$.

A symmetric, positive definite operator T on $\mathcal{H}$ has a generalized discrete spectrum if every set bounded in the energetic norm is compact in the norm of $\mathcal{H}$ (cf. Michlin [1]; Zeidler [2]). The eigenelements are also dense in the space $\mathcal{H}_T$ (cf. also Section 2.2.1).

Given a positive definite operator T on $\mathcal{H}$, an operator $G: \mathcal{H} \to \mathcal{H}$ can be defined as follows: For $f \in \mathcal{H}$ and $u \in \mathcal{H}_T$ a linear bounded functional on $\mathcal{H}_T$ is introduced by $(\!(u, f)\!)$. Using the Riesz lemma there exists a unique element $u_0 \in \mathcal{H}_R$ where

$$(\!(u, f)\!) = (\!(u, u_0)\!)_T \quad \text{for all} \quad u \in \mathcal{H}_T .$$

Then the equation $u_0 = Gf$ defines a linear operator on $\mathcal{H}$. This operator G is is symmetric, positive, and bounded and we have $G^{-1} = \tilde{T}$. G is compact if T has the property that every set bounded in the energetic norm is compact in the norm of $\mathcal{H}$. Then the eigenvalues of G are computed as the inverses of the generalized eigenvalues of T. The connection between G and T is used for the proof of the following theorem.

Theorem 1.8. *Let $T(\omega)$ be a positive definite, symmetric, random operator on a separable Hilbert space $\mathcal{H}$, $T(\omega)$: $\Omega \times \mathrm{D}(T) \to \mathcal{H}$. Let $\langle T(\omega)\rangle$ exist for all $x \in \mathrm{D}(T)$ and $\mathcal{H}_{T(\omega)} = \mathcal{H}_{\langle T\rangle}$ almost surely. Furthermore, every set bounded in the energetic norm $||.||_{\langle T\rangle}$ is supposed to be compact in the norm $||.||$ of the Hilbert space $\mathcal{H}$. Then the operator $T(\omega)$ possesses almost surely a generalized, discrete, random spectrum.*

Proof. First, it follows from the condition that for almost all ω every set bounded in the energetic norm $||.||_{T(\omega)}$ is compact in the norm $||.||$. Hence, $T(\omega)$ possesses almost surely a generalized discrete spectrum. The measurability of the eigenvalues $\lambda_k(\omega)$ must still be shown. The operator $G(\omega)$ associated with $T(\omega)$ is almost surely compact, and because $G^{-1}(\omega) = \tilde{T}(\omega)$ it is a random operator on $\mathcal{H}$ (cf. Hanš [3]). Using Theorem 1.6 its eigenvalues $1/\lambda_g(\omega)$ are measurable and hence also $\lambda_g(\omega)$. ◀

The measurability of eigenelements depends on parameters as in case of the eigenvectors of a random matrix. Let $\{u_g(\omega)\}$ be a system of measurable eigenelements of $T(\omega)$, and let $\varkappa_g(\omega)$ be scalar functions which are not measurable. Then $\{\varkappa_g(\omega)\, u_g(\omega)\}$ denotes a system of eigenelements, but this system is in generally not measurable. Using special conditions a system of measurable eigenelements can be found.

Theorem 1.9. *Let the random operator $T(\omega)$ satisfy the conditions of Theorem 1.8. The generalized eigenvalues $\lambda_g(\omega)$ are assumed to be simple. Then a complete system $\{u_i\}$ of generalized eigenelements can be found where the $u_g(\omega)$ are random variables with values in $\mathcal{H}$.*

Proof. We use the fact that the linear random operator defined above, associated with the operator $T(\omega)$ is compact, symmetric, and positive definite. This operator $G(\omega)$ possesses the same eigenelements as $T(\omega)$. Hence, the eigenvalue problem

$$G(\omega)\, u = \mu u \tag{1.14}$$

is considered and we will establish the measurability of the eigenelements $u_i(\omega)$. The problem (1.14) is equivalent to the variational problem

$$\mu_g(\omega) = \max_{u \in \mathcal{X}_g} \langle\!\langle G(\omega)\, u, u \rangle\!\rangle , \tag{1.15}$$

where

$$\mathcal{X}_g = \{v \in \mathcal{H} : ||v|| = 1 , \langle\!\langle v, u_i \rangle\!\rangle = 0 \text{ for } i = 1, 2, \dots , g-1\} .$$

We use the method of Ritz (cf. Section 2.2.1). Let $\{w_k\}$ be a system of elements of $\mathcal{H}$ with the properties:

(1) Every finite subset of the system $\{w_g\}$ is always independent.
(2) $\mathcal{H} = \overline{\bigcup \mathcal{H}_k}$ where $\mathcal{H}_k = \text{lin}\, \{w_1, \dots , w_k\}$.

The system $\{w_g\}$ is assumed to be orthonormal. We now put

$$^n u = \sum_{k=1}^{n} {}^n c_k w_k , \qquad \langle\!\langle {}^n u, {}^n u \rangle\!\rangle = 1 , \tag{1.16}$$

and obtain the Ritz' eigenvalue problem

$$\sum_{k=1}^{n} \left(\langle\!\langle G(\omega)\, w_j, w_k \rangle\!\rangle - {}^n\mu \delta_{jk}\right) {}^n c_k = 0 , \qquad j = 1, 2, \dots , n . \tag{1.17}$$

The eigenvalues of this problem are denoted by ${}^n\mu_g$, and the eigenvectors associated with the eigenvalues by $({}^n c_{gk})^{\mathsf{T}}_{1 \leq k \leq n}$. By means of Theorem 1.7 the components ${}^n c_{gk}(\omega)$ of the eigenvectors are measurable. Hence, the elements

$$^n u_g(\omega) = \sum_{k=1}^{n} {}^n c_{gk} w_k$$

yield random elements with values in $\mathcal{H}$, since for all $v \in \mathcal{H}$

$$\langle\!\langle {}^n u_g(\omega), v \rangle\!\rangle = \sum_{k=1}^{n} {}^n c_{gk}(\omega) \langle\!\langle w_k, v \rangle\!\rangle$$

is a random variable. For the approximations ${}^n\mu_g(\omega)$ of the eigenvalues the convergence

$$\lim_{n\to\infty} {}^n\mu_g(\omega) = \mu_g(\omega) \quad \text{a.s.}$$

can be proved (cf. Zeidler [2]). The measurability of the eigenvalues $\mu_g(\omega)$ follows from this convergence with probability one so that we have a second proof for Theorem 1.8. Since $\mu_g(\omega)$ was assumed to be simple, let the unique eigensolution of (1.14) be $u_g(\omega)$. Assume $({}^n u_g(\omega))$ is the sequence of Ritz approximate solution of (1.16), (1.17) belonging to $({}^n\mu_g)$.

We now consider an arbitrary subsequence $({}^{n'} u_g(\omega))$ of the sequence $({}^n u_g(\omega))$, and let ω be a fixed element of Ω. Because $||{}^{n'} u_g(\omega)|| = 1$, a subsequence $({}^{n''} u_g)$ exists which converges weakly to an element u. It can be proved that u is an eigenelement associated with the eigenvalue μ_g. Hence, we have the weak convergence to u_g; that is,

$$^{n''} u_g(\omega) \rightharpoonup u = u_g ,$$

since the eigenvalue μ_g was assumed to be simple. Because of the compactness of $G(\omega)$ the strong convergence

$$G(\omega)\, {}^{n''}u_g(\omega) \to G(\omega)\, u_g(\omega) = \mu_g(\omega)\, u_g(\omega)$$

follows and then

$${}^{n''}u_g(\omega) \to u_g(\omega)$$

(cf. ZEIDLER [2]). We remark that the subsequence $({}^{n''}u_g)$ can be dependent on ω.

Hence, from every subsequence of $\left({}^{n}u_g(\omega)\right)$ a subsequence can be found which converges in $\mathcal{H}$ to $u_g(\omega)$. Then we have

$$\lim_{n\to\infty} \|{}^{n}u_g(\omega) - u_g(\omega)\| = 0 \quad \text{a.s.},$$

and $u_g(\omega)$ is measurable in $\mathcal{H}$ if we use a theorem of HANŠ [2]. ◀

We will now give an example of a random operator which satisfies the conditions of Theorem 1.8 and Theorem 1.9. Consider the random Sturm-Liouville operator

$$T(\omega)\, u \doteq -\frac{\mathrm{d}}{\mathrm{d}x}\left(p(x, \omega)\frac{\mathrm{d}u}{\mathrm{d}x}\right) + q(x, \omega)\, u$$

for $a \leqq x \leqq b$, with boundary conditions

$$u(a) = \alpha u'(b) + \beta u(b) = 0\,.$$

In the above $p(x, \omega)$ and $q(x, \omega)$ are stochastic processes. The derivatives in the definition of the operator $T(\omega)$ are understood to be sample functions derivatives. For $p(x, \omega)$ and $q(x, \omega)$ we suppose the following conditions:

(1) $p(x, \omega) \geqq p_0(\omega) > 0$, $q(x, \omega) \geqq 0$ a.s.; $\alpha, \beta > 0$.

In case of $q(x, \omega) \geqq -c(\omega)$ a.s. we can consider the eigenvalue problem

$$\tilde{T}(\omega)\, u \doteq T(\omega)\, u + c(\omega)\, u = \left(\lambda + c(\omega)\right) u \doteq \tilde{\lambda} u\,.$$

(2) $p(x, \omega)$, $p'(x, \omega)$, $q(x, \omega)$ are almost surely sample continuous.

(3) $\langle p(x, \omega)\rangle'$, $\langle q(x, \omega)\rangle$ are continuous.

(4) The correlation matrix $R(x, x) \doteq \langle r^{\mathrm{T}}(x)\, r(x)\rangle$ exists and the elements are continuous functions where

$$r(x, \omega) \doteq \left(p(x, \omega), p'(x, \omega), q(x, \omega)\right).$$

Thus

$$\mathrm{D}\left(T(\omega)\right) = \{u \in \mathbf{C}^2[a, b] \colon u(a) = \alpha u'(b) + \beta u(b) = 0\}$$

and we obtain

$$T(\omega) \colon \Omega \times \mathrm{D}(T) \to \mathbf{L}_2(a, b)$$

since the condition (2) implies the relation $T(\omega)\, u \in \mathbf{L}_2(a, b)$ for all $u \in \mathrm{D}(T)$. From condition (4) it follows that

$$\langle \|T(\omega)\, u\|\rangle \leqq [\int_a^b \langle \left(T(\omega)\, u\right)^2\rangle\, \mathrm{d}x]^{1/2}$$

for every $u \in D(T)$, so that the averaged operator $\langle T(\omega)\rangle$ exists whith $D(\langle T\rangle) = D(T)$. With the aid of $\langle\langle\!\langle T(\omega)\, u, v\rangle\!\rangle\rangle = \langle\!\langle\langle T(\omega)\rangle\, u, v\rangle\!\rangle$ we observe that

$$\langle\!\langle \langle T(\omega)\rangle\, u, v\rangle\!\rangle = \left\langle \langle\!\langle pu', v'\rangle\!\rangle + \langle\!\langle qu, v\rangle\!\rangle + p(b, \omega)\frac{\beta}{\alpha}\, u(b)\, v(b)\right\rangle$$
$$= \langle\!\langle -(\langle p\rangle\, u')' + \langle q\rangle\, u, v\rangle\!\rangle$$

for $u, v \in D(T)$; and from this we have

$$\langle T(\omega)\rangle\, u = -\frac{\mathrm{d}}{\mathrm{d}x}\left(\langle p(x, \omega)\rangle \frac{\mathrm{d}u}{\mathrm{d}x}\right) + \langle q(x, \omega)\rangle\, u\,.$$

The symmetry of the operator $T(\omega)$ is clear, and from the estimate

$$||u||^2_{T(\omega)} = \int_a^b \big(p(x, \omega)\, u'(x)^2 + q(x, \omega)\, u^2(x)\big)\, \mathrm{d}x + p(b, \omega)\frac{\beta}{\alpha}\, u^2(b)$$
$$\geqq p_0(\omega) \int_a^b u'(x)^2\, \mathrm{d}x \geqq \frac{p_0(\omega)}{(b-a)^2}\, ||u||^2$$

it follows that $T(\omega)$ is positive definite. Hence we have $\mathcal{H}_{T(\omega)} = \mathcal{H}_{\langle T\rangle}$ (cf. MICHLIN [2]).

Let $\mathcal{M}$ denote a set which is bounded in the energetic metric of $\langle T\rangle$, that is

$$\langle p_0(\omega)\rangle \int_a^b u'(x)^2\, \mathrm{d}x \leqq ||u||^2_{\langle T\rangle} < C$$

for all $u \in \mathcal{M}$. Because

$$u(x) = \int_a^b K(x, t)\, u'(t)\, \mathrm{d}t$$

where

$$K(x, t) = \begin{cases} 1 & \text{for} \quad a \leqq t \leqq x\,, \\ 0 & \text{for} \quad x < t \leqq b\,, \end{cases}$$

and

$$\int_a^b\int_a^b K^2(x, t)\, \mathrm{d}t\, \mathrm{d}x < \infty$$

the set $\mathcal{M}$ is compact in the norm of $L_2(a, b)$.

Hence, applying Theorem 1.8 the operator $T(\omega)$ has a generalized discrete spectrum and the eigenvalues are random variables. These eigenvalues can be shown to be simple; and because of Theorem 1.9 a complete system of generalized eigenelements $\{u_g\}$ can be found which are random variables with values in $L_2(a, b)$.

1.3. Calculation of the moments for eigenvalues and eigenvectors of random matrices

Every random matrix $A(\omega) = \big(\alpha_{ij}(\omega)\big)_{1\leqq i,j\leqq n}$ can be written as the sum of a deterministic matrix A_0 and a random matrix $B(\omega)$ with $\langle B(\omega)\rangle = 0$ in the following form:

$$A(\omega) = A_0 + B(\omega)$$

where

$$A_0 \doteq \langle A(\omega) \rangle = (\langle \alpha_{ij} \rangle)_{1 \leqq i, j \leqq n} \doteq (a_{ij})_{1 \leqq i, j \leqq n}$$

and

$$B(\omega) \doteq A(\omega) - \langle A(\omega) \rangle = (\alpha_{ij} - \langle \alpha_{ij} \rangle)_{1 \leqq i, j \leqq n} \doteq \big(b_{ij}(\omega)\big)_{1 \leqq i, j \leqq n} .$$

In this section we will assume that the matrix $B(\omega)$ represents a random fluctuation or random perturbation of A_0 and we require

$$|B(\omega)| = \Big(\sum_{i,j=1}^{n} b_{ij}^2(\omega) \Big)^{1/2} < \varepsilon \quad \text{a.s.,} \tag{1.18}$$

where $\varepsilon > 0$ is sufficiently small.

1.3.1. Randomly perturbed symmetric matrices with simple eigenvalues of the unperturbed matrix

Consider a symmetric matrix A_0 which has the simple eigenvalues

$$\mu_1 < \mu_2 < \dots < \mu_n .$$

The matrix A_0 is perturbed by a random matrix $B(\omega)$ where $\langle B(\omega) \rangle = 0$ and the condition (1.18) is satisfied for a suitable $\varepsilon > 0$. Let

$$\lambda_1(\omega) < \lambda_2(\omega) < \dots < \lambda_n(\omega)$$

be the eigenvalues of $A_0 + B(\omega)$, and ${}^g x(\omega)$, $g = 1, 2, \dots, n$, the orthonormal eigenvectors.

We will consider the approximate calculation of the expectation and the correlation relations of the eigenvalues $\lambda_g(\omega)$ and the eigenvectors ${}^g x(\omega)$. First, we notice that for an almost surely bounded random operator $T(\omega)$ the existence of the k-th moment of the eigenvalues $\lambda_g(\omega)$ follows from the existence of the k-th moment of $||T(\omega)||$. This statement follows from the inequality

$$|\lambda_g(\omega)| \leqq ||T(\omega)|| \quad \text{a.s.}$$

In particular, from

$$|A(\omega)| = \Big(\sum_{i,j=1}^{n} a_{ij}^2(\omega) \Big)^{1/2}$$

and

$$\Big\langle \Big(\sum_{i,j=1}^{n} a_{ij}^2(\omega) \Big)^{1/2} \Big\rangle \leqq \Big(\sum_{i,j=1}^{n} \langle a_{ij}^2(\omega) \rangle \Big)^{1/2}$$

for a random matrix $A(\omega) = \big(a_{ij}(\omega)\big)_{1 \leqq i, j \leqq n}$ it follows that the existence of the first two moments of the eigenvalues $\lambda_g(\omega)$ of the random matrix $A(\omega)$ is implied by the existence of the first two moments of the a_{ij}.

For further considerations the transformation of A_0 to a diagonal matrix turns out to be advantageous. Let $C = (c_{ij})_{1 \leqq i, j \leqq n}$ denote the matrix which produces this transformation, that is,

$$C A_0 C^{\mathsf{T}} \doteq \hat{A}_0 = (\mu_i \delta_{ij})_{1 \leqq i, j \leqq n} \,.$$

Put

$$C B(\omega)\, C^{\mathsf{T}} \doteq \hat{B}(\omega)\,.$$

Then the mean of $\hat{B}$ is zero and for the elements $\hat{b}_{ij}$ we have

$$\hat{b}_{ij}(\omega) = \sum_{r, s=1}^{n} c_{ir} c_{js} b_{rs}(\omega) \doteq c_i^r c_j^s b_{rs}(\omega)\,. \tag{1.19}$$

The matrix $\hat{A}_0 + \hat{B}(\omega)$ has the same eigenvalues $\lambda_l(\omega)$ as the matrix $A_0 + B(\omega)$. If the normalized eigenvectors of $\hat{A}_0 + \hat{B}(\omega)$ are denoted by ${}^g y(\omega)$ then these vectors can be computed by the normalized eigenvectors ${}^g x(\omega)$ of $A_0 + B(\omega)$; that is

$${}^g y(\omega) = C\, {}^g x(\omega)\,, \qquad {}^g x(\omega) = C^{\mathsf{T}}\, {}^g y(\omega)\,.$$

The eigenvalues $\lambda_g(\omega)$ and the eigenvectors ${}^g y(\omega)$ can be calculated from the system of random linear equations

$$(\hat{A}_0 + \hat{B}(\omega) - \lambda_g I)\, {}^g y = 0 \tag{1.20}$$

in which I denotes the identity matrix. The eigenvectors ${}^g y$ are normalized by the equation $(\!({}^g y, {}^g y)\!) = 1$. For these eigenvalues $\lambda_g(\omega)$ and the normalized eigenvectors ${}^g y(\omega)$ we assume that the expansions

$$\lambda_g(\omega) = \mu_g - \sum_{k=1}^{\infty} \lambda_{gk}(\omega)\,, \tag{1.21}$$

$${}^g y(\omega) = \sum_{k=0}^{\infty} {}^g y_k(\omega)$$

exist, where the μ_g are the eigenvalues of the matrix A_0 and ${}^g y_0 = (0, \dots, 0, \overset{g}{1}, 0, \dots, 0)^{\mathsf{T}}$ the normalized eigenvectors associated with μ_g. $\lambda_{gk}(\omega)$, ${}^g y_k(\omega)$ denote the terms of λ_g and ${}^g y$, respectively, which are homogeneous of the k-th order in the perturbations $\hat{b}_{ij}$ for $k = 1, 2, \dots$

From the analytical dependence of λ_g and ${}^g y$ on the elements $\hat{b}_{ij}$, it can be shown that the expansions (1.21) converge at least for $\hat{b}_{ij}$ sufficiently small (cf. Section 2.2.2).

We now state a theorem which deals with the calculation of the homogeneous terms λ_{gk} and ${}^g y_k$ for $k = 1, 2, 3, 4$.

Theorem 1.10. *The homogeneous terms $\lambda_{gk}(\omega)$ and ${}^g y_k(\omega)$ up to the fourth order can be given in the following form:*

$$\lambda_{g1} = -\hat{b}_{gg}\,,$$

$$\lambda_{g2} = \sum_{\substack{i=1 \\ i \neq g}}^{n} \frac{\hat{b}_{gi} \hat{b}_{ig}}{\mu_{ig}}\,,$$

$$\lambda_{g3} = - \sum_{\substack{i,j=1\\ i,j\neq g}}^{n} \frac{\hat{b}_{gi}\hat{b}_{ij}\hat{b}_{jg}}{\mu_{ig}\mu_{jg}} + \hat{b}_{gg} \sum_{\substack{i=1\\ i\neq g}}^{n} \frac{\hat{b}_{gi}\hat{b}_{ig}}{\mu_{ig}^2},$$

$$\lambda_{g4} = \sum_{\substack{i,j,k=1\\ i,j,k\neq g}}^{n} \frac{\hat{b}_{gi}\hat{b}_{ij}\hat{b}_{jk}\hat{b}_{kg}}{\mu_{ig}\mu_{jg}\mu_{kg}} - \sum_{\substack{i,j=1\\ i,j\neq g}}^{n} \frac{\hat{b}_{gi}\hat{b}_{ig}\hat{b}_{gj}\hat{b}_{jg}}{\mu_{ig}^2\mu_{jg}}$$

$$- \hat{b}_{gg} \sum_{\substack{i,j=1\\ i,j\neq g}}^{n} \frac{1}{\mu_{ig}\mu_{jg}}\left(\frac{1}{\mu_{ig}} + \frac{1}{\mu_{jg}}\right)\hat{b}_{gi}\hat{b}_{ij}\hat{b}_{jg} + \hat{b}_{gg}^2 \sum_{\substack{i=1\\ i\neq g}}^{n} \frac{\hat{b}_{gi}\hat{b}_{ig}}{\mu_{ig}^3},$$

$${}^g y_1 = {}^g z_1 ,$$

$${}^g y_k = {}^g z_k - \tfrac{1}{2} \sum_{q=2}^{k} {}^g z_{k-q} \left[\sum_{r=1}^{q-1} (\!({}^g y_r, {}^g y_{q-r})\!) \right], \qquad \text{for} \quad k = 2, 3, 4 ,$$

where $\mu_{ig} \doteq \mu_i - \mu_g$. *The vectors* ${}^g z_k = ({}^g z_{k1}, {}^g z_{k2}, \ldots, {}^g z_{kn})^{\mathsf{T}} = ({}^g z_{ki})^{\mathsf{T}}_{1\leq i\leq n}$ *can be determined by the perturbations* $\hat{b}_{ij}$:

$${}^g z_1 \colon {}^g z_{1i} = -\frac{\hat{b}_{ig}}{\mu_{ig}} \quad \textit{for} \quad i \neq g; \qquad {}^g z_{1g} = 0 ,$$

$${}^g z_2 \colon {}^g z_{2i} = \frac{1}{\mu_{ig}}\left[\sum_{\substack{j=1\\ j\neq g}}^{n} \frac{\hat{b}_{ij}\hat{b}_{jg}}{\mu_{jg}} - \frac{\hat{b}_{gg}\hat{b}_{ig}}{\mu_{ig}} \right] \quad \textit{for} \quad i \neq g; \qquad {}^g z_{2g} = 0 ,$$

$${}^g z_3 \colon {}^g z_{3i} = \frac{1}{\mu_{ig}}\left[- \sum_{\substack{j,k=1\\ j,k\neq g}}^{n} \frac{\hat{b}_{ij}\hat{b}_{jk}\hat{b}_{kg}}{\mu_{jg}\mu_{kg}} + \sum_{\substack{j=1\\ j\neq g}}^{n} \frac{\hat{b}_{gj}\hat{b}_{jg}\hat{b}_{ig}}{\mu_{ig}\mu_{jg}} - \frac{\hat{b}_{gg}^2\hat{b}_{ig}}{\mu_{ig}^2} \right.$$

$$\left. + \hat{b}_{gg} \sum_{\substack{j=1\\ j\neq g}}^{n} \frac{1}{\mu_{jg}}\left(\frac{1}{\mu_{ig}} + \frac{1}{\mu_{jg}}\right)\hat{b}_{ij}\hat{b}_{jg} \right] \quad \textit{for} \quad i \neq g; \qquad {}^g z_{3g} = 0 ,$$

$${}^g z_4 \colon {}^g z_{4i} = \frac{1}{\mu_{ig}}\left[\sum_{\substack{j,k,m=1\\ j,k,m\neq g}}^{n} \frac{\hat{b}_{ij}\hat{b}_{jk}\hat{b}_{km}\hat{b}_{mg}}{\mu_{jg}\mu_{kg}\mu_{mg}} - \sum_{\substack{j,k=1\\ j,k\neq g}}^{n} \left(\frac{1}{\mu_{jg}} + \frac{1}{\mu_{ig}}\right) \frac{\hat{b}_{ij}\hat{b}_{jg}\hat{b}_{gk}\hat{b}_{kg}}{\mu_{jg}\mu_{kg}} \right.$$

$$- \hat{b}_{ig} \sum_{\substack{j,k=1\\ j,k\neq g}}^{n} \frac{\hat{b}_{gj}\hat{b}_{jk}\hat{b}_{kg}}{\mu_{ig}\mu_{jg}\mu_{kg}} - \hat{b}_{gg} \sum_{\substack{j,k=1\\ j,k\neq g}}^{n} \left(\frac{1}{\mu_{jg}} + \frac{1}{\mu_{kg}} + \frac{1}{\mu_{ig}}\right) \frac{\hat{b}_{ij}\hat{b}_{jk}\hat{b}_{kg}}{\mu_{jg}\mu_{kg}}$$

$$+ \hat{b}_{ig}\hat{b}_{gg} \sum_{\substack{j=1\\ j\neq g}}^{n} \left(\frac{1}{\mu_{jg}} + \frac{2}{\mu_{ig}}\right) \frac{\hat{b}_{gj}\hat{b}_{jg}}{\mu_{ig}\mu_{jg}} - \frac{\hat{b}_{gg}^3\hat{b}_{ig}}{\mu_{ig}^3}$$

$$\left. + \hat{b}_{gg}^2 \sum_{\substack{j=1\\ j\neq g}}^{n} \left(\frac{1}{\mu_{jg}^2} + \frac{1}{\mu_{jg}\mu_{ig}} + \frac{1}{\mu_{ig}^2}\right) \frac{\hat{b}_{ij}\hat{b}_{jg}}{\mu_{jg}} \right] \quad \textit{for} \quad i \neq g; \quad {}^g z_{4q} = 0 .$$

Proof. To obtain the homogeneous terms we substitute these expressions (1.21) into the eigenvalue problem (1.20); that is

$$(\hat{A}_0 - \mu_g I)\, {}^g y_0 + [(\hat{A}_0 - \mu_g I)\, {}^g y_1 + (\hat{B} + \lambda_{g1} I)\, {}^g y_0]$$
$$+ \sum_{k=2}^{\infty} [(\hat{A}_0 - \mu_g I)\, {}^g y_k + (\hat{B} + \lambda_{g1} I)\, {}^g y_{k-1} + \sum_{s=2}^{k} \lambda_{gs}\, {}^g y_{k-s}] = 0 ,$$

and pick out the homogeneous terms, which are supposed to vanish separately. Then we obtain the following equations:

$$k = 0: \qquad (\hat{A}_0 - \mu_g I)\, {}^g y_0 = 0 , \tag{1.22}$$
$$k = 1: \qquad (\hat{A}_0 - \mu_g I)\, {}^g y_1 = -(\hat{B} + \lambda_{g1})\, {}^g y_0 \tag{1.23}$$
$$k = 2, 3, \ldots: \qquad (\hat{A}_0 - \mu_g I)\, {}^g y_k = -(\hat{B} + \lambda_{g1} I)\, {}^g y_{k-1} - \sum_{s=2}^{k} \lambda_{gs}\, {}^g y_{k-s} . \tag{1.24}$$

Equation (1.22) is satisfied by the given ${}^g y_0$. From the condition of solvability of (1.23) we obtain

$$\langle\!\langle (\hat{B} + \lambda_{g1} I)\, {}^g y_0, {}^g y_0 \rangle\!\rangle = 0 \quad \text{or} \quad \lambda_{g1} = -\hat{b}_{gg} .$$

The scalar product of $\mathbb{R}^n$ is denoted by $\langle\!\langle .\,, . \rangle\!\rangle$ and the norm by $|.|$. If we put the arbitrary parameter in the g-th component of ${}^g y_1$ equal to zero, then by means of the solution of (1.23) we obtain the vector ${}^g y_1$. Now we determine solutions of (1.24) for $k = 2, 3, \ldots$ having the property that the g-th component vanishes. These solutions are denoted by ${}^g z_k$ for $k = 0, 1, 2, \ldots$ where ${}^g z_0 = {}^g y_0$, ${}^g z_1 = {}^g y_1$. λ_{gk} can be obtained by forming the scalar product of the right-hand side of (1.24) with ${}^g y_0$. This results in

$$\lambda_{gk} = - \sum_{j=1}^{n} \hat{b}_{gj}\, {}^g z_{k-1, j} \tag{1.25}$$

and hence the solution ${}^g z_k$ of (1.24) gives

$$\begin{aligned} {}^g z_{ki} &= -\frac{1}{\mu_{ig}} [\sum_{j=1}^{n} \hat{b}_{ij}\, {}^g z_{k-1, j} + \sum_{j=0}^{k-1} \lambda_{g, k-j}\, {}^g z_{ji}] \quad \text{for} \quad i \neq g , \\ {}^g z_{kg} &= 0 \end{aligned} \tag{1.26}$$

if ${}^g z_{pg} = 0$ for $p \leqq k$ is assumed. Then the expressions for λ_{gk} and ${}^g z_k$ can be computed by the successive application of (1.25) and (1.26).

The eigenvectors ${}^g z = \sum_{k=0}^{\infty} {}^g z_k$ are normalized by the condition $\langle\!\langle {}^g z, {}^g z_0 \rangle\!\rangle = 1$, because

$$\langle\!\langle {}^g z_0, {}^g z_0 \rangle\!\rangle = 1 \quad \text{and} \quad \langle\!\langle {}^g z_k, {}^g z_0 \rangle\!\rangle = 0 \quad \text{for} \quad k \geqq 1 .$$

To obtain ${}^g y = \sum_{k=0}^{\infty} {}^g y_k$ from ${}^g z$ where ${}^g y$ is normalized by $\langle\!\langle {}^g y, {}^g y \rangle\!\rangle = 1$ we put

$${}^g y_k = \sum_{p=0}^{k} a_{k-p}\, {}^g z_p . \tag{1.27}$$

These vectors ${}^g y_k$ for $k = 0, 1, 2, \dots$, satisfy (1.22), (1.23), and (1.24), respectively. The values a_p are determined so that

$$1 = \langle\!\langle {}^g y, {}^g y \rangle\!\rangle = \langle\!\langle \sum_{k=0}^{\infty} {}^g y_k, \sum_{k=0}^{\infty} {}^g y_k \rangle\!\rangle = \sum_{k=0}^{\infty} \sum_{p=0}^{k} \langle\!\langle {}^g y_p, {}^g y_{k-p} \rangle\!\rangle ;$$

and we obtain

$$\langle\!\langle {}^g y_0, {}^g y_0 \rangle\!\rangle = 1$$

and

$$\sum_{p=0}^{k} \langle\!\langle {}^g y_p, {}^g y_{k-p} \rangle\!\rangle = 0 \quad \text{for} \quad k \geqq 1 .$$

From these equations the values a_k can be calculated successively. The equation

$$\langle\!\langle {}^g y_k, {}^g y_0 \rangle\!\rangle = -\tfrac{1}{2} \sum_{p=1}^{k-1} \langle\!\langle {}^g y_p, {}^g y_{k-p} \rangle\!\rangle$$

follows from $\sum_{p=0}^{k} \langle\!\langle {}^g y_p, {}^g y_{k-p} \rangle\!\rangle = 0$. Because of (1.27), and the normalization of ${}^g z$ we compute

$$\langle\!\langle {}^g y_k, {}^g y_0 \rangle\!\rangle = a_k \langle\!\langle {}^g z_0, {}^g z_0 \rangle\!\rangle + \sum_{p=1}^{k} a_{k-p} \langle\!\langle {}^g z_p, {}^g z_0 \rangle\!\rangle = a_k$$

and

$$a_1 = 0 , \qquad a_k = -\tfrac{1}{2} \sum_{p=1}^{k-1} \langle\!\langle {}^g y_p, {}^g y_{k-p} \rangle\!\rangle \quad \text{for} \quad k \geqq 2 . \blacktriangleleft$$

In particular, we calculate

$$\begin{aligned} {}^g y_2 &= {}^g z_2 + a_2 {}^g z_0 , \\ {}^g y_3 &= {}^g z_3 + a_3 {}^g z_0 + a_2 {}^g z_1 , \\ {}^g y_4 &= {}^g z_4 + a_4 {}^g z_0 + a_3 {}^g z_1 + a_2 {}^g z_2 , \end{aligned}$$

where

$$\begin{aligned} a_2 &= -\tfrac{1}{2} \langle\!\langle {}^g y_1, {}^g y_1 \rangle\!\rangle = -\tfrac{1}{2} \langle\!\langle {}^g z_1, {}^g z_1 \rangle\!\rangle , \\ a_3 &= -\langle\!\langle {}^g y_1, {}^g y_2 \rangle\!\rangle = -\langle\!\langle {}^g z_1, {}^g z_2 \rangle\!\rangle , \\ a_4 &= -\tfrac{1}{2} [2\langle\!\langle {}^g y_1, {}^g y_3 \rangle\!\rangle + \langle\!\langle {}^g y_2, {}^g y_2 \rangle\!\rangle] = -\tfrac{1}{2} [2\langle\!\langle {}^g z_1, {}^g z_3 \rangle\!\rangle + \langle\!\langle {}^g z_2, {}^g z_2 \rangle\!\rangle \\ &\quad - \tfrac{3}{4} \langle\!\langle {}^g z_1, {}^g z_1 \rangle\!\rangle^2] . \end{aligned}$$

The m-th approximations

$$\lambda_g^{(m)} = \mu_g - \sum_{k=1}^{m} \lambda_{gk}$$

and

$${}^g y^{(m)} = \sum_{k=0}^{m} {}^g y_k$$

satisfy the equation

$$(\hat{A}_0 + \hat{B} - \lambda_g I) \, {}^g y = 0$$

and the normalization condition $\langle\!\langle {}^g y, {}^g y \rangle\!\rangle = 1$, up to terms of the $(m + 1)$-th order in the perturbations $\hat{b}_{ij}$.

By means of the expressions given in Theorem 1.10 the expectations and the correlation relations of the eigenvalues and the eigenvectors can be determined approximately. For the means we obtain

$$\langle \lambda_g(\omega) \rangle = \mu_g - \langle \lambda_{g2} \rangle - \langle \lambda_{g3} \rangle - \langle \lambda_{g4} \rangle - \dots$$

and

$$\langle {}^g y(\omega) \rangle = {}^g y_0 + \langle {}^g y_2 \rangle + \langle {}^g y_3 \rangle + \langle {}^g y_4 \rangle + \dots ,$$

because $\langle \hat{b}_{ij} \rangle = 0$. If the correlations between the $\hat{b}_{ij}$ are only given then the approximations up to terms of second order can be calculated:

$$\langle \lambda_g(\omega) \rangle = \mu_g - \sum_{\substack{j=1 \\ j \neq g}}^{n} \frac{\langle \hat{b}_{jg} \hat{b}_{gj} \rangle}{\mu_{jg}} , \tag{1.28}$$

$$\langle {}^g y(\omega) \rangle = (1 + \langle a_2 \rangle)\, {}^g y_0 + \langle {}^g z_2 \rangle . \tag{1.29}$$

Put ${}^g y \doteq ({}^g Y_1, {}^g Y_2, \dots, {}^g Y_n)^{\mathsf{T}}$ then the components are determined by

$$\langle {}^g Y_i \rangle = \frac{1}{\mu_{ig}} \left[\sum_{\substack{j=1 \\ j \neq g}}^{n} \frac{\langle \hat{b}_{ij} \hat{b}_{jg} \rangle}{\mu_{jg}} - \frac{\langle \hat{b}_{gg} \hat{b}_{ig} \rangle}{\mu_{ig}} \right] \quad \text{for} \quad i \neq g ,$$

$$\langle {}^g Y_g \rangle = 1 - \frac{1}{2} \sum_{\substack{j=1 \\ j \neq g}}^{n} \frac{\langle \hat{b}_{jg}^2 \rangle}{\mu_{jg}^2} . \tag{1.30}$$

Using (1.28) and (1.30) we can also compute an approximate solution of the so-called averaging problem for the eigenvalues and the eigenvectors up to terms of the second order in the perturbations. This is a problem of importance in the theory of random equations. If the random quantities in a given random problem (algebraic equations, operator equations, eigenvalue problems, and others) are substituted by their expected values then we get the averaged problem. In our case because $\langle \hat{B} \rangle = 0$, the averaged problem is $(\hat{A}_0 - \mu I)\, y = 0$ where μ_g denote the eigenvalues and ${}^g y_0$ denotes the eigenvectors. The averaging problem concerns the calculation of the difference between the expected solution of the eigenvalue problem and the deterministic solution of the averaged problem. Hence, we must compute the differences

$$\langle \lambda_g \rangle - \mu_g , \qquad \langle {}^g y \rangle - {}^g y_0 ,$$

which follow immediately from (1.28) and (1.30) up to terms of second order. If the moments of the $\hat{b}_{ij}$ up to the fourth order are assumed to exist, then from Theorem 1.10 we can determine the means of λ_g and ${}^g y$ up to the fourth order; and also the solution of the averaging problem up to this order.

For the smallest eigenvalue λ_1 and the greatest eigenvalue λ_n an interesting inequality is obtained from (1.28) for symmetric perturbation matrices. Since the term

$$\sum_{\substack{j=1 \\ j \neq g}}^{n} \frac{\langle \hat{b}_{jg} \hat{b}_{gj} \rangle}{\mu_{jg}}$$

dominates the terms of a higher order, and

$$\langle \lambda_1 \rangle \approx \mu_1 - \sum_{j=2}^{n} \frac{\langle \hat{b}_{j1}^2 \rangle}{\mu_{j1}} \quad \text{where} \quad \mu_{j1} = \mu_j - \mu_1 > 0 \, ,$$

$$\langle \lambda_n \rangle \approx \mu_n - \sum_{j=1}^{n-1} \frac{\langle \hat{b}_{jn}^2 \rangle}{\mu_{jn}} \quad \text{where} \quad \mu_{jn} = \mu_j - \mu_n < 0$$

hold for a symmetric matrix $\hat{B}$, we can derive

$$\langle \lambda_1 \rangle \leqq \mu_1 \, , \tag{1.31}$$

$$\langle \lambda_n \rangle \geqq \mu_n \, . \tag{1.32}$$

The inequalities (1.31), (1.32) are proved more generally for compact symmetric operators by means of Courant's principle and Fatou's lemma without complications.

Using the expansions of Theorem 1.10, the correlation relations of the eigenvalues and the eigenvectors can be calculated approximately. We obtain the correlation relations of the eigenvalues and the eigenvectors up to the $(k+1)$-th order in the perturbations $\hat{b}_{ij}$ if the expansions are used up to the k-th order, $k = 1, 2, 3, 4$. Thus, the correlations R_{gh} of the eigenvalues are given by

$$\begin{aligned} R_{gh} &\doteq \langle \big(\lambda_g(\omega) - \langle \lambda_g(\omega) \rangle\big) \big(\lambda_h(\omega) - \langle \lambda_h(\omega) \rangle\big) \rangle \\ &= \langle (-\lambda_{g1} - \lambda_{g2} - \lambda_{g3} + \langle \lambda_{g2} \rangle + \langle \lambda_{g3} \rangle) \\ &\quad \times (-\lambda_{h1} - \lambda_{h2} - \lambda_{h3} + \langle \lambda_{h2} \rangle + \langle \lambda_{h3} \rangle) \rangle \end{aligned}$$

if the expansions are considered up to third order. We notice that $\langle \hat{b}_{ij} \rangle = 0$, and we only write terms up to fourth order so that we can calculate

$$\begin{aligned} R_{gh} &= \langle \lambda_{g1} \lambda_{h1} \rangle + \langle \lambda_{g2} \lambda_{h1} + \lambda_{h2} \lambda_{g1} \rangle + \langle \lambda_{g3} \lambda_{h1} + \lambda_{h3} \lambda_{g1} \rangle \\ &\qquad + \langle (\lambda_{g2} - \langle \lambda_{g2} \rangle)(\lambda_{h2} - \langle \lambda_{h2} \rangle) \rangle \\ &= \langle \hat{b}_{gg} \hat{b}_{hh} \rangle - \sum_{\substack{p=1 \\ p \neq g}}^{n} \frac{1}{\mu_{pg}} \langle \hat{b}_{pg} \hat{b}_{gp} \hat{b}_{hh} \rangle - \sum_{\substack{p=1 \\ p \neq h}}^{n} \frac{1}{\mu_{ph}} \langle \hat{b}_{ph} \hat{b}_{hp} \hat{b}_{gg} \rangle \\ &\qquad - \sum_{\substack{p=1 \\ p \neq g}}^{n} \frac{1}{\mu_{pg}^2} \langle \hat{b}_{gp} \hat{b}_{pg} \hat{b}_{gg} \hat{b}_{hh} \rangle - \sum_{\substack{p=1 \\ p \neq h}}^{n} \frac{1}{\mu_{ph}^2} \langle \hat{b}_{hp} \hat{b}_{ph} \hat{b}_{gg} \hat{b}_{hh} \rangle \\ &\qquad + \sum_{\substack{p,\, q=1 \\ p,\, q \neq g}}^{n} \frac{1}{\mu_{pg} \mu_{qg}} \langle \hat{b}_{gp} \hat{b}_{pq} \hat{b}_{qg} \hat{b}_{hh} \rangle + \sum_{\substack{p,\, q=1 \\ p,\, q \neq h}}^{n} \frac{1}{\mu_{ph} \mu_{qh}} \langle \hat{b}_{hp} \hat{b}_{pq} \hat{b}_{qh} \hat{b}_{gg} \rangle \\ &\qquad + \sum_{\substack{p,\, q=1 \\ p \neq g,\, q \neq h}}^{n} \frac{1}{\mu_{pg} \mu_{qh}} \langle (\hat{b}_{pg} \hat{b}_{gp} - \langle \hat{b}_{pg} \hat{b}_{gp} \rangle)(\hat{b}_{qh} \hat{b}_{hq} - \langle \hat{b}_{qh} \hat{b}_{hq} \rangle) \rangle \, , \end{aligned}$$

$$R_{gh} = \langle \hat{b}_{gg}\hat{b}_{hh}\rangle - \sum_{\substack{p=1\\p\neq g}}^{n} \frac{1}{\mu_{pg}} \left\langle \hat{b}_{pg}\hat{b}_{gp}\hat{b}_{hh}\left(1 + \frac{\hat{b}_{gg}}{\mu_{pg}}\right)\right\rangle$$
$$- \sum_{\substack{p=1\\p\neq h}}^{n} \frac{1}{\mu_{ph}} \left\langle \hat{b}_{ph}\hat{b}_{hp}\hat{b}_{gg}\left(1 + \frac{\hat{b}_{hh}}{\mu_{ph}}\right)\right\rangle$$
$$+ \sum_{\substack{p,q=1\\p,q\neq g}}^{n} \frac{1}{\mu_{pg}\mu_{qg}} \langle \hat{b}_{gp}\hat{b}_{pq}\hat{b}_{qg}\hat{b}_{hh}\rangle + \sum_{\substack{p,q=1\\p,q\neq h}}^{n} \frac{1}{\mu_{ph}\mu_{qh}} \langle \hat{b}_{hp}\hat{b}_{pq}\hat{b}_{qh}\hat{b}_{gg}\rangle$$
$$+ \sum_{\substack{p,q=1\\p\neq g,\, q\neq h}}^{n} \frac{1}{\mu_{pg}\mu_{qh}} (\langle \hat{b}_{pg}\hat{b}_{gp}\hat{b}_{qh}\hat{b}_{hq}\rangle - \langle \hat{b}_{pg}\hat{b}_{gp}\rangle \langle \hat{b}_{qh}\hat{b}_{hq}\rangle)\,. \tag{1.33}$$

In particular, for the variance of $\lambda_g(\omega)$ we have

$$\operatorname{var}\lambda_g(\omega) = R_{gg} = \langle \hat{b}_{gg}^2\rangle - 2 \sum_{\substack{p=1\\p\neq g}}^{n} \frac{1}{\mu_{pg}} \left\langle \hat{b}_{pg}\hat{b}_{gp}\hat{b}_{gg}\left(1 + \frac{\hat{b}_{gg}}{\mu_{pg}}\right)\right\rangle$$
$$+ 2 \sum_{\substack{p,q=1\\p,q\neq g}}^{n} \frac{1}{\mu_{pg}\mu_{qg}} \left[\langle \hat{b}_{gp}\hat{b}_{pq}\hat{b}_{qg}\hat{b}_{gg}\rangle + \frac{1}{2}\langle \hat{b}_{pg}\hat{b}_{gp}\hat{b}_{qg}\hat{b}_{gq}\rangle \right.$$
$$\left. - \frac{1}{2}\langle \hat{b}_{pg}\hat{b}_{gp}\rangle \langle \hat{b}_{qg}\hat{b}_{gq}\rangle\right]. \tag{1.34}$$

Equations (1.33) and (1.34) are exact up to terms of fourth order in the perturbations $\hat{b}_{ij}$.

For the correlation matrix of the g-th and the h-th eigenvector we can derive

$$R^{gh} \doteq \langle ({}^g y - \langle {}^g y\rangle)\, ({}^h y - \langle {}^h y\rangle)^{\mathsf{T}}\rangle$$
$$= \left((1 - \delta_{ig})\,(1 - \delta_{jh})\,[\langle {}^g z_{1i}\, {}^h z_{1j}\rangle + \langle {}^g z_{1i}\, {}^h z_{2j}\rangle + \langle {}^g z_{2i}\, {}^h z_{1j}\rangle]\right)_{1\leqq i,\, j\leqq n}$$
$$+ (\delta_{jh}\langle {}^g z_{1i}\, {}^h a_2\rangle + \delta_{ig}\langle {}^h z_{1j}\, {}^g a_2\rangle)_{1\leqq i,\, j\leqq n}$$
$$= \left((1 - \delta_{ig})\,(1 - \delta_{jh})\,\frac{1}{\mu_{ig}\mu_{jh}} \left[\langle \hat{b}_{ig}\hat{b}_{jh}\rangle + \frac{\langle \hat{b}_{gg}\hat{b}_{ig}\hat{b}_{jh}\rangle}{\mu_{ig}} + \frac{\langle \hat{b}_{hh}\hat{b}_{jh}\hat{b}_{ig}\rangle}{\mu_{jh}}\right.\right.$$
$$\left.\left. - \sum_{\substack{p=1\\p\neq g,\, h}}^{n} \left(\frac{\langle \hat{b}_{ip}\hat{b}_{pg}\hat{b}_{jh}\rangle}{\mu_{pg}} + \frac{\langle \hat{b}_{jp}\hat{b}_{ph}\hat{b}_{ig}\rangle}{\mu_{ph}}\right)\right]\right)_{1\leqq i,\, j\leqq n}$$
$$+ \left(\frac{1}{2}\,\delta_{hj}(1 - \delta_{ig}) \sum_{\substack{p=1\\p\neq h}}^{n} \frac{\langle \hat{b}_{ph}^2\hat{b}_{ig}\rangle}{\mu_{ph}^2\mu_{ig}}\right.$$
$$\left. + \frac{1}{2}\,\delta_{gi}(1 - \delta_{jh}) \sum_{\substack{p=1\\p\neq g}}^{n} \frac{\langle \hat{b}_{pg}^2\hat{b}_{jh}\rangle}{\mu_{pg}^2\mu_{jh}}\right)_{1\leqq i,\, j\leqq n} \tag{1.35}$$

up to terms of third order. In particular, the correlation quantities of components of the g-th eigenvector emerge from this by

$$R_{ij}^{gg} \doteq \langle ({}^gY_i - \langle {}^gY_i\rangle)({}^gY_j - \langle {}^gY_j\rangle)\rangle$$
$$= (1-\delta_{ig})(1-\delta_{jg})\frac{1}{\mu_{ig}\mu_{jg}}\Bigg[\langle \hat{b}_{ig}\hat{b}_{jg}\rangle + \langle \hat{b}_{gg}\hat{b}_{jg}\hat{b}_{ig}\rangle\left(\frac{1}{\mu_{ig}}+\frac{1}{\mu_{jg}}\right) - \sum_{\substack{p=1\\p\neq g}}^{n}\frac{1}{\mu_{pg}}\langle \hat{b}_{pg}(\hat{b}_{ip}\hat{b}_{jg}+\hat{b}_{jp}\hat{b}_{ig})\rangle\Bigg]$$
$$+\frac{1}{2}\delta_{gj}(1-\delta_{ig})\sum_{\substack{p=1\\p\neq g}}^{n}\frac{\langle \hat{b}_{pg}^2\hat{b}_{ig}\rangle}{\mu_{pg}^2\mu_{ig}} + \frac{1}{2}\delta_{gi}(1-\delta_{jg})\sum_{\substack{p=1\\p\neq g}}^{n}\frac{\langle \hat{b}_{pg}^2\hat{b}_{jg}\rangle}{\mu_{pg}^2\mu_{jg}}\,; \quad (1.36)$$

and the variances of the components of the g-th eigenvector are given by

$$\operatorname{var}({}^gY_i) = R_{ii}^{gg} = \langle ({}^gY_i - \langle {}^gY_i\rangle)^2\rangle$$
$$= \frac{1}{\mu_{ig}^2}(1-\delta_{ig})\left[\langle \hat{b}_{ig}^2\rangle + 2\frac{\langle \hat{b}_{ig}^2\hat{b}_{gg}\rangle}{\mu_{ig}} - 2\sum_{\substack{p=1\\p\neq g}}^{n}\frac{\langle \hat{b}_{pg}\hat{b}_{ip}\hat{b}_{ig}\rangle}{\mu_{pg}}\right] \quad (1.37)$$

up to third order in each case.

To solve the original problem the given expansions for the $\lambda_g(\omega)$ and ${}^gy(\omega)$, respectively, we have to go back to the perturbations b_{ij} using (1.19). For instance, we can calculate, up to terms of second order,

$$\lambda_g(\omega) = \mu_g - c_g^p c_g^q b_{pq} + \sum_{\substack{j=1\\j\neq g}}^{n}\frac{1}{\mu_{jg}}c_g^p c_j^q c_j^r c_g^s b_{pq}b_{rs}\,,$$

$${}^gY_i(\omega) = -\frac{1}{\mu_{ig}}\left[c_i^p c_g^q b_{pq} + \sum_{\substack{j=1\\j\neq g}}^{n}\frac{1}{\mu_{jg}}c_i^p c_j^q c_j^r c_g^s b_{pq}b_{rs} - c_g^p c_g^q c_i^r c_g^s b_{pq}b_{rs}\right] \quad \text{for} \quad i \neq g\,,$$

$${}^gY_g(\omega) = 1 - \frac{1}{2}\sum_{\substack{j=1\\j\neq g}}^{n}\frac{1}{\mu_{jg}^2}c_j^p c_g^q c_j^r c_g^s b_{pq}b_{rs}\,.$$

Applying $\langle b_{ij}\rangle = 0$ and $\langle b_{pq}b_{rs}\rangle \doteq R_{pqrs}$ we have

$$\langle \lambda_g(\omega)\rangle = \sum_{\substack{j=1\\j\neq g}}^{n}\frac{1}{\mu_{jg}}c_g^p c_j^q c_j^r c_g^s R_{pqrs}\,,$$

$$\langle {}^gY_i(\omega)\rangle = -\frac{1}{\mu_{ig}}\left[\sum_{\substack{j=1\\j\neq g}}^{n}\frac{1}{\mu_{jg}}c_i^p c_j^q c_j^r c_g^s - c_g^p c_g^q c_i^r c_g^s\right]R_{pqrs} \quad \text{for} \quad i \neq g\,,$$

$$\langle {}^gY_g(\omega)\rangle = 1 - \frac{1}{2}\sum_{\substack{j=1\\j\neq 1}}^{n}\frac{1}{\mu_{jg}^2}c_j^p c_g^q c_j^r c_g^s R_{pqrs}\,,$$

in each case up to terms of second order.

For instance, if the perturbations b_{ij} are independent with $\langle b_{ij}^2\rangle = \sigma^2$ then the expressions

$$\langle\lambda_g(\omega)\rangle = \mu_g + \sigma^2 \sum_{\substack{j=1\\ j\neq g}}^{n} \frac{1}{\mu_{jg}} A_{gj}^2 ,$$

$$\langle{}^gY_i(\omega)\rangle = -\frac{\sigma^2}{\mu_{ig}}\left[\sum_{\substack{j=1\\ j\neq g}}^{n} \frac{1}{\mu_{jg}} A_{ij}A_{jg} - A_{gi}A_{gg}\right] \quad \text{for} \quad i \neq g ,$$

$$\langle{}^gY_g(\omega)\rangle = 1 - \frac{\sigma^2}{2} \sum_{\substack{j=1\\ j\neq g}}^{n} \frac{1}{\mu_{jg}^2} A_{jj}A_{gg}$$

are obtained since $R_{pqrs} = \sigma^2\delta_{pr}\delta_{qs}$ where

$$A_{ij} \doteq \sum_{p=1}^{n} c_i^p c_j^p .$$

The mean of the g-th eigenvector ${}^gx = ({}^gX_1, \dots, {}^gX_n)^\mathsf{T}$ of the matrix $A_0 + B(\omega)$ follows from $\langle{}^gx\rangle = C^\mathsf{T}\langle{}^gy\rangle$.

We will now consider the more general random eigenvalue problem

$$\big(A_0 + B(\omega)\big)\, x = \lambda\big(C_0 + D(\omega)\big)\, x , \tag{1.38}$$

where the eigenvectors are normalized by the condition

$$\langle\!\langle \big(C_0 + D(\omega)\big)\, x, x \rangle\!\rangle = 1 . \tag{1.39}$$

Let the unperturbed (deterministic) problem

$$A_0 x = \mu C_0 x$$

possess the simple eigenvalue μ_g. The matrices A_0 and C_0 are supposed to be symmetric and the eigenvectors ${}^lx_0$ to be orthonormal relative to C_0; that is

$$\langle\!\langle C_0\, {}^lx_0, {}^mx_0 \rangle\!\rangle = \delta_{lm} . \tag{1.40}$$

$B(\omega)$ and $D(\omega)$ are random matrices which satisfy condition (1.18) for a sufficiently small ε and $\langle B\rangle = \langle D\rangle = 0$.

The following expressions for the g-th eigenvalue $\lambda_g(\omega)$ and the g-th eigenvector ${}^gx(\omega)$

$$\begin{aligned} \lambda_g(\omega) &= \mu_g - \sum_{k=1}^{\infty} \lambda_{gk}(\omega) , \\ {}^gx(\omega) &= \sum_{k=0}^{\infty} {}^gx_k(\omega) \end{aligned} \tag{1.41}$$

are assumed, respectively, where ${}^gx_k(\omega)$ and $\lambda_{gk}(\omega)$ denote the homogeneous terms of the k-th order in the random perturbations $b_{ij}(\omega)$ and $d_{ij}(\omega)$. We substitute (1.41) in (1.38) and then it is possible to satisfy the system of linear equations (1.38) if the expressions of the k-th order vanish for $k = 0, 1, 2, \dots$

This leads to the following system:

$$k = 0: \qquad (A_0 - \mu_g C_0)\, {}^g x_0 = 0\,, \tag{1.42}$$

$$k = 1: \qquad (A_0 - \mu_g C_0)\, {}^g x_1 = (\mu_g D - B)\, {}^g x_0 - \lambda_{g1} C_0\, {}^g x_0\,, \tag{1.43}$$

$$k = 2, 3, \ldots: (A_0 - \mu_g C_0)\, {}^g x_k = (\mu_g D - B)\, {}^g x_{k-1} - \lambda_{gk} C_0\, {}^g x_0 - \sum_{j=1}^{k-1} \lambda_{gj} (C_0\, {}^g x_{k-j} + D\, {}^g x_{k-j-1})\,. \tag{1.44}$$

For $k = 0$ we have the unperturbed problem with the eigenvalues μ_l and the eigenvectors ${}^l x_0$. Now we solve (1.43). The condition of solvability is

$$⟪(\mu_g D - B)\, {}^g x_0 - \lambda_{g1} C_0\, {}^g x_0, {}^g x_0⟫ = 0\,,$$

and by means of $⟪C_0\, {}^g x_0, {}^g x_0⟫ = 1$ we get

$$\lambda_{g1} = ⟪(\mu_g D - B)\, {}^g x_0, {}^g x_0⟫\,. \tag{1.45}$$

We now determine the solution ${}^g x_1$ of (1.43). For this the operator T_g is defined by

$$T_g x = \sum_{\substack{k=1\\k\neq g}}^{n} \frac{⟪x, {}^k x_0⟫}{\mu_{kg}}\, {}^k x_0\,, \tag{1.46}$$

where this operator has the properties

$$T_g (A_0 - \mu_g C_0)\, x = x - ⟪x, C_0\, {}^g x_0⟫\, {}^g x_0\,,$$

$$(A_0 - \mu_g C_0)\, T_g\, x = x - ⟪x, {}^g x_0⟫\, C_0\, {}^g x_0$$

for all $x \in \mathbb{R}^n$.

If the operator T_g acts on (1.43) so that

$${}^g x_1 = T_g [(\mu_g D - B) - \lambda_{g1} C_0]\, {}^g x_0 + a_1\, {}^g x_0\,, \tag{1.47}$$

where a_1 is an arbitrary constant. This constant can be computed from the normalization condition given by (1.39). We substitute the expansion (1.41) for ${}^g x$ in (1.39) and obtain

$$⟪(C_0 + D) \sum_{k=0}^{\infty} {}^g x_k, \sum_{k=0}^{\infty} {}^g x_k⟫ = 1\,;$$

that is, arranged according to the homogeneous terms,

$$⟪C_0\, {}^g x_0, {}^g x_0⟫ + \sum_{k=1}^{\infty} \Big[\sum_{j=0}^{k} ⟪C_0\, {}^g x_{k-j}, {}^g x_j⟫ + \sum_{j=0}^{k-1} ⟪D\, {}^g x_{k-j-1}, {}^g x_j⟫\Big] = 1\,.$$

Thus

$$⟪C_0\, {}^g x_0, {}^g x_0⟫ = 1\,,$$

$$\sum_{j=0}^{k} ⟪C_0\, {}^g x_{k-j}, {}^g x_j⟫ + \sum_{j=0}^{k-1} ⟪D\, {}^g x_{k-j-1}, {}^g x_j⟫ = 0 \quad \text{for} \quad k = 1, 2, \ldots\,; \tag{1.48}$$

and for $k = 1$

$$⟪C_0\, {}^g x_1, {}^g x_0⟫ = -\tfrac{1}{2} ⟪D\, {}^g x_0, {}^g x_0⟫\,.$$

With the aid of (1.47) the constant a_1 can be calculated as

$$a_1 = -\langle\!\langle C_0\,{}^g x_0, T_g[\mu_g D - B - \lambda_{g1} C_0]\,{}^g x_0\rangle\!\rangle - \tfrac{1}{2}\langle\!\langle D\,{}^g x_0, {}^g x_0\rangle\!\rangle . \quad (1.49)$$

The matrix C_0 is symmetric and so

$$a_1 = -\langle\!\langle C_0 T_g[\mu_g D - B - \lambda_{g1} C_0]\,{}^g x_0, {}^g x_0\rangle\!\rangle - \tfrac{1}{2}\langle\!\langle D\,{}^g x_0, {}^g x_0\rangle\!\rangle .$$

For an arbitrary matrix F we can derive

$$\langle\!\langle C_0 T_g F\,{}^g x_0, {}^g x_0\rangle\!\rangle = \sum_{\substack{k=1\\k\neq g}}^{n} \frac{\langle\!\langle F\,{}^g x_0, {}^k x_0\rangle\!\rangle}{\mu_{kg}} \langle\!\langle C_0\,{}^g x_0, {}^g x_0\rangle\!\rangle = 0 \quad (1.50)$$

and

$$\langle\!\langle F T_g C_0\,{}^g x_0, {}^g x_0\rangle\!\rangle = \sum_{\substack{k=1\\k\neq g}}^{n} \frac{\langle\!\langle C_0\,{}^g x_0, {}^k x_0\rangle\!\rangle}{\mu_{kg}} \langle\!\langle F\,{}^k x_0, {}^g x_0\rangle\!\rangle = 0\,, \quad (1.51)$$

since the eigenvectors ${}^l x_0$ are orthonormal relative to C_0 (cf. (1.40)). Hence, we obtain

$$a_1 = -\tfrac{1}{2}\langle\!\langle D\,{}^g x_0, {}^g x_0\rangle\!\rangle . \quad (1.52)$$

We now suppose that the solutions $\mu_g, \lambda_{g1}, \dots, \lambda_{g,k-1}, {}^g x_0, {}^g x_1, \dots, {}^g x_{k-1}$ have been computed from (1.42), (1.43), (1.44) and the constants $a_1, a_2, \dots, a_{k-1}$ from (1.48). We will now determine the quantities λ_{gk}, ${}^g x_k$ and a_k.

The condition of solvability for (1.44) yields

$$\langle\!\langle (\mu_g D - B)\,{}^g x_{k-1} - \sum_{j=1}^{k-1} \lambda_{gj}(C_0\,{}^g x_{k-j} + D\,{}^g x_{k-j-1}), {}^g x_0\rangle\!\rangle$$
$$- \lambda_{gk}\langle\!\langle C_0\,{}^g x_0, {}^g x_0\rangle\!\rangle = 0$$

and then

$$\lambda_{gk} = \langle\!\langle (\mu_g D - B)\,{}^g x_{k-1}, {}^g x_0\rangle\!\rangle - \sum_{j=1}^{k-1} \lambda_{gj}\langle\!\langle C_0\,{}^g x_{k-j} + D\,{}^g x_{k-j-1}, {}^g x_0\rangle\!\rangle . \quad (1.53)$$

Put

$${}^g y_{k-1} \doteq (\mu_g D - B)\,{}^g x_{k-1} - \sum_{j=1}^{k-1} \lambda_{gj}(C_0\,{}^g x_{k-j} + D\,{}^g x_{k-j-1}) - \lambda_{gk} C_0\,{}^g x_0$$

with λ_{gk} from (1.53). Thus

$${}^g x_k = T_g\,{}^g y_{k-1} + a_k\,{}^g x_0 \quad (1.54)$$

from (1.44). The constant a_k can be calculated using (1.48), from which we obtain

$$\langle\!\langle C_0\,{}^g x_0, {}^g x_k\rangle\!\rangle = -\tfrac{1}{2}\Big(\sum_{j=1}^{k-1} \langle\!\langle C_0\,{}^g x_{k-j}, {}^g x_j\rangle\!\rangle + \sum_{j=0}^{k-1} \langle\!\langle D\,{}^g x_{k-j-1}, {}^g x_j\rangle\!\rangle\Big)$$

and

$$a_k = -\langle\!\langle C_0\,{}^g x_0, T_g\,{}^g y_{k-1}\rangle\!\rangle - \tfrac{1}{2}\Big(\sum_{j=1}^{k-1} \langle\!\langle C_0\,{}^g x_{k-j}, {}^g x_j\rangle\!\rangle$$
$$+ \sum_{j=0}^{k-1} \langle\!\langle D\,{}^g x_{k-j-1}, {}^g x_j\rangle\!\rangle\Big) \quad (1.55)$$

if relation (1.54) is used. On the right-hand side of (1.55) we can only find terms which were previously computed. Thus we have determined the desired quantities λ_{gk}, ${}^g x_k$, and a_k.

With the aid of (1.46) for T_g it is possible, in principle, that $\lambda_{gk}(\omega)$ and ${}^g x_k(\omega)$ can be expressed by ${}^l x_0$, λ_{l0} and the perturbation matrices $B(\omega)$, $D(\omega)$. The calculations are very extensive. As an example we will determine the mean of $\lambda_g(\omega)$ up to terms of second order.

Because $\langle \lambda_{g1} \rangle = 0$ we obtain

$$\langle \lambda_g(\omega) \rangle = \mu_g - \langle \lambda_{g2}(\omega) \rangle ,$$

and we now compute $\langle \lambda_{g2}(\omega) \rangle$. From (1.53), for $k = 2$, we have

$$\begin{aligned}\lambda_{g2} &= (\!((\mu_g D - B)\, {}^g x_1, {}^g x_0)\!) - \lambda_{g1} (\!(C_0\, {}^g x_1 + D\, {}^g x_0, {}^g x_0)\!) \\ &= (\!((\mu_g D - B - \lambda_{g1} C_0)\, {}^g x_1, {}^g x_0)\!) - \lambda_{g1} (\!(D\, {}^g x_0, {}^g x_0)\!) \\ &= (\!((\mu_g D - B - \lambda_{g1} C_0)\, T_g (\mu_g D - B - \lambda_{g1} C_0)\, {}^g x_0, {}^g x_0)\!) \\ &\quad + a_1 (\!((\mu_g D - B - \lambda_{g1} C_0)\, {}^g x_0, {}^g x_0)\!) - \lambda_{g1} (\!(D\, {}^g x_0, {}^g x_0)\!) .\end{aligned}$$

Using (1.45), (1.50), and (1.51) we have

$$\begin{aligned}\lambda_{g2} &= \mu_g^2 (\!(D T_g D\, {}^g x_0, {}^g x_0)\!) - \mu_g (\!((D T_g B + B T_g D)\, {}^g x_0, {}^g x_0)\!) \\ &\quad + (\!(B T_g B\, {}^g x_0, {}^g x_0)\!) - \mu_g (\!(D\, {}^g x_0, {}^g x_0)\!)^2 + (\!(B\, {}^g x_0, {}^g x_0)\!)(\!(D\, {}^g x_0, {}^g x_0)\!)\end{aligned}$$

and because of (1.46)

$$\begin{aligned}\lambda_{g2} = \sum_{\substack{k=1\\k \neq g}}^{n} \frac{1}{\mu_{kg}} &\left[\mu_g^2 (\!(D\, {}^g x_0, {}^k x_0)\!)(\!(D\, {}^k x_0, {}^g x_0)\!) + (\!(B\, {}^g x_0, {}^k x_0)\!)(\!(B\, {}^k x_0, {}^g x_0)\!) \right. \\ &\left. - \mu_g \{(\!(B\, {}^k x_0, {}^g x_0)\!)(\!(D\, {}^g x_0, {}^k x_0)\!) + (\!(B\, {}^g x_0, {}^k x_0)\!)(\!(D\, {}^k x_0, {}^g x_0)\!)\}\right] \\ &+ (\!(B\, {}^g x_0, {}^g x_0)\!)(\!(D\, {}^g x_0, {}^g x_0)\!) - \mu_g (\!(D\, {}^g x_0, {}^g x_0)\!)^2 .\end{aligned}$$

Put ${}^m x_0 = ({}^m X_1, {}^m X_2, \ldots , {}^m X_n)^{\mathsf{T}}$. Then we obtain

$$(\!(F\, {}^k x_0, {}^g x_0)\!)(\!(G\, {}^g x_0, {}^k x_0)\!) = \sum_{i,j,r,s=1}^{n} {}^k X_j\, {}^g X_i\, {}^g X_s\, {}^k X_r\, f_{ij} g_{rs}$$

and

$$\begin{aligned}\langle \lambda_{g2} \rangle = \sum_{\substack{k=1\\k \neq g}}^{n} \frac{1}{\mu_{kg}} &\left[\sum_{i,j,r,s=1}^{n} {}^k X_j\, {}^g X_i\, {}^g X_s\, {}^k X_r\, \{\mu_g^2 \langle d_{ij} d_{rs} \rangle - \mu_g \{\langle b_{ij} d_{rs} \rangle \right. \\ &\left. + \langle d_{ij} b_{rs} \rangle\} + \langle b_{ij} b_{rs} \rangle\}\right] \\ + \sum_{i,j,r,s=1}^{n} &{}^g X_j\, {}^g X_i\, {}^g X_s\, {}^g X_r\, \{\langle b_{ij} d_{rs} \rangle - \mu_g \langle d_{ij} d_{rs} \rangle\} .\end{aligned}$$

We further remark that the derived expansions are also possible if the eigenvalue problem $A_0 x_0 = \mu C_0 x_0$ has no simple eigenvalues. We only have to assume that the investigated eigenvalue μ_g is simple. For instance, the case $g = 1$ is of interest in many technological problems and the other eigenvalues are often not of great interest. In these cases the simplicity of μ_1 is sufficient.

1.3.2. Randomly perturbed symmetric matrices with multiple eigenvalues of the unperturbed matrix

Consider the random eigenvalue problem

$$(A_0 + B(\omega))\, x = \lambda x\,.$$

Let $A_0 = (a_{ij})_{1 \leqq i, j \leqq n}$ be a symmetric deterministic matrix with eigenvalues

$$\mu_1 \leqq \mu_2 \leqq \dots \leqq \mu_n\,.$$

and let $B(\omega) = \big(b_{ij}(\omega)\big)_{1 \leqq i, j \leqq n}$ a random perturbation matrix with $\langle B(\omega) \rangle = 0$ and $|B(\omega)|$ sufficiently small. The eigenvalues of the matrix $A_0 + B(\omega)$ are denoted by $\lambda_l(\omega)$. These eigenvalues are random variables whose first two moments exist if the first two moments of the random variables b_{ij} exist.

We put

$$\lambda_g(\omega) = -\sum_{p=0}^{\infty} \lambda_{gp}(\omega)\,,$$

where the quantity $\lambda_{gk}(\omega)$ denotes the homogeneous terms of the k-th order in the $b_{ij}(\omega)$. Thus $\lambda_{g0}(\omega) = -\mu_g$.

The eigenvalue $\lambda_g(\omega)$ can be calculated from the characteristic equation

$$\det\big(a_{ij} + b_{ij}(\omega) - \lambda \delta_{ij}\big) = 0$$

which may also be written in the form

$$\sum_{k=0}^{n} (-1)^{n-k} \lambda^{n-k}(\omega)\, J_k(\omega) = 0 \tag{1.56}$$

where $J_k(\omega)$ is the k-th invariant of $A_0 + B(\omega)$ for $k \geqq 1$ and $J_0 \doteq 1$. We now give the expansion of $J_k(\omega)$ in homogeneous terms of the $b_{ij}(\omega)$:

$$J_k(\omega) = \sum_{p=0}^{k} J_{pk}(\omega)\,, \tag{1.57}$$

where J_{0k} and $J_{kk}(\omega)$ denote the k-th invariant of A_0 and $B(\omega)$, respectively. The expansion of $\lambda_g(\omega)$ and (1.57) are substituted in (1.56), and we put

$$\Big(\sum_{p=0}^{\infty} \lambda_{gp}\Big)^{n-k} = \sum_{p=0}^{\infty} \Lambda_p^{n-k}\,,$$

where

$$\Lambda_p^t \doteq \sum_{\Sigma r_s = p} \lambda_{gr_1} \lambda_{gr_2} \dots \lambda_{gr_t}$$

and

$$\Lambda_0^0 \doteq 1\,, \qquad \Lambda_k^0 \doteq 0 \quad \text{for} \quad k > 0\,.$$

Then we obtain

$$\sum_{k=0}^{n} (-\lambda_g)^{n-k} J_k(\omega) = \sum_{p=0}^{\infty} \Big(\sum_{k=0}^{n} \sum_{i=0}^{p} \Lambda_i^{n-k} J_{p-i,\,k}\Big) = 0$$

where $J_{pk} = 0$ for $p > k$. The expression

$$\sum_{k=0}^{n} \sum_{i=0}^{p} \Lambda_i^{n-k} J_{p-i,\,k}$$

denotes the term of the characteristic equation which is homogeneous in the b_{ij} of the p-th order. The terms λ_{gp} are determined so that the homogeneous parts vanish separately, that is, from the equations

$$\sum_{k=1}^{n} \sum_{i=\max\{0,\,p-k\}}^{p-1} \Lambda_i^{n-k} J_{p-i,\,k} + \sum_{k=0}^{n} \hat{\Lambda}_p^{n-k} J_{0k} + \lambda_{gp} \sum_{k=0}^{n} (n-k)\,\lambda_{g0}^{n-k-1} J_{0k} = 0 \tag{1.58}$$

for $p = 1, 2, 3, \ldots$ where

$$\hat{\Lambda}_s^{n-k} \doteq \Lambda_s^{n-k} - (n-k)\,\lambda_{gs}\lambda_{g0}^{n-k-1}\,, \quad \text{for} \quad s \geqq 1\,,$$

and $J_{ip} = 0$ for $i > p$.

Let μ_g be an eigenvalue of A_0 of multiplicity h; that is, the eigenvalues of A_0 are

$$\mu_1 \leqq \mu_2 \leqq \ldots \leqq \mu_{g-1} < \mu_g = \mu_{g+1} = \ldots = \mu_{g+h-1} < \mu_{g+h} \leqq \ldots \leqq \mu_n$$

where $h \geqq 2$. We introduce the matrix $\tilde{A}_0 \doteq A_0 - \mu_g I$ and get $\tilde{\lambda}_l = \lambda_l - \mu_g$ if $\tilde{\lambda}_l(\omega)$ denote the eigenvalues of the matrix $\tilde{A}_0 + B(\omega)$. By means of

$$\tilde{\lambda}_g = -\sum_{p=1}^{\infty} \tilde{\lambda}_{gp}(\omega)$$

and $\tilde{\lambda}_{g0} = -\tilde{\mu}_g = 0$, $\tilde{J}_{0,\,n-1} = 0$, from (1.58) it follows that

$$\sum_{k=1}^{n} \sum_{i=\max\{0,\,p-k\}}^{p-1} \tilde{\Lambda}_i^{n-k} \tilde{J}_{p-i,\,k} + \sum_{k=0}^{n} \tilde{\hat{\Lambda}}_p^{n-k} \tilde{J}_{0k} = 0 \tag{1.59}$$

for $p = 1, 2, 3, \ldots$ The sign "$\sim$" describes the corresponding quantities relative to $\tilde{A}_0 + B(\omega)$. Thus $\tilde{J}_{q,\,n-p} = 0$ for $p + q < h$ and $\tilde{\Lambda}_i^t = 0$ for $t > i$ since $\tilde{\mu}_g = \tilde{\mu}_{g+1} = \ldots = \tilde{\mu}_{g+h-1} = 0$. Hence, (1.59) is always satisfied for $p = 1, 2, \ldots, h-1$. For $p = h$ we obtain from (1.59)

$$\sum_{t=0}^{h-1} \sum_{i=\max\{t,\,p-n+t\}}^{t+p-h} \tilde{\Lambda}_i^t \tilde{J}_{p-i,\,n-t} + \sum_{t=h}^{n-1} \sum_{i=\max\{t,p-n+t\}}^{p-1} \tilde{\Lambda}_i^t \tilde{J}_{p-i,\,n-t} + \sum_{t=h}^{\min\{n,\,p\}} \tilde{\Lambda}_p^t \tilde{J}_{0,\,n-t} = 0\,. \tag{1.60}$$

In particular, for $p = h$ it follows that

$$\sum_{t=0}^{h} \tilde{\lambda}_{g1}^t \tilde{J}_{h-t,\,n-t} = 0 \qquad \text{a.s.} \tag{1.61}$$

as an equation for $\tilde{\lambda}_{g1}$. We write (1.60) in the form

$$\tilde{\lambda}_{g,\,p-h+1} \sum_{t=1}^{h} t\tilde{\lambda}_{g1}^{t-1} \tilde{J}_{h-t,\,n-t} = F_{p-h+1}(\tilde{\lambda}_{g1}, \ldots, \tilde{\lambda}_{g,\,p-h}) \quad \text{a.s.}, \tag{1.62}$$

where all terms which only contain $\tilde{\lambda}_{g1}, \dots, \tilde{\lambda}_{g,p-h}$ are written on the right-hand side.

To calculate the homogeneous terms $\tilde{\lambda}_{gp}$, we first solve (1.61) for $\tilde{\lambda}_{g1}$ and obtain h roots $\tilde{\lambda}_{g1;0}, \tilde{\lambda}_{g1;1}, \dots, \tilde{\lambda}_{g1;h-1}$ of this algebraic equation. For further considerations we assume that these roots $\tilde{\lambda}_{g1;s}$ of equation (1.61) are simple. If the roots are not simple we obtain the terms of the expansion of $\lambda_g(\omega)$ in the same manner employed above. Assuming that the zeros $\tilde{\lambda}_{g1;s}$ are simple, we observe that

$$\frac{\mathrm{d}}{\mathrm{d}x}\Big(\sum_{t=0}^{h} x^t \tilde{J}_{h-t,n-t}\Big)\Big|_{x=\tilde{\lambda}_{g1;s}} = \sum_{t=1}^{h} t\tilde{\lambda}_{g1;s}^{t-1}\tilde{J}_{h-t,n-t} \neq 0$$

for $s = 0, 1, \dots, h-1$. Then, further terms of the expansion of $\lambda_g(\omega)$ for a fixed $\tilde{\lambda}_{g1;s}$ can be calculated successively from (1.62) by

$$\tilde{\lambda}_{gr;s} = \frac{F_r(\tilde{\lambda}_{g1;s}, \dots, \tilde{\lambda}_{g,r-1;s})}{\sum_{t=1}^{h} t\tilde{\lambda}_{g1;s}^{t-1}\tilde{J}_{h-t,n-t}} \quad \text{for} \quad r = 2, 3, \dots$$

Remark 1.1. *If we proceed from the transformed characteristic equation*

$$\det\big((\mu_{ig} - \tilde{\lambda}_g)\,\delta_{ij} + \hat{b}_{ij}(\omega)\big) = 0\,,$$

where $\hat{B}(\omega) = CBC^{\mathsf{T}} = \big(\hat{b}_{ij}(\omega)\big)_{1\leqq i,j\leqq n}$, $\mu_{ig} = \mu_i - \mu_g$, *then* (1.61) *can be written as*

$$\tilde{\lambda}_{g1}^h + \sum_{t=0}^{h} \tilde{\lambda}_{g1}^t \frac{\hat{\tilde{J}}_{h-t,n-t}}{\hat{\tilde{J}}_{0,n-h}} = 0\,.$$

The quantities $\hat{\tilde{J}}_{ki}$ *are the homogeneous terms in the* $\hat{b}_{pq}(\omega)$ *of the invariant* $\hat{\tilde{J}}_i$ *of the determinant* $\det(\mu_{ig}\delta_{ij} + \hat{b}_{ij})$. *We have*

$$\frac{\hat{\tilde{J}}_{h-t,n-t}}{\hat{\tilde{J}}_{0,n-h}} \doteq I_{h-t}$$

where I_k *denotes the* k-th *invariant of the determinant of the matrix* $\hat{B}_{hg}(\omega) = \big(\hat{b}_{ij}(\omega)\big)_{g\leqq i,j\leqq g+h-1}$. *Hence, the values* $-\tilde{\lambda}_{g1;s}$ *are just the eigenvalues of the matrix* $\hat{B}_{hg}(\omega)$.

Rellich [1, 2, 3, 4] has investigated problems of the convergence of the perturbation series of symmetric matrices. He proved the following estimate:

$$|\lambda_{g+s}(\omega) - \big(\mu_g - \sum_{p=1}^{m} \varepsilon^p \lambda_{gp;s}(\omega)\big)| \leqq \tfrac{1}{4}\, d_g\big(\varepsilon C_{gs}(\omega)\big)^{m+1} \quad \text{a.s.}$$

for $m = 1, 2, \dots$, which is expressed here for the random case. This estimate is true for the $(g+s)$-th eigenvalue of $A_0 + \varepsilon B(\omega)$, for $s = 0, 1, \dots, h-1$, if μ_g

denotes an eigenvalue of A_0 of multiplicity h. Applying the notation

$$d_g \doteq \min \{\mu_g - \mu_{g-1}, \mu_{g+h} - \mu_g\} \quad \text{and} \quad e_{gp}(\omega) \doteq \min_{\substack{p \neq q \\ p, q = 0, \ldots, h-1}} \{|\lambda_{g1;p} - \lambda_{g1;q}|\}$$

we have

$$C_{gs}(\omega) \doteq 24 \left(\frac{|B(\omega)|}{d_g} + 2 \right) \frac{|B(\omega)|}{d_g} \left(1 + \frac{(h-1)\, d_g}{e_{gp}} \right).$$

The above estimate yields almost sure convergence of the expansion $\sum_{p=1}^{\infty} \varepsilon^p \lambda_{gp;s}$ for $|\varepsilon| < C_{gs}^{-1}(\omega)$.

For our case we use the convergence of the expansion for small b_{ij} and $\varepsilon = 1$. This does not follow from Rellich's estimate since the term $|B(\omega)|/e_{gp}$ is of the 0-th order in the b_{ij}. Therefore, we will give another convergence condition where the proof differs only in some places from the proof of Rellich.

Theorem 1.11. *For the expansion*

$$\lambda_{g+s}(\omega) = \mu_g - \sum_{i=1}^{\infty} \tilde{\lambda}_{gi;s}(\omega)$$

of the eigenvalues of the matrix $\tilde{A}_0 + B(\omega)$ *where the terms* $\tilde{\lambda}_{gi;s}$ *are determined by* (1.61) *and* (1.62) *the estimate*

$$|\lambda_{g+s}(\omega) - (\mu_g - \sum_{i=1}^{m} \tilde{\lambda}_{gi;s}(\omega))| \leqq \frac{d_g}{4} \frac{1}{1 + \dfrac{(h-1)}{e_{gs}} |B|} D_{gs}^{m+1}(\omega)$$

is obtained where

$$D_{gs}(\omega) \doteq \frac{24}{d_g} |B(\omega)| \left\{ \frac{h-1}{e_{gs}} |B(\omega)| + 2 \right\};$$

that is, the expansion converges almost surely if $D_{gs}(\omega) < 1$ *almost surely. The values* e_{gs} *and* d_g *are defined above.*

Remark 1.2. *For the expansion of the eigenvector* ${}^{g+s}x(\omega)$ *which is normalized by the condition* $(\!({}^{g+s}x, {}^{g+s}x)\!) = 1$, *we have the estimate*

$$|{}^{g+s}x(\omega) - \sum_{k=0}^{m} {}^{g}x_{k;s}(\omega)| \leqq \tfrac{1}{2} D_{gs}^{m+1}(\omega),$$

where the quantity D_{gs} *is defined in Theorem* 1.11.

Proof. Substituting the expansions

$$\lambda_g(\omega) = \mu_g - \sum_{k=1}^{\infty} \lambda_{gk}(\omega),$$

$${}^{g}x(\omega) = \sum_{k=0}^{\infty} {}^{g}x_k(\omega)$$

in the eigenvalue problem $(A_0 + B(\omega))\,x = \lambda x$ we obtain the following equations

$$(A_0 - \mu_g I)\,{}^g x_0 = 0\,,$$
$$(A_0 - \mu_g I)\,{}^g x_m = -\,B\,{}^g x_{m-1} - \sum_{s=1}^{m} \lambda_{gs}\,{}^g x_{m-s}\,, \quad \text{for} \quad m = 1, 2, \ldots \tag{1.63}$$

The operator T_g is defined analogously to (1.46) by

$$T_g u \doteq \sum_k{}' \frac{\langle\!\langle u, {}^k x_0 \rangle\!\rangle}{\mu_{kg}}\,{}^k x_0\,,$$

where the sum is taken over all orthonormal eigenvectors ${}^k x_0$ of A_0 which are associated with eigenvalues $\mu_k \neq \mu_g$. This operator has the following properties:

$$T_g(A_0 - \mu_g I)\,u = (A_0 - \mu_g I)\,T_g u = \sum_k{}' \langle\!\langle u, {}^k x_0 \rangle\!\rangle\,{}^k x_0\,,$$
$$|T_g u| \leq \frac{1}{d_g}\,|u|\,.$$

Let the h vectors v_i for $i = 0, 1, \ldots, h-1$ constitute the eigenspace of the eigenvector μ_g. These vectors v_i can be found such that

$$\langle\!\langle v_p, v_q \rangle\!\rangle = \delta_{pq} \quad \text{and} \quad \langle\!\langle Bv_p, v_q \rangle\!\rangle = -\,r_p \delta_{pq} \quad \text{for} \quad p, q = 0, 1, \ldots, h-1.$$

From (1.63) using the relation

$$x_0 = \sum_{p=0}^{h-1} c_{0p} v_p\,,$$

the condition of solvability

$$-\lambda_{g1}\langle\!\langle {}^g x_0, v_p \rangle\!\rangle - \langle\!\langle B\,{}^g x_0, v_p \rangle\!\rangle = -\sum_{q=0}^{h-1} c_{0q}\,[\langle\!\langle Bv_q, v_p \rangle\!\rangle + \lambda_{g1}\delta_{pq}] = 0$$

follows from the case where $m = 1$, i.e. the values $r_0, r_1, \ldots, r_{h-1}$ are the h approximations of the first order of the perturbed eigenvalue. Put $\lambda_{g1} \doteq \lambda_{g1;\,p} = r_p$ and $c_{0q} = \delta_{pq}$, because $\langle\!\langle {}^g x_0, {}^g g_0 \rangle\!\rangle = 1$. Furthermore, from (1.63) for $m = 1$ we have

$${}^{g+p}x_1 = \sum_{q=0}^{h-1} {}^{g+p}c_{1q} v_q - T_g B\,{}^{g+p}x_0\,. \tag{1.64}$$

For further calculations we omit the number of the eigenvalue and retain the index which denotes the order of the approximation, i.e. we write $\bar{x}_1$ instead of ${}^{g+p}x_1$, and $\bar{\lambda}_1$ instead of $\lambda_{g+p,\,1}$ and so on. The condition of solvability of the $(m+1)$-th equation of (1.63) is

$$\langle\!\langle B\bar{x}_m, v_q \rangle\!\rangle + \sum_{s=1}^{m+1} \bar{\lambda}_s \langle\!\langle \bar{x}_{m+1-s}, v_q \rangle\!\rangle = \bar{\lambda}_{m+1}\delta_{pq} + \sum_{s=1}^{m} \bar{\lambda}_s \langle\!\langle x_{m+1-s}, v_q \rangle\!\rangle + \langle\!\langle B\bar{x}_m, v_q \rangle\!\rangle = 0 \tag{1.65}$$

for $q = 0, 1, \ldots, h-1$; and from the m-th equation of (1.63) we have

$$\bar{x} = \sum_{s=0}^{h-1} \bar{c}_{ms} v_s - T_g B\bar{x}_{m-1} - \sum_{s=1}^{m-1} \bar{\lambda}_s T_g \bar{x}_{m-s}\,. \tag{1.66}$$

Using (1.65) for $p = q$ and (1.66) we have

$$\bar{\lambda}_{m+1} = -\sum_{s=2}^{m} \bar{\lambda}_s (\!(\bar{x}_{m+1-s}, \bar{x}_0)\!) + \sum_{s=1}^{m-1} \bar{\lambda}_s (\!(BT_g \bar{x}_{m-s}, \bar{x}_0)\!) + (\!(BT_g B\bar{x}_{m-1}, \bar{x}_0)\!) . \tag{1.67}$$

By means of (1.65), for $p \neq q$, we obtain

$$\bar{c}_{mq} = \frac{1}{\bar{\lambda}_1 - \lambda_{g1;q}} [(\!(BT_g B\bar{x}_{m-1}, v_q)\!) + \sum_{s=1}^{m-1} \bar{\lambda}_s (\!(BT_g \bar{x}_{m-s}, v_q)\!) - \sum_{s=2}^{m} \bar{\lambda}_s (\!(\bar{x}_{m+1-s}, v_q)\!)] .$$

If we use the condition $(\!({}^{g+p}x, {}^{g+p}x)\!) = 1$, then for $m \geqq 1$

$$\sum_{k=0}^{m} (\!(\bar{x}_k, \bar{x}_{m-k})\!) = 0 .$$

From (1.66) it follows that $\bar{c}_{mp} = (\!(\bar{x}_m, \bar{x}_0)\!)$ and

$$\bar{c}_{mp} = -\tfrac{1}{2} \sum_{k=1}^{m-1} (\!(\bar{x}_k, \bar{x}_{m-k})\!) .$$

Hence, we have

$$\begin{aligned} &\bar{\lambda}_1 = -(\!(B\bar{x}_0, \bar{x}_0)\!) , \\ &\bar{\lambda}_2 = (\!(BT_g B\bar{x}_0, \bar{x}_0)\!) , \\ &\bar{\lambda}_{m+1} \ \text{ from (1.67) for } m = 2, 3, \dots , \\ &\bar{x}_1 = \sum_{\substack{s=0 \\ s \neq p}}^{h-1} \frac{(\!(TB_g B\bar{x}_0, v_s)\!)}{\bar{\lambda}_1 - \lambda_{g1;s}} - T_g B\bar{x}_0 , \\ &\bar{x}_m = \sum_{\substack{s=0 \\ s \neq p}}^{h-1} \frac{1}{\bar{\lambda}_1 - \lambda_{g1;s}} [(\!(BT_g B\bar{x}_{m-1}, v_s)\!) + \sum_{t=1}^{m-1} \bar{\lambda}_t (\!(BT_g \bar{x}_{m-t}, v_s)\!) - \sum_{t=2}^{m} \bar{\lambda}_t (\!(\bar{x}_{m+1-t}, v_s)\!)] \\ &\qquad - \frac{1}{2} \sum_{k=1}^{m-1} (\!(\bar{x}_k, \bar{x}_{m-k})\!) \bar{x}_0 - T_g B\bar{x}_{m-1} - \sum_{s=1}^{m-1} \bar{\lambda}_s T_g \bar{x}_{m-s} \quad \text{for } m = 2, 3, \dots \end{aligned} \tag{1.68}$$

For the estimation of the rate of convergence of the computed expansion from (1.67), we find

$$\Lambda_{m+1} \leqq \sum_{s=2}^{m} |\bar{x}_{m+1-s}| (\Lambda_s + \alpha \Lambda_{s-1}) + \alpha^2 |\bar{x}_{m-1}|$$

for $m \geqq 2$, where

$$\alpha \doteq \frac{|B|}{d_g} \quad \text{and} \quad \frac{1}{d_g} |\bar{\lambda}_p| \doteq \Lambda_p \quad \text{for} \quad p \geqq 1 .$$

By addition of $\alpha \Lambda_m$ we obtain

$$w_{m+1} \leqq (|\bar{x}_1| + \alpha)\, w_m + \sum_{s=2}^{m-1} |\bar{x}_{m-s+1}|\, w_s \tag{1.69}$$

for $m \geqq 2$ if we define

$$w_1 \doteq \Lambda_1 + 2\alpha\,, \qquad w_2 \doteq \Lambda_2 + \alpha \Lambda_1 + \alpha^2\,, \qquad w_k \doteq \Lambda_k + \alpha \Lambda_{k-1} \quad \text{for } k \geqq 3$$

and apply the inequality $\Lambda_k \leqq w_k$. Put

$$\frac{1}{2}\,\eta_s \doteq \frac{(h-1)\, d_g}{e_{gp}}\, w_{s+1} + w_s$$

for $s \geqq 1$. Furthermore, if we multiply (1.69) by $(h-1)\, d_g/e_{gp}$ and add the inequality (1.69) for $m-1$ to this inequality then we have

$$\eta_m \leqq (|\bar{x}_1| + \alpha)\, \eta_{m-1} + \sum_{s=2}^{m-1} |\bar{x}_{m-s+1}|\, \eta_{s-1} \tag{1.70}$$

for $m \geqq 3$. This inequality is also valid for $m = 2$, since from

$$\Lambda_1 \doteq \frac{1}{d_g} |\bar{\lambda}_1| \leqq \alpha$$

and

$$\Lambda_2 \doteq \frac{1}{d_g} |\bar{\lambda}_2| \leqq \alpha^2$$

(cf. (1.68)) the inequality

$$w_2 \leqq (|\bar{x}_1| + \alpha)\, w_1$$

follows. By the use of (1.69) for $m = 2$, (1.70) is also valid for $m = 2$.

From (1.68) a simple estimation implies

$$|\bar{x}_m| \leqq \tfrac{1}{2} \sum_{s=2}^{m} \eta_{s-1} |\bar{x}_{m-s+1}| + \tfrac{1}{2} \sum_{s=1}^{m-1} |\bar{x}_{m-s}|\, |\bar{x}_s| \tag{1.71}$$

if we apply the inequalities $\Lambda_1 + \alpha \leqq w_1$ and $\Lambda_s \leqq w_s$. Put

$$|\bar{x}_s| = q_s \quad \text{for} \quad s \geqq 2 \quad \text{and} \quad |\bar{x}_1| + \alpha = q_1;$$

then because of (1.70) and (1.71) we verify that

$$\eta_m \leqq \sum_{s=1}^{m-1} q_{m-s} \eta_s\,,$$

$$q_m \leqq \tfrac{1}{2} \sum_{s=1}^{m-1} q_{m-s} \eta_s + \tfrac{1}{2} \sum_{s=1}^{m-1} q_{m-s} q_s$$

for $m = 2, 3, \ldots$ For q_1, η_1 we have

$$q_1 \leqq \alpha \left(2 + \frac{(h-1)\, d_g}{e_{gp}}\, \alpha\right),$$

$$\eta_1 \leqq 6\alpha \left(2 + \frac{(h-1)\, d_g}{e_{gp}}\, \alpha\right),$$

and with

$$D_{gp} \doteq 24\alpha\left(2 + \frac{(h-1)\,d_g}{e_{gp}}\,\alpha\right)$$

then $q_1 \leqq \frac{1}{4} D_{gp}$, $\eta_1 \leqq \frac{1}{4} D_{gp}$.

The sequences (u_m), (y_m) defined by

$$u_1 = y_1 = \tfrac{1}{4} D_{gp}\,,$$

$$y_m = \sum_{s=1}^{m-1} y_s u_{m-s}\,, \qquad u_m = \tfrac{1}{2}\sum_{s=1}^{m-1} y_s u_{m-s} + \tfrac{1}{2}\sum_{s=1}^{m-1} u_s u_{m-s}$$

satisfy the relations

$$u_m \geqq q_m\,, \qquad y_m \geqq q_m \quad \text{and} \quad u_m = y_m \quad \text{for} \quad m = 1, 2, \ldots\,.$$

For the function $f(z)$ given by

$$f(z) = \sum_{s=1}^{\infty} y_s z^s$$

we calculate

$$f^2(z) = \sum_{s=2}^{\infty}\Big(\sum_{t=1}^{s-1} y_t y_{s-t}\Big) z^s = f(z) - \tfrac{1}{4} D_{gp} z$$

from which we have

$$f(z) = \tfrac{1}{2}\left(1 - \sqrt{1 - D_{gp} z}\right) = \tfrac{1}{2}\sum_{s=1}^{\infty}\left|\binom{1/2}{s}\right| D_{gp}^s z^s\,.$$

Thus

$$y_s = \tfrac{1}{2}\left|\binom{1/2}{s}\right| D_{gp}^s$$

and

$$|x_s| \leqq \tfrac{1}{2}\left|\binom{1/2}{s}\right| D_{gp}^s\,,$$

$$|\bar{\lambda}_s| \leqq \tfrac{1}{2}\, d_g\left|\binom{1/2}{s}\right| D_{gp}^s (1 + (h-1)\, d_g/e_{gp})^{-1}\,.$$

These estimates lead to the statement of the above theorem and to the remark since the expansion for λ_{g+p} deduced in the beginning of Section 1.3.2 coincides with the expansion given in this proof because of uniqueness. ◀

1.3.3. Applications

1.3.3.1. Random matrices with independent almost surely bounded elements

In the following we deal with the special case of a symmetric matrix $A_0 + B(\omega)$. Assume that the elements $b_{ij}(\omega)$ of the matrix $B(\omega)$ are independent, almost surely bounded, random variables where the assumed almost sure symmetry $b_{ij}(\omega) = b_{ji}(\omega)$ is required. Furthermore, let these random variables $b_{ij}(\omega)$

possess the same symmetric distribution function $F(x)$. This case is interesting if all elements of the matrix A_0 have small independent errors. For instance, we can assume that the $b_{ij}(\omega)$ are uniformly distributed on $[-d, d]$.

In this important case the expansions of the expectation of the eigenvalues and their correlation relations up to the fourth order in the perturbations can be calculated relatively simply. We now assume that the transformation matrix $C = (c_{ij})_{1 \leq i, j \leq n}$ is known which transforms A_0 to a diagonal matrix.

Theorem 1.12. *Let A_0 be a symmetric n by n matrix with the simple eigenvalue μ_g. This matrix A_0 is perturbed by an almost surely symmetric random matrix $B(\omega)$ satisfying condition* (1.18). *Let the $b_{ij}(\omega)$ be independent random variables with $b_{ij} = b_{ji}$ almost surely which have the same symmetric distribution function $F(x)$ where*

$$\sigma^2 = \int_{-\infty}^{\infty} x^2 \, dF(x) , \qquad \sigma_4 = \int_{-\infty}^{\infty} x^4 \, dF(x)$$

and the remainder is $r = \sigma_4 - 3\sigma^4$. Then the expectation of the eigenvalue $\lambda_g(\omega)$ of the matrix $A_0 + B(\omega)$ can be given up to terms of the fourth order in the $b_{ij}(\omega)$; namely

$$\langle \lambda_g(\omega) \rangle = \mu_g - \langle \lambda_{g2}(\omega) \rangle - \langle \lambda_{g4}(\omega) \rangle ,$$

where

$$\langle \lambda_{g2} \rangle = \sigma^2 \sum_{\substack{p=1 \\ p \neq g}}^{n} \frac{1}{\mu_{pg}} (1 - C_{ppg}) , \tag{1.72}$$

$$\begin{aligned} \langle \lambda_{g4} \rangle = & \sum_{\substack{p, q, s=1 \\ p, q, s \neq g}}^{n} \frac{1}{\mu_{pg}\mu_{qg}\mu_{sg}} \{(\sigma^4 + r) D_{gspq}^2 + (\sigma^4 + 3r) [C_{pqg}C_{pqs} + C_{gps}C_{gpq}] \\ & \qquad + r[C_{ggq}C_{ssp} - 7E_{gpqs}]\} \\ & - \sum_{\substack{p, q=1 \\ p, q \neq g}}^{n} \frac{1}{\mu_{pg}^2\mu_{qg}} \{(4\sigma^4 + 6r) C_{pqg}^2 + (2\sigma^4 + 6r) C_{pgg}C_{gpq} \\ & \qquad + (\sigma^4 + C_{ggg}r) C_{ppq} + (\sigma^4 + 5r) C_{ppg}C_{qqg} \\ & \qquad + (2\sigma^4 + 6r) C_{gpq}C_{qgg} - \sigma^4 C_{qqg} - 21rE_{ggpq}\} \\ & + \sum_{\substack{p=1 \\ p \neq g}}^{n} \frac{1}{\mu_{pg}^3} \{(1 - C_{ggg}) \sigma^4 + \big(3\sigma^4 + C_{ggg}(\sigma^4 + 4r)\big) C_{ppg} \\ & \qquad + (2\sigma^4 + 4r) \; C_{pgg}^2 - 7rE_{gggp}\} . \end{aligned} \tag{1.73}$$

The following expressions have been used above:

$$D_{ijks} \doteq \sum_{p=1}^{n} c_{ip}c_{jp}c_{kp}c_{sp} ,$$

$$E_{ijks} \doteq \sum_{p=1}^{n} c_{ip}^2 c_{jp}^2 c_{kp}^2 c_{sp}^2 ,$$

$$C_{ijk} \doteq D_{ijkk} .$$

Estimates for the terms of second and fourth order are given by

$$|\langle\lambda_{g2}\rangle| \leqq \sigma^2 A_{01}\,, \tag{1.74}$$

$$\begin{aligned}|\langle\lambda_{g4}\rangle| \leqq{}& A_{21}^2\left[\frac{1}{\bar{\mu}_g}|\sigma^4+r| + (|\sigma^4+3r|+|r|)\,A_{01} + 7\,|r|\,A_{21}\right]\\ &+ A_{21}\,[c_g^2\,|\sigma^4+3r|\,A_{01}^2 + (|4\sigma^4+6r| + |\sigma^4+5r| + 21c_g^2\,|r|)\,A_{22}\\ &\qquad + (|2\sigma^4+6r|\,c_g^2+\sigma^4)\,A_{02}]\\ &+ A_{22}A_{01}\,(|\sigma^4+C_{ggg}\,r| + c_g^2\,|2\sigma^4+6r|) + (1-C_{ggg})\,\sigma^4\,A_{03}\\ &+ A_{23}(c_g^2\,|2\sigma^4+4r| + |3\sigma^4+C_{ggg}(\sigma^4+4r)| + 7\,|r|\,c_g^4)\,, \end{aligned} \tag{1.75}$$

where

$$c_i \doteq \max_p |c_{ip}|\,, \qquad \bar{\mu}_g \doteq \min_{p \neq g} |\mu_{pg}|$$

and

$$A_{st} \doteq \sum_{\substack{p=1\\p\neq g}}^{n} \frac{c_p^s}{|\mu_{pg}|^t}\,.$$

Proof. From the conditions on the $b_{ij}(\omega)$, we can derive the relations

$$\langle b_{i_1j_1} b_{i_2j_2}\rangle = \sigma^2[\delta_{i_1i_2}\delta_{j_1j_2} + \delta_{i_1j_2}\delta_{i_2j_1} - \delta_{i_1i_2j_1j_2}]\,,$$

$$\langle b_{i_1j_1} b_{i_2j_2} b_{i_3j_3}\rangle = 0\,,$$

$$\begin{aligned}&\langle b_{i_1j_1} b_{i_2j_2} b_{i_3j_3} b_{i_4j_4}\rangle\\ &= \langle b_{i_1j_1} b_{i_2j_2}\rangle\,\langle b_{i_3j_3} b_{i_4j_4}\rangle + \langle b_{i_1j_1} b_{i_3j_3}\rangle\,\langle b_{i_2j_2} b_{i_4j_4}\rangle + \langle b_{i_1j_1} b_{i_4j_4}\rangle\,\langle b_{i_2j_2} b_{i_3j_3}\rangle\\ &\quad + r[\delta_{i_1i_2i_3i_4}\delta_{j_1j_2j_3j_4} + \delta_{i_1i_2i_3j_4}\delta_{j_1j_2j_3i_4} + \delta_{i_1i_2j_3i_4}\delta_{j_1j_2i_3j_4} + \delta_{i_1j_2i_3i_4}\delta_{j_1i_2j_3j_4}\\ &\qquad + \delta_{j_1i_2i_3i_4}\delta_{i_1j_2j_3j_4} + \delta_{i_1i_2j_3j_4}\delta_{j_1j_2i_3i_4} + \delta_{i_1j_2i_3j_4}\delta_{j_1i_2j_3i_4} + \delta_{j_1i_2i_3j_4}\delta_{i_1j_2j_3i_4}\\ &\qquad - 7\delta_{i_1i_2i_3i_4j_1j_2j_3j_4}]\,,\end{aligned}$$

where

$$\delta_{k_1k_2\ldots k_p} = \begin{cases} 1 & \text{for } k_1 = k_2 = \ldots = k_p\,,\\ 0 & \text{otherwise}\,.\end{cases}$$

Because $CBC^{\mathsf{T}} = \hat{B}$ it follows that

$$\hat{b}_{ij} = \sum_{p,\,q=1}^{n} c_{ip}c_{jq}b_{pq} \doteq c_i^p c_j^q b_{pq}\,;$$

and we then have

$$\langle \hat{b}_{p_1q_1}\hat{b}_{p_2q_2}\ldots\hat{b}_{p_kq_k}\rangle = c_{p_1}^{i_1}c_{q_1}^{j_1}\ldots c_{p_k}^{i_k}c_{q_k}^{j_k}\,\langle b_{i_1j_1}\ldots b_{i_kj_k}\rangle\,.$$

Put

$$D_{i_1\ldots i_k} \doteq \sum_{p=1}^{n} c_{i_1p}c_{i_2p}\ldots c_{i_kp}\,.$$

Using $\sum_{p=1}^{n} c_{ip}c_{jp} = \delta_{ij}$ for $i, j = 1, 2, \ldots, n$ (because of $CC^{\mathsf{T}} = I$) we obtain

$$\langle \hat{b}_{p_1q_1}\hat{b}_{p_2q_2}\rangle = \sigma^2[\delta_{p_1q_2}\delta_{p_2q_1} + \delta_{p_1p_2}\delta_{q_1q_2} - D_{p_1p_2q_1q_2}]\,,$$

$$\langle \hat{b}_{p_1q_1}\hat{b}_{p_2q_2}\hat{b}_{p_3q_3}\rangle = 0\,,$$

and

$$\begin{aligned}
&\langle \hat{b}_{p_1q_1}\hat{b}_{p_2q_2}\hat{b}_{p_3q_3}\hat{b}_{p_4q_4}\rangle \\
&= \sigma^4[(\delta_{p_1p_2}\delta_{q_1q_2} + \delta_{p_1q_2}\delta_{p_2q_1} - D_{p_1p_2q_1q_2})(\delta_{p_3p_4}\delta_{q_3q_4} + \delta_{p_3q_4}\delta_{p_4q_3} - D_{p_3p_4q_3q_4}) \\
&\quad + (\delta_{p_1p_3}\delta_{q_1q_3} + \delta_{p_1q_3}\delta_{p_3q_1} - D_{p_1p_3q_1q_3})(\delta_{p_2p_4}\delta_{q_2q_4} + \delta_{p_2q_4}\delta_{p_4q_2} - D_{p_2p_4q_2q_4}) \\
&\quad + (\delta_{p_1p_4}\delta_{q_1q_4} + \delta_{p_1q_4}\delta_{p_4q_1} - D_{p_1p_4q_1q_4})(\delta_{p_2p_3}\delta_{q_2q_3} + \delta_{p_2q_3}\delta_{p_3q_2} - D_{p_2p_3q_2q_3})] \\
&\quad + r[D_{p_1p_2p_3p_4}D_{q_1q_2q_3q_4} + D_{p_1p_2p_3q_4}D_{q_1q_2q_3p_4} + D_{p_1p_2q_3p_4}D_{q_1q_2p_3q_4} \\
&\qquad + D_{p_1q_2p_3p_4}D_{q_1p_2q_3q_4} + D_{q_1p_2p_3p_4}D_{p_1q_2q_3q_4} + D_{p_1p_2q_3q_4}D_{q_1q_2p_3p_4} \\
&\qquad + D_{p_1q_2p_3q_4}D_{q_1p_2q_3p_4} + D_{q_1p_2p_3q_4}D_{p_1q_2q_3p_4} - 7D_{p_1p_2p_3p_4q_1q_2q_3q_4}] .
\end{aligned}$$

Assuming that the indices p, q, s are different from g, we have

$$\begin{aligned}
\langle \hat{b}_{pg}^2\rangle &= \sigma^2(1 - C_{ppg}) , \\
\langle \hat{b}_{pg}^2\hat{b}_{qg}^2\rangle &= \sigma^4[(1 - C_{ppg})(1 - C_{qqg}) + 2(\delta_{pq} - C_{pqg})^2] \\
&\quad + r[C_{ggg}C_{ppq} + C_{ppg}C_{qqg} + 2C_{gpq}C_{qgg} + 2C_{gpq}C_{pgg} \\
&\qquad + 2C_{pqg}^2 - 7E_{pqgg}] , \\
\langle \hat{b}_{gg}^2\hat{b}_{gp}^2\rangle &= \sigma^4[(2 - C_{ggg})(1 - C_{ppg}) + 2C_{pgg}^2] \\
&\quad + r[4C_{ggg}C_{ppg} + 4C_{pgg}^2 - 7E_{gggp}] , \\
\langle \hat{b}_{gg}\hat{b}_{gp}\hat{b}_{pq}\hat{b}_{qg}\rangle &= \sigma^4[C_{pgg}C_{gpq} + C_{qgg}C_{gpq} - C_{pqg}(\delta_{pq} - C_{pqg})] \\
&\quad + r[2C_{pqg}^2 + 2C_{pgg}C_{gpq} + 2C_{qgg}C_{ppg} \\
&\qquad + 2C_{gqp}C_{qgg} - 7E_{ggpq}] , \\
\langle \hat{b}_{gp}\hat{b}_{pq}\hat{b}_{qs}\hat{b}_{sg}\rangle &= \sigma^4[C_{gpq}C_{gqs} + D_{gspq}^2 + (\delta_{sp} - C_{spg})(\delta_{pqs} + \delta_{ps} - C_{spq})] \\
&\quad + r[D_{gpqs}^2 + C_{pqg}C_{pqs} + C_{psg}C_{psq} + C_{qsg}C_{qsp} + C_{gps}C_{gpq} \\
&\qquad + C_{gsp}C_{gsq} + C_{gqs}C_{gqp} + C_{ggq}C_{ssp} - 7E_{gpqs}] .
\end{aligned}$$

By using the expressions from the expansions of Theorem 1.10, we can calculate $\langle \lambda_{g3}\rangle = 0$ and (1.72) and (1.73) for $\langle \lambda_{g2}\rangle$ and $\langle \lambda_{g4}\rangle$, respectively.

We see that from

$$1 = \sum_{p=1}^{n} c_{ip}^2 \sum_{q=1}^{n} c_{gq}^2 = \sum_{p=1}^{n} c_{ip}^2 c_{gp}^2 + \sum_{\substack{p=1 \\ p \neq q}}^{n} c_{ip}^2 c_{gq}^2$$

the relation

$$1 \geqq C_{iig} = \sum_{p=1}^{n} c_{ip}^2 c_{gp}^2 \geqq 0$$

follows; and then the estimate for $\langle \lambda_{g2}\rangle$ is immediately clear from (1.72). The estimate (1.75) is obtained by some simple calculations from

$$|C_{pqs}| = |\sum_{i=1}^{n} c_{pi}c_{qi}c_{si}^2| \leqq c_p c_q \sum_{i=1}^{n} c_{si}^2 = c_p c_q ,$$

$$E_{pqst} = \sum_{i=1}^{n} c_{pi}^2 c_{qi}^2 c_{si}^2 c_{ti}^2 \leqq c_p^2 c_q^2 c_s^2 \sum_{i=1}^{n} c_{ti}^2 = c_p^2 c_q^2 c_s^2 ,$$

and

$$\left| \sum_{\substack{p,q,s=1\\ p,q,s\neq g}}^{n} \frac{1}{\mu_{sg}\mu_{pg}\mu_{qg}} D^2_{gspq} \right| \leqq \frac{1}{\mu_g} \sum_{\substack{p,q=1\\ p,q\neq g}}^{n} \sum_{s=1}^{n} \frac{1}{|\mu_{pg}\mu_{qg}|} D^2_{gspq}$$

$$= \frac{1}{\mu_g} \sum_{i,j=1}^{n} \sum_{\substack{p,q=1\\ p,q\neq g}}^{n} \frac{c_{gi}c_{gj}c_{pi}c_{pj}c_{qi}c_{qj}}{|\mu_{pg}\mu_{qg}|} \sum_{s=1}^{n} c_{si}c_{sj}$$

$$\leqq \frac{1}{\mu_g} \sum_{i=1}^{n} c^2_{gi} \left(\sum_{\substack{p=1\\ p\neq g}}^{n} \frac{1}{|\mu_{pg}|} c^2_p \right)^2 = \frac{1}{\mu_g} \left(\sum_{\substack{p=1\\ p\neq g}}^{n} \frac{1}{|\mu_{pg}|} c^2_p \right)^2 . \quad ◀$$

Theorem 1.13. *Assume the same conditions and notation of Theorem* 1.12. *The elements of the correlation matrix of the eigenvalues are given up to terms of the fourth order by*

$$R_{gh} = [R_{gh}]_2 + [R_{gh}]_4 ,$$

where

$$[R_{gh}]_2 = \sigma^2(2\delta_{gh} - C_{ggh}) ,$$

$$[R_{gh}]_4 = 2\sigma^4 \left[\frac{1}{\mu^2_{gh}} (1 - \delta_{gh}) - \delta_{gh} \sum_{\substack{p-1\\ p\neq g}} \frac{1}{\mu^2_{pg}} \right] + \alpha_{gh} + \alpha_{hg} + \beta_{gh} + \beta_{hg} + \gamma_{gh} .$$

The quantities α_{gh}, β_{gh}, *and* γ_{gh} *are defined by*

$$\alpha_{gh} \doteq - \sum_{\substack{p=1\\ p\neq g}}^{n} \frac{1}{\mu^2_{pg}} \{\sigma^4[C_{pph} + C_{ggh}(C_{ggp} - 1) + 2C_{pgg}C_{gph}] + r[4C_{hgg}C_{gph} + 4C_{phg} - 7E_{ggph}]\} ,$$

$$\beta_{gh} \doteq \sum_{\substack{p,q=1\\ p,q\neq g}}^{n} \frac{1}{\mu_{pg}\mu_{qg}} \{\sigma^4[2C_{gqp}C_{gqh} + C_{pqg}C_{pqh}] + r[2D^2_{gpqh} + 4C_{phg}C_{phq} + 2C_{ghq}C_{ghp} - 7E_{gpqh}]\} ,$$

$$\gamma_{gh} \doteq \sum_{\substack{p,q=1\\ p\neq g\\ q\neq h}}^{n} \frac{1}{\mu_{pg}\mu_{qh}} \{2\sigma^4 D^2_{gpqh} + r[C_{ppq}C_{ggh} + 2C_{hqp}C_{hqg} + 2C_{pgq}C_{pgh} + 2D^2_{gpqh} + C_{pph}C_{qqg} - 7E_{gpqh}]\} .$$

The proof of Theorem 1.13 is based on (1.33) and uses the expressions for $\langle \hat{b}_{p_1q_1}\hat{b}_{p_2q_2}\rangle$, $\langle \hat{b}_{p_1q_1}\hat{b}_{p_2q_2}\hat{b}_{p_3q_3}\hat{b}_{p_4q_4}\rangle$ calculated in the proof of Theorem 1.12. We have simply indicated the proof; and not given all details.

Let us consider a numerical example. The matrix

$$A_0 = \begin{pmatrix} 0 & 2 & -2 & 0 \\ 2 & 1 & 0 & 0 \\ -2 & 0 & -1 & 0 \\ 0 & 0 & 0 & 1 \end{pmatrix} \tag{1.76}$$

is perturbed by a random matrix $B(\omega) = \big(b_{ij}(\omega)\big)_{1\leq i,j\leq 4}$ which satisfied the conditions of Theorem 1.12. In particular, let the realizations of the perturbations be sufficiently small. We can see that the matrix A_0 can be transformed by

$$C = \frac{1}{3}\begin{pmatrix} 2 & -1 & 2 & 0 \\ -1 & 2 & 2 & 0 \\ 0 & 0 & 0 & 3 \\ -2 & -2 & 1 & 0 \end{pmatrix}$$

to a diagonal matrix

$$CA_0C^{\mathsf{T}} = (\mu_i\delta_{ij})_{1\leq i,j\leq 4}\,,$$

where $\mu_1 = -3$, $\mu_2 = 0$, $\mu_3 = 1$, $\mu_4 = 3$. Then from (1.72) and (1.73) we can calculate

$$\begin{aligned}
\langle\lambda_1\rangle &= -3 - 0.6018518\sigma^2 - 0.0518\sigma^4 - 0.0038r\,,\\
\langle\lambda_2\rangle &= 0 - \sigma^2 + 0.4075\sigma^4 + 0.4074r\,,\\
\langle\lambda_3\rangle &= 1 + 0.7500000\sigma^2 - 0.3906\sigma^4 - 0.3904r\,,\\
\langle\lambda_4\rangle &= 3 + 0.8518518\sigma^2 + 0.0349\sigma^4 - 0.0132r
\end{aligned} \tag{1.77}$$

up to terms of the fourth order. In particular, if we use uniformly distributed random variables $b_{ij}(\omega)$ on $[-d, d]$ with $\sigma^2 = d^2/3$, $r = -2\,d^4/15$ then the following holds:

$$\begin{aligned}
\langle\lambda_1\rangle &= -3 - 0.2006173d^2 - 0.0052d^4\,,\\
\langle\lambda_2\rangle &= 0 - 0.3333333d^2 - 0.0090d^4\,,\\
\langle\lambda_3\rangle &= 1 + 0.2500000d^2 + 0.0087d^4\,,\\
\langle\lambda_4\rangle &= 3 + 0.2839506d^2 + 0.0056d^4\,.
\end{aligned}$$

The estimates of the expectations of the eigenvalues are given by

$$\begin{aligned}
|\langle\lambda_{12}\rangle| &\leqq 0.2500\,d^2, & |\langle\lambda_{14}\rangle| &\leqq 0.4825d^4\,,\\
|\langle\lambda_{22}\rangle| &\leqq 0.5556d^2\,, & |\langle\lambda_{24}\rangle| &\leqq 8.5133d^4\,,\\
|\langle\lambda_{32}\rangle| &\leqq 0.5833d^2\,, & |\langle\lambda_{34}\rangle| &\leqq 5.2630d^4\,,\\
|\langle\lambda_{42}\rangle| &\leqq 0.3333d^2\,, & |\langle\lambda_{44}\rangle| &\leqq 1.4371d^4
\end{aligned}$$

from (1.70) and (1.71) if we assume uniformly distributed random variables $b_{ij}(\omega)$.

From Theorem 1.13 we obtain the following correlation matrix of the eigenvalues up to terms of the fourth order in the perturbations

$$(R_{gh})_{1 \leqq g,\, h \leqq 4}$$
$$= \begin{pmatrix} 1.5926 & -0.2963 & 0 & -0.2963 \\ -0.2963 & 1.5926 & 0 & -0.2963 \\ 0 & 0 & 1 & 0 \\ -0.2963 & -0.2963 & 0 & 1.5926 \end{pmatrix} \sigma^2$$
$$+ \begin{pmatrix} -0.3012 & 0.5051 & -0.3333 & 0.1293 \\ 0.5051 & -1.5656 & 0.5000 & 0.5606 \\ -0.3333 & 0.5000 & 0 & -0.1667 \\ 0.1293 & 0.5606 & -0.1667 & -0.5232 \end{pmatrix} \sigma^4$$
$$+ \begin{pmatrix} 0.0307 & 0.0821 & -0.0625 & -0.0431 \\ 0.0821 & 0.3764 & -0.3333 & -0.1401 \\ -0.0625 & -0.3333 & 0.3125 & 0.0833 \\ -0.0431 & -0.1401 & 0.0833 & 0.1071 \end{pmatrix} r\,.$$

1.3.3.2. Random matrices with independent normally distributed elements

Consider a symmetric matrix A_0 with a simple eigenvalue μ_g, which is perturbed by a symmetric random matrix $B(\omega)$. Let the matrix B have normally distributed elements $b_{ij}(\omega)$ where $\langle b_{ij}(\omega)\rangle = 0$ and $\langle b_{ij}^2(\omega)\rangle = \sigma^2$. These elements are assumed to be independent with the symmetry property $b_{ij}(\omega) = b_{ji}(\omega)$ almost surely. The eigenvalues of the matrix $A_0 + B(\omega)$ are denoted by

$$\lambda_1(\omega) \leqq \lambda_2(\omega) \leqq \dots \leqq \lambda_n(\omega)\,.$$

For these eigenvalues, expansions can be formally written in the form

$$\lambda_g(\omega) = \mu_g - \sum_{k=1}^{\infty} \lambda_{gk}(\omega) \doteq \sum_{k=0}^{\infty} \lambda'_{gk}(\omega)\,, \tag{1.78}$$

where the terms $\lambda'_{gk}(\omega)$ are homogeneous of the k-th order in the $b_{ij}(\omega)$. These homogeneous terms can be determined by the methods given in Section 1.3.1. Assuming normally distributed perturbations the conditions for the convergence of the expansions are not satisfied with positive probability since the perturbations $b_{ij}(\omega)$ exceed every given number with positive probability. We will show that the expansions given in Section 1.3.3.1 may still be used as approximations, and we will estimate the error.

The expansion (1.78) converges for $|b_{ij}(\omega)| \leqq \varepsilon$; that is, for all $\omega \in A \subset \Omega$ with

$$A = \{\omega \in \Omega\colon |b_{ij}(\omega)| \leqq \varepsilon,\ i, j = 1, 2, \dots, n\,,\ i \leqq j\}\,.$$

Because of the independence of the elements b_{ij}, for $i \leqq j$, we have the relation

$$\mathsf{P}(A) = (1 - \eta)^N\,, \qquad N = \tfrac{1}{2}\,(n^2 + n)\,,$$

where $\eta \doteq \mathsf{P}\{|b_{ij}| > \varepsilon\}$. Then for $\bar{A} = \Omega \setminus A$ we have

$$\mathsf{P}(\bar{A}) = 1 - (1 - \eta)^N .$$

The set $\bar{A}$ is the set of all $\omega \in \Omega$ for which the expansion (1.78) is not necessarily convergent. Hence, for the eigenvalue $\lambda_g(\omega)$ we have

$$\lambda_g(\omega) = \begin{cases} \sum_{k=0}^{\infty} \lambda'_{gk}(\omega) & \text{for} \quad \omega \in A , \\ \lambda_g(\omega) & \text{for} \quad \omega \in \bar{A} ; \end{cases}$$

and the mean

$$\begin{aligned} \langle \lambda_g(\omega) \rangle &= \int_A \sum_{k=0}^{\infty} \lambda'_{gk}(\omega) \, \mathsf{P}(\mathrm{d}\omega) + \int_{\bar{A}} \lambda_g(\omega) \, \mathsf{P}(\mathrm{d}\omega) \\ &= \sum_{k=0}^{\infty} \int_A \lambda'_{gk}(\omega) \, \mathsf{P}(\mathrm{d}\omega) + \int_{\bar{A}} \lambda_g(\omega) \, \mathsf{P}(\mathrm{d}\omega) , \end{aligned} \tag{1.79}$$

because of the uniform convergence of the series (1.78) on A. From the inequality $|\lambda_g(\omega)| \leqq |A_0 + B(\omega)|$ it follows that

$$\begin{aligned} \Big| \int_{\bar{A}} \lambda_g(\omega) \, \mathsf{P}(\mathrm{d}\omega) \Big| &\leqq \int_{\bar{A}} |\lambda_g(\omega)| \, \mathsf{P}(\mathrm{d}\omega) \leqq \int_{\bar{A}} [\sum_{i,j=1}^{n} (a_{ij} + b_{ij}(\omega))^2]^{1/2} \, \mathsf{P}(\mathrm{d}\omega) \\ &\leqq [\sum_{i,j=1}^{n} \int_{\bar{A}} (a_{ij} + b_{ij}(\omega))^2 \, \mathsf{P}(\mathrm{d}\omega)]^{1/2} \end{aligned}$$

if we apply the inequality $\langle \sqrt{X} \rangle \leqq \sqrt{\langle X \rangle}$. Furthermore, we observe that

$$\begin{aligned} \int_{\bar{A}} (a_{ij} + b_{ij}(\omega))^2 \, \mathsf{P}(\mathrm{d}\omega) = a_{ij}^2 \, \mathsf{P}(\bar{A}) & \\ + 2a_{ij} [\int_{\bar{A}_{ij}^1} b_{ij}(\omega) \, \mathsf{P}(\mathrm{d}\omega) + \int_{\bar{A}_{ij}^2} b_{ij}(\omega) \, \mathsf{P}(\mathrm{d}\omega)] & \\ + \int_{\bar{A}_{ij}^1} b_{ij}^2(\omega) \, \mathsf{P}(\mathrm{d}\omega) + \int_{\bar{A}_{ij}^2} b_{ij}^2(\omega) \, \mathsf{P}(\mathrm{d}\omega) , & \end{aligned}$$

where

$$\bar{A}_{ij}^1 = \bar{A} \cap \{|b_{ij}| \leqq \varepsilon\} , \qquad \bar{A}_{ij}^2 = \bar{A} \cap \{|b_{ij}| > \varepsilon\} .$$

Thus

$$\Big| \int_{\bar{A}_{ij}^1} b_{ij}^k(\omega) \, \mathsf{P}(\mathrm{d}\omega) \Big| \leqq \varepsilon^k \mathsf{P}(\bar{A}) \quad \text{for} \quad k = 1, 2 ,$$

$$\begin{aligned} \left| \int_{\bar{A}_{ij}^2} b_{ij}(\omega) \, \mathsf{P}(\mathrm{d}\omega) \right| &\leqq \frac{1}{\sqrt{2\pi}\sigma} \int_{|x| > \varepsilon} |x| \exp\left(- \frac{x^2}{2\sigma^2} \right) \mathrm{d}x \\ &= \frac{2\sigma}{\sqrt{2\pi}} \exp\left(- \frac{\varepsilon^2}{2\sigma^2} \right) , \end{aligned} \tag{1.80}$$

$$\begin{aligned} \left| \int_{\bar{A}_{ij}^2} b_{ij}^2(\omega) \, \mathsf{P}(\mathrm{d}\omega) \right| &\leqq \frac{1}{\sqrt{2\pi}\sigma} \int_{|x| > \varepsilon} x^2 \exp\left(- \frac{x^2}{2\sigma^2} \right) \mathrm{d}x \\ &= \frac{2\sigma}{\sqrt{2\pi}} \varepsilon \exp\left(- \frac{\varepsilon^2}{2\sigma^2} \right) + \sigma^2 \eta \end{aligned}$$

and then

$$\int_{\bar{A}} (a_{ij} + b_{ij})^2 \, \mathsf{P}(\mathrm{d}\omega) \leqq (|a_{ij}| + \varepsilon)^2 \, \mathsf{P}(\bar{A}) + \sigma^2 \eta + \frac{2\sigma}{\sqrt{2\pi}} \exp\left(-\frac{\varepsilon^2}{2\sigma^2}\right) [2\,|a_{ij}| + \varepsilon] .$$

Hence, we obtain

$$\left| \int_{\bar{A}} \lambda_g(\omega) \, \mathsf{P}(\mathrm{d}\omega) \right| \leqq \left[\mathsf{P}(\bar{A}) \sum_{i,j=1}^{n} (|a_{ij}| + \varepsilon)^2 + n^2\sigma^2\eta + \{2 \sum_{i,j=1}^{n} |a_{ij}| + n^2\varepsilon\} \frac{2\sigma}{\sqrt{2\pi}} \exp\left(-\frac{\varepsilon^2}{2\sigma^2}\right)\right]^{1/2}$$
$$\leqq \left[\mathsf{P}(\bar{A}) \, (|A_0| + n\varepsilon)^2 + n^2\sigma^2\eta + \{2n\,|A_0| + n^2\varepsilon\} \times \frac{2\sigma}{\sqrt{2\pi}} \exp\left(-\frac{\varepsilon^2}{2\sigma^2}\right)\right]^{1/2} \doteq \beta_{21} . \tag{1.81}$$

Let β_1^{gm} denote the difference between the series (1.78) and the sum up to the m-th term, that is,

$$\Big| \sum_{k=m+1}^{\infty} \int_{A} \lambda'_{gk}(\omega) \, \mathsf{P}(\mathrm{d}\omega) \Big| \leqq \beta_1^{gm} .$$

Then from (1.78) it follows that

$$|\langle \lambda_g \rangle - \sum_{k=0}^{m} \int_{A} \lambda'_{gk}(\omega) \, \mathsf{P}(\mathrm{d}\omega)| \leqq \beta_1^{gm} + \beta_{21} ;$$

and finally

$$|\langle \lambda_g \rangle - \sum_{k=0}^{m} \langle \lambda'_{gk} \rangle| \leqq |\sum_{k=0}^{m} \int_{\bar{A}} \lambda'_{gk}(\omega) \, \mathsf{P}(\mathrm{d}\omega)| + \beta_1^{gm} + \beta_{21} .$$

Every summand $\int_{\bar{A}} \lambda'_{gk}(\omega) \, \mathsf{P}(\mathrm{d}\omega)$ is of the form

$$c_{p_1 \dots p_k} \int_{\bar{A}} X_{p_1} \dots X_{p_k} \, \mathsf{P}(\mathrm{d}\omega) ,$$

where the $c_{p_1 \dots p_k}$ are real numbers and the $X_{p_1}, \dots, X_{p_k}$ are k parts of the $b_{ij}(\omega)$. Because of the independence of the b_{ij}, the term $\int_{\bar{A}} X_{p_1} \dots X_{p_k} \, \mathsf{P}(\mathrm{d}\omega)$ can be decomposed into factors of the form $\int_{\bar{A}} X_p^r \, \mathsf{P}(\mathrm{d}\omega)$, where

$$\left| \int_{\bar{A}} X_p^r \, \mathsf{P}(\mathrm{d}\omega) \right| \leqq \varepsilon^r \, \mathsf{P}(\bar{A}) + \frac{2}{\sqrt{2\pi}\sigma} \int_{\varepsilon}^{\infty} x^r \exp\left(-\frac{x^2}{2\sigma^2}\right) \mathrm{d}x .$$

Thus

$$\Big| \int_{\bar{A}} \lambda'_{gk} \, \mathsf{P}(\mathrm{d}\omega) \Big| \leqq d_{gk}(\varepsilon, \sigma)$$

with

$$\lim_{\sigma\to 0} d_{gk}(\varepsilon, \sigma) = 0 \quad \text{and} \quad \lim_{\varepsilon\to\infty} d_{gk}(\varepsilon, \sigma) = 0$$

since

$$0 \leqq \lim_{\varepsilon\to\infty} \varepsilon^r \eta \leqq \lim_{\varepsilon\to\infty} \frac{2\varepsilon^r}{\sqrt{2\pi}\sigma} \int_{\varepsilon}^{\infty} x \exp\left(-\frac{x^2}{2\sigma^2}\right) dx$$

$$= \lim_{\varepsilon\to\infty} \frac{2\sigma}{\sqrt{2\pi}} \varepsilon^r \exp\left(-\frac{\varepsilon^2}{2\sigma^2}\right) = 0 .$$

Hence, we have

$$|\langle\lambda_g\rangle - \sum_{k=0}^{m} \langle\lambda'_{gk}\rangle| \leqq \beta_1^{gm} + \beta_{21} + \beta_{22}^{gm} ,$$

where

$$\beta_{22}^{gm} \doteq \sum_{k=0}^{m} d_{gk}(\varepsilon, \sigma)$$

so that

$$\lim_{\varepsilon/\sigma\to\infty} [\beta_{21} + \beta_{22}^{gm}] = 0 .$$

These results can be collected in the following theorem.

Theorem 1.14. *Let A_0 be a symmetric matrix with the simple eigenvalue μ_g and let $B(\omega)$ be a symmetric random matrix whose elements $b_{ij}(\omega)$ are independent and normally distributed with $\langle b_{ij}\rangle = 0$ and $\langle b_{ij}^2\rangle = \sigma^2$. Then for the mean of the eigenvalue $\lambda_g(\omega)$ of the matrix $A_0 + B(\omega)$ the expression*

$$\langle\lambda_g(\omega)\rangle = \mu_g - \sum_{k=1}^{m} \langle\lambda_{gk}(\omega)\rangle + F_{gm}$$

can be obtained, where λ_{gk} results from perturbation theory. The error F_{gm} is composed of the error following from the substitution of the series (1.78) by the sum of its first m terms and the error which is obtained by the fact that the series does not converge for $\omega \in \bar{A}$. This last error converges to zero as $\varepsilon/\sigma \to \infty$.

We now wish to give a more exact estimate for $\sum_{k=0}^{2} \int_{\bar{A}} \lambda'_{gk} \mathsf{P}(d\omega)$. First, we put

$$a \doteq \int_{\bar{A}} b_{ij} \mathsf{P}(d\omega) \quad \text{and} \quad b \doteq \int_{\bar{A}} b_{ij}^2 \mathsf{P}(d\omega) ,$$

where these quantities are independent of the indices i and j. The expressions

$$\lambda'_{g1} = -\lambda_{g1} = \hat{b}_{gg} , \qquad \lambda'_{g2} = -\lambda_{g2} = \sum_{\substack{k=1\\k\neq g}}^{n} \frac{\hat{b}_{gk}^2}{\mu_{gk}}$$

from Theorem 1.10 are used where

$$\hat{B} = (\hat{b}_{ij})_{1\leqq i,j\leqq n} = CBC^\mathsf{T} \quad \text{and} \quad CA_0C^\mathsf{T} = (\mu_i\delta_{ij})_{1\leqq i,j\leqq n} .$$

From this we obtain

$$\int_{\bar{A}} \lambda'_{g1} \, \mathsf{P}(\mathrm{d}\omega) = \int_{\bar{A}} \hat{b}_{gg} \, \mathsf{P}(\mathrm{d}\omega) = \sum_{i,\, j=1}^{n} c_{gi} c_{gj} \int_{\bar{A}} b_{ij} \, \mathsf{P}(\mathrm{d}\omega) = a\Big(\sum_{i=1}^{n} c_{gi}\Big)^2$$

and because of the independence of the random variables b_{ij}, $i, j = 1, 2, \dots, n$, $i \leqq j$,

$$\int_{\bar{A}} b_{ij} b_{rs} \, \mathsf{P}(\mathrm{d}\omega) = b(\delta_{ir}\delta_{js} + \delta_{is}\delta_{jr} - \delta_{ijrs}) + a^2(1 - \delta_{ir}\delta_{js} - \delta_{is}\delta_{jr} + \delta_{ijrs}) \,.$$

For $k \neq g$ it follows that

$$\int_{\bar{A}} \hat{b}^2_{gk} \, \mathsf{P}(\mathrm{d}\omega) = \sum_{i,\, j,\, r,\, s=1}^{n} c_{gi} c_{kj} c_{gr} c_{ks} \int_{\bar{A}} b_{ij} b_{rs} \, \mathsf{P}(\mathrm{d}\omega)$$

$$= b(1 - C_{gk}) + a^2 \Big[\Big(\sum_{i=1}^{n} c_{gi}\Big)^2 \Big(\sum_{j=1}^{n} c_{kj}\Big)^2 - 1 + C_{gk}\Big]$$

where $C_{gk} \doteq \sum_{i=1}^{n} c_{gi}^2 c_{ki}^2$. Finally, we compute

$$\int_{\bar{A}} \lambda'_{g2} \, \mathsf{P}(\mathrm{d}\omega) = \sum_{\substack{k=1 \\ k \neq g}}^{n} \frac{1}{\mu_{gk}} \int_{\bar{A}} \hat{b}^2_{gk} \, \mathsf{P}(\mathrm{d}\omega)$$

$$= b \sum_{\substack{k=1 \\ k \neq g}}^{n} \frac{1 - C_{gk}}{\mu_{gk}} + a^2 \Bigg[\Big(\sum_{i=1}^{n} c_{gi}\Big)^2 \sum_{\substack{k=1 \\ k \neq g}}^{n} \frac{1}{\mu_{gk}} \Big(\sum_{j=1}^{n} c_{kj}\Big)^2 - \sum_{\substack{k=1 \\ k \neq g}}^{n} \frac{1 - C_{gk}}{\mu_{gk}}\Bigg].$$

Using the estimates for a and b from (1.80), namely

$$|a| \leqq \varepsilon \, \mathsf{P}(\bar{A}) + \frac{2\sigma}{\sqrt{2\pi}} \exp\left(-\frac{\varepsilon^2}{2\sigma^2}\right) \doteq a_1 \,,$$

$$b \leqq \varepsilon^2 \, \mathsf{P}(\bar{A}) + \frac{2\sigma\varepsilon}{\sqrt{2\pi}} \exp\left(-\frac{\varepsilon^2}{2\sigma^2}\right) + \sigma^2 \eta \doteq b_1$$

we have

$$\Big|\sum_{k=0}^{2} \int_{\bar{A}} \lambda'_{gk} \, \mathsf{P}(\mathrm{d}\omega)\Big| \leqq \sum_{k=0}^{2} d_{gk}(\varepsilon; \sigma) \,, \tag{1.82}$$

where

$$d_{g0}(\varepsilon, \sigma) \leqq |\mu_g| \, \mathsf{P}(\bar{A}) \,,$$

$$d_{g1}(\varepsilon, \sigma) \leqq a_1 \Big(\sum_{i=1}^{n} c_{gi}\Big)^2 \,,$$

$$d_{g2}(\varepsilon, \sigma) \leqq b_1 \left|\sum_{\substack{k=1 \\ k \neq g}}^{n} \frac{1 - C_{gk}}{\mu_{gk}}\right| + a_1^2 \left|\Big(\sum_{i=1}^{n} c_{gi}\Big)^2 \sum_{\substack{k=1 \\ k \neq g}}^{n} \frac{1}{\mu_{gk}} \Big(\sum_{j=1}^{n} c_{kj}\Big)^2 - \sum_{\substack{k=1 \\ k \neq g}}^{n} \frac{1 - C_{gk}}{\mu_{gk}}\right|.$$

As a numerical example we consider the matrix

$$A_0 = \begin{pmatrix} -2 & 4 \\ 4 & 4 \end{pmatrix}$$

with eigenvalues $\mu_1 = -4$, $\mu_2 = 6$ which is perturbed by the random matrix

$$B(\omega) = \begin{pmatrix} b_{11}(\omega) & b_{12}(\omega) \\ b_{21}(\omega) & b_{22}(\omega) \end{pmatrix}.$$

The eigenvalues of the random matrix $A_0 + B(\omega)$ are calculated by

$$\lambda_{1/2}(\omega) = 1 + \tfrac{1}{2}\left(b_{11}(\omega) + b_{22}(\omega)\right) \mp 5\sqrt{1 + X(\omega)}$$

where

$$X(\omega) = \tfrac{1}{25}\left(3(b_{22} - b_{11}) + 8b_{12} + \tfrac{1}{4}(b_{11} - b_{22})^2 + b_{12}^2\right).$$

Furthermore, the expansion of these eigenvalues up to terms of second order follows from

$$5\sqrt{1 + X(\omega)} = 5 + \tfrac{3}{10}(b_{22} - b_{11}) + \tfrac{4}{5}b_{12} + \tfrac{2}{125}(b_{22} - b_{11})^2 + \tfrac{9}{250}b_{12}^2 + \dots$$

This series converges for $\omega \in A$, where

$$A = \{\omega : |b_{ij}(\omega)| \leqq 1.47433,\ i, j = 1, 2\}$$

since $|X(\omega)| < 1$ holds for $\omega \in A$. Assuming that the b_{ij} are zero mean independent random variables with variance $\langle b_{ij}^2\rangle = \sigma^2$, than we have for the mean of the eigenvalues up to terms of seccond order,

$$\langle\lambda_1\rangle = -4 - \tfrac{17}{250}\sigma^2, \qquad \langle\lambda_2\rangle = 6 + \tfrac{17}{250}\sigma^2.$$

The same result can be obtained from (1.72) using

$$C = \frac{1}{\sqrt{5}}\begin{pmatrix} -2 & 1 \\ 1 & 2 \end{pmatrix}.$$

Let us now consider normally distributed random variables with $\sigma = 1/50$. For an estimate of the error, we note that from

$$\sum_{k=3}^{\infty} \lambda'_{gk} = \pm \tfrac{5}{8}\gamma \pm \tfrac{5}{16}X^3(1 + \vartheta X)^{-5/2},$$

where $0 < \vartheta(\omega) < 1$, and

$$\gamma = \tfrac{3}{1250}(b_{22} - b_{11})^3 + \tfrac{6}{625}b_{12}^2(b_{22} - b_{11}) + \tfrac{4}{625}b_{12}(b_{11} - b_{22})^2 + \tfrac{16}{625}b_{12}^3 + \tfrac{1}{625}\left(\tfrac{1}{4}(b_{11} - b_{22})^2 + b_{12}^2\right)^2,$$

it follows that

$$\int_{\bar{A}} \sum_{k=3}^{\infty} \lambda'_{gk} \, \mathsf{P}(\mathrm{d}\omega) \leqq \tfrac{5}{8} \int_{\Omega} \gamma(\omega) \, \mathsf{P}(\mathrm{d}\omega) + \tfrac{5}{16} (1-d)^{-5/2} \int_{\Omega} |X|^3 \, \mathsf{P}(\mathrm{d}\omega) \, .$$

The error can be estimated which is due to the substitution of the series by the first terms of the series. We note that

$$|X| < d \, , \qquad d = \tfrac{14}{25} \varepsilon + \tfrac{2}{25} \varepsilon^2 \, .$$

By the aid of (1.81) for β_{21} and the estimate (1.82) for $\sum\limits_{k=0}^{2} \int\limits_{\bar{A}} \lambda'_{gk} \, \mathsf{P}(\mathrm{d}\omega)$, and because of

$$\left| \sum_{\substack{k=1 \\ k \neq g}}^{2} \frac{1 - C_{gk}}{\mu_{kg}} \right| = 0.068$$

and

$$\left| \left(\sum_{k=1}^{2} c_{gk} \right)^2 \sum_{\substack{k=1 \\ k \neq g}}^{2} \frac{1}{\mu_{kg}} \left(\sum_{r=1}^{2} c_{kr} \right)^2 - \sum_{\substack{k=1 \\ k \neq g}}^{2} \frac{1 - C_{kg}}{\mu_{kg}} \right| = 0.032$$

Table 1.1 is obtained. It can be seen that the error caused by the nonconvergence of the expansion for $\omega \in \bar{A}$ decreases very quickly with increasing quotient ε/σ.

Table 1.1

σ	ε	ε/σ	β_1^{g2}	β_{21}	β_{22}^{g2}
1/50	3/50	3	$1.2138 \cdot 10^{-6}$	$4.7173 \cdot 10^{-1}$	$2.5025 \cdot 10^{-2}$
1/50	5/50	5	$1.2432 \cdot 10^{-6}$	$4.5053 \cdot 10^{-3}$	$5.4810 \cdot 10^{-5}$
1/50	8/50	8	$1.2930 \cdot 10^{-6}$	$1.3135 \cdot 10^{-6}$	$1.8290 \cdot 10^{-13}$
1/50	15/50	15	$1.4446 \cdot 10^{-6}$	$4.5070 \cdot 10^{-24}$	$2.1763 \cdot 10^{-48}$

We will compare the results obtained by perturbation expansions for normally distributed perturbations and the results achieved by Monte Carlo simulation. These investigations were considered for the matrix A_0 of (1.76) which is perturbed by a symmetric random matrix $B(\omega)$ with independent elements $b_{ij}(\omega)$ for $i \leqq j$.

We will consider normally distributed perturbations with $\sigma^2 = 1/100$ and uniformly distributed perturbations on $[-d, d]$ with the same variance, that is, with $d = \sqrt{3}/100$. We calculate the expectations of the four eigenvalues from (1.77) up to terms of fourth order, where $r = 0$ is taken for the normally distributed case. The values of $\langle \lambda_g \rangle$ for $g = 1, 2, 3, 4$ obtained by simulation of N realizations are denoted by L_g^N. The results are given in Table 1.2. We observe that for normally distributed perturbations the simulation results coincide better with $\langle \lambda_g \rangle_{\text{perturb}}$ than the results for uniformly distributed perturbations with the same variance.

Table 1.2

	Uniformly distributed perturbations on $[-\sqrt{3}/100, \sqrt{3}/100]$		Normally distributed perturbations with $\sigma^2 = 1/100$	
g	$\langle\lambda_g\rangle_{\text{perturb}}$	L_g^{2000}	$\langle\lambda_g\rangle_{\text{perturb}}$	L_g^{2000}
1	−3.00602	−3.00194	−3.00602	−3.00342
2	−0.01000	−0.00700	−0.01000	−0.01247
3	1.00751	1.00230	1.00751	1.00910
4	3.00852	3.00363	3.00852	3.00429

1.3.3.3. Probabilistic characteristics of eigenvalues of multiplicity two

An important special case of the expansion given in Section 1.3.2 is obtained for $h = 2$. In this case the terms $\tilde{\lambda}_{g1;\,0}, \tilde{\lambda}_{g1;\,1}$ can be written explicitly, namely

$$\tilde{\lambda}_{g1;\,0/1} = -\frac{\tilde{J}_{1,\,n-1}}{2\tilde{J}_{0,\,n-1}} \pm \frac{1}{2}\sqrt{\left(\frac{\tilde{J}_{1,\,n-1}}{\tilde{J}_{0,\,n-2}}\right)^2 - 4\frac{\tilde{J}_{2n}}{\tilde{J}_{0,\,n-2}}}\,.$$

In the term $\tilde{\lambda}_{..;\,0/1}$ the 0 belongs to the upper sign and the 1 to the lower sign. For $p = 3$ (1.60) yields

$$\tilde{\lambda}_{g2;\,0/1} = \frac{\mp 1}{\tilde{J}_{0,\,n-2}\sqrt{\left(\frac{\tilde{J}_{1,\,n-1}}{\tilde{J}_{0,\,n-2}}\right)^2 - 4\frac{\tilde{J}_{2n}}{\tilde{J}_{0,\,n-2}}}} \times \{\tilde{J}_{3n} + \tilde{\lambda}_{g1;\,0/1}\tilde{J}_{2,\,n-1} + \tilde{\lambda}^2_{g1;\,0/1}\tilde{J}_{1,\,n-2} + \tilde{\lambda}^3_{g1;\,0/1}\tilde{J}_{0,\,n-3}\}\,.$$

Hence, we obtain the expansion

$$\lambda_{g+p}(\omega) = \mu_g - \tilde{\lambda}_{g1;\,p} - \tilde{\lambda}_{g2;\,p} - \ldots$$

for $p = 0, 1$, from which some statistical characteristics can be computed. The first approximation of $\lambda_{g+p}(\omega)$ is defined by

$$\lambda^{\mathrm{I}}_{g+p}(\omega) \doteq \mu_g - \lambda_{g1;\,p}(\omega)\,.$$

It then follows that

$$\langle\lambda_{g+p}(\omega)\rangle \approx \langle\lambda^{\mathrm{I}}_{g+p}(\omega)\rangle = \mu_g \mp \frac{1}{2\,|\tilde{J}_{0,\,n-2}|}\langle\sqrt{\tilde{J}^2_{1,\,n-1} - 4\tilde{J}_{2n}\tilde{J}_{0,\,n-2}}\rangle\,.$$

Using $X \doteq \tilde{J}^2_{1,\,n-1} - 4\tilde{J}_{2n}\tilde{J}_{0,\,n-2} \geqq 0$, almost surely the mean of the first approximation is calculated by

$$\langle\lambda^{\mathrm{I}}_{g+p}\rangle = \mu_g \mp \frac{1}{2\,|\tilde{J}_{0,\,n-2}|}\sqrt{\langle X\rangle}\left[\sum_{\nu=0}^{m}\binom{1/2}{\nu}\left\langle\left(\frac{X-\langle X\rangle}{\langle X\rangle}\right)^{\nu}\right\rangle + \binom{1/2}{m+1}\left\langle\left(\frac{X-\langle X\rangle}{\langle X\rangle}\right)^{m+1}\left\{1+\vartheta(\omega)\frac{X-\langle X\rangle}{\langle X\rangle}\right\}^{-(2m+1)/2}\right\rangle\right]$$

where $0 < \vartheta(\omega) < 1$ almost surely. For m odd we have

$$\binom{1/2}{m+1} < 0$$

and

$$\left(\frac{X - \langle X \rangle}{\langle X \rangle}\right)^{m+1} \left\{1 + \vartheta(\omega) \frac{X - \langle X \rangle}{\langle X \rangle}\right\}^{-(2m+1)/2} \geqq 0 \quad \text{almost surely;}$$

and the estimate for the mean of the first approximation

$$\langle \lambda_g^{\mathrm{I}} \rangle \geqq \mu_g - A_{gk}\,, \qquad \langle \lambda_{g+1}^{\mathrm{I}} \rangle \leqq \mu_g + A_{gk}$$

can be derived, where

$$A_{gk} \doteq \frac{1}{2\,|\tilde{J}_{0,\,n-2}|} \sqrt{\langle X \rangle} \sum_{\nu=0}^{2k+1} \binom{1/2}{\nu} \left\langle \left(\frac{X - \langle X \rangle}{\langle X \rangle}\right)^{\nu} \right\rangle$$

for $k = 0, 1, 2, \ldots$ The right-hand side of the inequality above converges to the mean of λ_g^{I} or $\lambda_{g+1}^{\mathrm{I}}$, respectively, if $|X - \langle X \rangle| < \langle X \rangle$ almost surely. For example this is satisfied if $X \in [0, 1]$ almost surely and $\langle X \rangle \geqq 1/2$.

For the calculation of the eigenvalues we assumed that A_0 was transformed into a diagonal matrix,

$$C\big(\tilde{A}_0 + B(\omega)\big)\, C^{\mathsf{T}} = \big((\mu_i - \mu_g)\, \delta_{ij}\big)_{1 \leqq i,\, j \leqq n} + \hat{B}(\omega)$$

where

$$\hat{B}(\omega) = \big(\hat{b}_{ij}(\omega)\big)_{1 \leqq i,\, j \leqq n}\,, \qquad C = (c_{ij})_{1 \leqq i,\, j \leqq n}$$

and

$$\hat{b}_{ij}(\omega) = \sum_{s,\, t=1}^{n} c_{is} c_{jt} b_{st}(\omega)\,.$$

Then the first approximations of the eigenvalues λ_{g+p} for $p = 0, 1$ are given by

$$\tilde{\lambda}_{g1;\,0/1}(\omega) = -\frac{1}{2}\,(\hat{b}_{gg} + \hat{b}_{g+1g+1}) \pm \frac{1}{2} \sqrt{(\hat{b}_{gg} - \hat{b}_{g+1g+1})^2 + 4\hat{b}_{gg+1}^2}\,,$$

$$\begin{aligned} \tilde{\lambda}_{g2;\,0/1}(\omega) = {} & \frac{1}{2} \sum_{\substack{p=1 \\ p \neq g,\, g+1}}^{n} \frac{1}{\mu_{pg}} (\hat{b}_{pg+1}^2 + \hat{b}_{pg}^2) \\ & \mp \frac{2}{\sqrt{(\hat{b}_{gg} - \hat{b}_{g+1g+1})^2 + 4\hat{b}_{gg+1}^2}} \\ & \times \Bigg\{ \hat{b}_{gg+1} \sum_{\substack{p=1 \\ p \neq g,\, g+1}}^{n} \frac{1}{\mu_{pg}} \hat{b}_{pg} \hat{b}_{pg+1} + \frac{1}{4}\,(\hat{b}_{g+1g+1} - \hat{b}_{gg}) \\ & \times \sum_{\substack{p=1 \\ p \neq g,\, g+1}}^{n} \frac{1}{\mu_{pg}} (b_{pg+1}^2 - b_{pg}^2) \Bigg\}\,. \end{aligned} \tag{1.83}$$

Problems of the distribution of $\lambda_{g1;\,p}$ as a root of a random polynomial are treated by BHARUCHA-REID [2, 4], HAMBLEN [1] and HAMMERSLEY [1]. The results given in these books and papers are more of theoretical interest and only in some cases one can give numerical calculations of the mean and other statistical characteristics of the roots.

As a numerical example for this class of problems, consider the diagonal matrix

$$A_0 = \begin{pmatrix} -1 & 0 & 0 \\ 0 & 1 & 0 \\ 0 & 0 & 1 \end{pmatrix}$$

and the perturbation matrix

$$B = \begin{pmatrix} 0 & b_{12} & 0 \\ b_{21} & b_{22} & b_{23} \\ 0 & b_{32} & b_{33} \end{pmatrix}.$$

Let the b_{ij} be independent, and let

$$b_{ij} = \pm\eta \ (\eta > 0) , \quad \text{where} \quad \mathsf{P}(b_{ij} = \eta) = \mathsf{P}(b_{ij} = -\eta) = \tfrac{1}{2} .$$

The matrix A_0 has the eigenvalues $\mu_1 = -1$, $\mu_2 = \mu_3 = 1$. Furthermore, the expansions

$$\lambda_{2+p}(\omega) = \mu_2 - \sum_{k=1}^{\infty} \lambda_{2k,\,p}(\omega) , \tag{1.84}$$

for $p = 0, 1$, converge for all η for which $D_{2;0/1} = 6\sqrt{6}(\sqrt{6} + 4)\,\eta < 1$, that is, for all η with

$$0 < \eta < 0.01054 ,$$

because

$$e_{20} = e_{21} = |\lambda_{21;\,0} - \lambda_{21;\,1}| = \sqrt{(b_{22} - b_{33})^2 + 4b_{23}^2} \geqq 2\eta .$$

$$d_2 = |\mu_2 - \mu_1| = 2 \quad \text{and} \quad |B| \leqq \sqrt{6}\eta .$$

Since the estimates in the proof of Theorem 1.11 are crude, the domain of convergence for the expansions (1.84) will be larger in general than the given in Theorem 1.11. The substance of this theorem is the convergence of the expansions given for small values of the $b_{ij}(\omega)$. Some numerical calculations for the example given above will underline these remarks, since in this case for comparison we can also compute the exact mean of the eigenvalues by calculation of the realizations.

For our example, from (1.83) it follows that

$$\begin{aligned} \langle\lambda_{21;\,0/1}\rangle &= \pm\tfrac{1}{2}\,\langle\sqrt{(b_{22} - b_{33})^2 + 4b_{23}^2}\rangle , \\ \langle\lambda_{22;\,0/1}\rangle &= -\tfrac{1}{4}\,\langle b_{12}^2\rangle , \end{aligned} \tag{1.85}$$

because of the independence of the b_{ij}, and from Theorem 1.10 for the first eigenvalue

$$\langle\lambda_{11}\rangle = 0 , \qquad \langle\lambda_{12}\rangle = \tfrac{1}{2}\,\langle b_{12}^2\rangle .$$

Put

$$\langle \lambda_g^{\mathrm{I}} \rangle \doteq \mu_g - \langle \lambda_{g1} \rangle ,$$

$$\langle \lambda_g^{\mathrm{II}} \rangle \doteq \mu_g - \langle \lambda_{g1} \rangle - \langle \lambda_{g2} \rangle .$$

Table 1.3

	g	$\langle \lambda_g \rangle$	$\langle \lambda_g^{\mathrm{I}} \rangle$	$\langle \lambda_g^{\mathrm{II}} \rangle$
$\eta = 0.01$	1	−1.00005	−1.00000	−1.00005
	2	0.98793	0.98791	0.98794
	3	1.01209	1.01207	1.01209
$\eta = 0.1$	1	−1.00501	−1.00000	−1.00500
	2	0.88193	0.87928	0.88178
	3	1.12308	1.12071	1.12321
$\eta = 0.2$	1	−1.02019	−1.00000	−1.02000
	2	0.76972	0.75857	0.76587
	3	1.25048	1.24142	1.25142

The comparison in Table 1.3 shows the great correspondence between the exact values $\langle \lambda_g \rangle$ and $\langle \lambda_g^{\mathrm{I}} \rangle$ or $\langle \lambda_g^{\mathrm{II}} \rangle$, respectively, also for such values of η which are not contained in the domain of convergence given by Theorem 1.11.

We now consider the example given above for the case when the perturbations b_{ij} take the values $\pm\eta$ and 0 with probabilities

$$\mathsf{P}(b_{ij} = \eta) = \mathsf{P}(b_{ij} = -\eta) = \mathsf{P}(b_{ij} = 0) = 1/3 .$$

With the aid of the expansions of the perturbation theory we obtain very good results, although in this case exact statements relative to the convergence were not be proved. The reason is that $e_{20} = e_{21} = 0$ for the realizations with $b_{22} = b_{33}$, $b_{23} = 0$. For $\eta = 0, 1$ the results are given in Table 1.4. We note that for very simple distributions the calculation of the mean of a root of a random variable as in (1.85) is difficult, and we must therefore calculate this term approximately. Thus

$$\langle \sqrt{X} \rangle \leqq \sqrt{\langle X \rangle} \sum_{\nu=0}^{2k+1} \binom{1/2}{\nu} \left\langle \left(\frac{X - \langle X \rangle}{\langle X \rangle} \right)^{\nu} \right\rangle .$$

Table 1.4

g	$\langle \lambda_g \rangle$	$\langle \lambda_g^{\mathrm{I}} \rangle$	$\langle \lambda_g^{\mathrm{II}} \rangle$
1	−1.00334	−1.00000	−1.00333
2	0.91044	0.90888	0.91055
3	1.09290	1.09111	1.09278

For $X = (b_{22} - b_{33})^2 + 4b_{23}^2$ we have

$$\langle \lambda_2^{\mathrm{I}} \rangle \geqq \mu_2 - A_{2k} , \qquad \langle \lambda_3^{\mathrm{I}} \rangle \leqq \mu_2 + A_{2k} ,$$

where

$$A_{2k} = \frac{1}{2} \sqrt{\langle X \rangle} \sum_{\nu=0}^{2k+1} \binom{1/2}{\nu} \left\langle \left(\frac{X - \langle X \rangle}{\langle X \rangle} \right)^{\nu} \right\rangle .$$

Since for our example we have $\langle X\rangle = 6\eta^2$ and $|(X - \langle X\rangle)/\langle X\rangle| \leqq 1/3$, the term $\mu_2 - A_{2k}$ converges to $\langle\lambda_2^{\mathrm{I}}\rangle$ as $k \to \infty$, and $\mu_2 + A_{2k}$ to $\langle\lambda_3^{\mathrm{I}}\rangle$. Since

$$\left\langle\left(\frac{X - \langle X\rangle}{\langle X\rangle}\right)^{\nu}\right\rangle = \begin{cases}(1/3)^{\nu} & \text{for } \nu \text{ even},\\ 0 & \text{for } \nu \text{ odd}\end{cases}$$

it follows that

$$\langle\lambda_{2/3}^{\mathrm{I}}\rangle = \mu_2 \mp \frac{\sqrt{6}}{2}\eta \sum_{\nu=0}^{\infty}\binom{1/2}{2\nu}\left(\frac{1}{3}\right)^{2\nu}.$$

As a second example we consider continuously distributed perturbations of the same matrix A_0. Let the b_{ij} be independent, and let b_{22}, b_{33} and b_{12} be uniformly distributed on $[-d, d]$, b_{23} uniformly distributed on $I_r = [-d, -d/r] \cup [d/r, d]$, where $r > 1$. Using these conditions we can show convergence of the expansions

$$\lambda_{g+p}(\omega) = \mu_g - \sum_{k=1}^{\infty}\lambda_{gk;\,p}(\omega) \quad \text{almost surely}$$

for $p = 0, 1$. These series converge for those d for which the inequality $D_{2p} < 12\sqrt{6}\; d(\frac{1}{2}\sqrt{6r} + 2) < 1$ holds, hence, for such d such that $d < [6\sqrt{6}\,(\sqrt{6r} + 4)]^{-1}$ we can apply the inequalities

$$e_{2p} = \left((b_{22} - b_{33})^2 + 4b_{23}^2\right)^{1/2} \geqq 2\frac{d}{r} \quad \text{and} \quad |B| \leqq \sqrt{6}d\,.$$

Hence,

$$\langle\lambda_{2+p}(\omega)\rangle = \mu_2 - \sum_{k=1}^{\infty}\langle\lambda_{2k;\,p}(\omega)\rangle\,,$$

since $\langle D_{2p}^{m+1}\rangle = D_{2p}^{m+1}$ converges to zero as $m \to \infty$. By means of (1.83) it follows that

$$\langle\lambda_{2+p}(\omega)\rangle = \mu_2 \mp \frac{1}{2}\langle\sqrt{(b_{22} - b_{33})^2 + 4b_{23}^2}\rangle + \frac{1}{4}\langle b_{12}^2\rangle$$
$$\pm\left\langle\frac{b_{12}^2(b_{33} - b_{22})}{4\sqrt{(b_{22} - b_{33})^2 + 4b_{23}^2}}\right\rangle \pm \ldots$$

where the upper sign is written for $p = 0$ and the lower sign for $p = 1$. Furthermore, with the aid of the conditions made on the b_{ij} the means are computed by

$$\langle\sqrt{(b_{22} - b_{33})^2 + 4b_{23}^2}\rangle$$
$$= \frac{1}{8d^3(1 - 1/r)}\int_{-d}^{d}\int_{-d}^{d}\int_{I_r}\sqrt{(x_1 - x_2)^2 + 4x_3^2}\,\mathrm{d}x_3\,\mathrm{d}x_2\,\mathrm{d}x_1\,,$$

$$\langle b_{12}^2\rangle = \frac{1}{2d}\int_{-d}^{d}x^2\,\mathrm{d}x = \frac{1}{3}d^2\,,$$

$$\left\langle\frac{b_{12}^2(b_{33} - b_{22})}{4\sqrt{(b_{22} - b_{33})^2 + 4b_{23}^2}}\right\rangle = 0\,.$$

To estimate the expectation of $\sqrt{X}$ with $X = (b_{22} - b_{33})^2 + 4b_{23}^2$ we apply the expansion of the expectation of a root given above. The moments of the random variable X can be determined by

$$\langle X^k \rangle = \frac{1}{8d^3(1 - 1/r)} \int_{-d}^{d} \int_{-d}^{d} \int_{I_r} [(x_1 - x_2)^2 + 4x_3^2]^k \, dx_3 \, dx_2 \, dx_1$$

$$= \frac{1}{8d^3(1 - 1/r)} \sum_{p=0}^{k} \binom{k}{p} 4^{k-p} \int_{-d}^{d} \int_{-d}^{d} \int_{I_r} (x_1 - x_2)^{2p} x_3^{2(k-p)} \, dx_3 \, dx_2 \, dx_1$$

$$= \frac{2r}{r - 1} (2d)^{2k} C_k$$

where

$$C_k \doteq \sum_{p=0}^{k} \binom{k}{p} \frac{1}{(2k - 2p + 1)(2p + 1)(2p + 2)} (1 - r^{2p-2k-1}) .$$

Table 1.5

k	Upper bounds for $\langle \sqrt{X} \rangle$
0	$1.4652d$
1	$1.4010d$
2	$1.3936d$
3	$1.4024d$
4	$1.4441d$

For $\langle \sqrt{X} \rangle$ upper bounds are given in Table 1.5. Hence, we obtain $\langle \sqrt{X} \rangle \leqq 1.3936d$ and then

$$\langle \lambda_2^{\mathrm{II}} \rangle \geqq 1 - 0.6968d + 0.8333d^2,$$

$$\langle \lambda_3^{\mathrm{II}} \rangle \leqq 1 + 0.6968d + 0.8333d^2.$$

The variance of $\lambda_2(\omega)$ and $\lambda_3(\omega)$ can be calculated up to terms of second order from

$$\operatorname{var} \lambda_{2+p}(\omega) = \langle (\lambda_{2+p} - \langle \lambda_{2+p} \rangle)^2 \rangle = \langle (\lambda_{21;p} - \langle \lambda_{21;p} \rangle)^2 \rangle$$

$$= \langle \lambda_{21;p}^2 \rangle - \langle \lambda_{21;p} \rangle^2 ,$$

where

$$\langle \lambda_{21;p}^2 \rangle = \tfrac{1}{4} \langle (b_{22} + b_{33})^2 \rangle \mp \tfrac{1}{2} \langle (b_{22} + b_{33}) \sqrt{X} \rangle + \tfrac{1}{4} \langle X \rangle = 0.7033d^2$$

because $\langle (b_{22} + b_{33}) \sqrt{X} \rangle = 0$. The variance of λ_2 and λ_3 satisfies the estimate

$$\operatorname{var} \lambda_2 = \operatorname{var} \lambda_3 \geqq 0.2177d^2$$

up to terms of second order.

1.4. Calculation of the moments of eigenvalues and eigenfunctions for stochastic differential operators

1.4.1. General case

We will consider stochastic eigenvalue problems of the form

$$\begin{aligned} &\big(M_0 + M_1(\omega)\big)\, u = \lambda\big(r(x) + N_1(\omega)\big)\, u\,, \qquad 0 \leqq x \leqq 1\,, \\ &U_i[u] \doteq U_{i0}[u] + U_{i1}(\omega)[u] = 0\,, \qquad i = 1, 2, \ldots, 2m\,, \end{aligned} \tag{1.86}$$

where M_0 denotes a deterministic differential operator of the order $2m$,

$$M_0 u(x) \doteq \sum_{k=0}^{m} (-1)^k \left[f_k(x)\, u^{(k)}\right]^{(k)}, \tag{1.87}$$

where $f_k \in \mathbf{C}^k[0, 1]$, and $f_n(x) \neq 0$. $r(x)$ is a positive continuous deterministic function. Let $M_1(\omega)$ and $N_1(\omega)$ be stochastic differential operators of the order p and q, respectively, with $p, q \leqq 2m$; that is

$$M_1(\omega)\, u(x) \doteq \sum_{k=0}^{p} h_k(x, \omega)\, u^{(k)}(x)\,,$$

$$N_1(\omega)\, u(x) \doteq \sum_{k=0}^{q} g_k(x, \omega)\, u^{(k)}(x)$$

where $h_k(x, \omega)$ and $g_k(x, \omega)$ are sufficiently often a.s. differentiable random processes. The $2m$ linearly independent boundary conditions $U_i[u] = 0$ contain the values of the function u and its derivatives $u^{(k)}$, $k = 0, 1, \ldots, 2m - 1$, at the points $x = 0$ and $x = 1$ of the interval in which the solution is sought. The boundary conditions are independent of the parameter λ. The deterministic and stochastic parts of the boundary conditions are denoted by $U_{i0}[u]$ and $U_{i1}[u]$, respectively. We have

$$\begin{aligned} &U_{i0}[u] \doteq \sum_{k=0}^{2m-1} \big(a_{ik} u^{(k)}(0) + b_{ik} u^{(k)}(1)\big)\,, \\ &U_{i1}(\omega)\,[u] \doteq \sum_{k=0}^{2m-1} \big(\alpha_{ik}(\omega)\, u^{(k)}(0) + \beta_{ik}(\omega)\, u^{(k)}(1)\big)\,. \end{aligned} \tag{1.88}$$

In the following we assume that the coefficients $h_k(x, \omega)$, $g_k(x, \omega)$ of the operators $M_1(\omega)$, $N_1(\omega)$ and the coefficients $\alpha_{ik}(\omega)$, $\beta_{ik}(\omega)$ in $U_{i1}[u]$ are sufficiently small so that the conditions of the perturbation theory are satisfied, i.e. the eigenvalues and eigenfunctions can be developed in terms homogeneous in the coefficients h_k, g_k, α_{ik}, β_{ik}. Rellich [1, 2, 3, 4] gives conditions for this. In Section 1.4.2 we deal with such conditions in connection with an estimation of the error for an important special case of the eigenvalue problem (1.86). Furthermore, we assume the unperturbed problem of (1.86)

$$M_0 w = \mu r w\,, \qquad U_{i0}[w] = 0\,, \quad i = 1, 2, \ldots, 2m\,, \tag{1.89}$$

to be self-adjoint and positive definite. Let all eigenvalues μ_l of (1.89) be simple and let $w_l(x)$ denote the eigenfunction corresponding to the eigenvalue μ_l.

These eigenfunctions are assumed to be orthonormal in $\mathbf{L}_2^r(0, 1)$, i.e. we have for $i, j = 1, 2, \ldots$

$$\langle\!\langle w_i, w_j \rangle\!\rangle_r \doteq \int_0^1 w_i(x)\, w_j(x)\, r(x)\, \mathrm{d}x = \delta_{ij}\,.$$

For the eigenvalues $\lambda_g(\omega)$ and the eigenfunctions $u_g(x, \omega)$ normalized by

$$\langle\!\langle u_g, u_g \rangle\!\rangle_r = 1 \tag{1.90}$$

we put

$$\begin{aligned} u_g(x, \omega) &= \sum_{k=0}^{\infty} u_{gk}(x, \omega)\,, \\ \lambda_g(\omega) &= \sum_{k=0}^{\infty} \lambda_{gk}(\omega) \end{aligned} \tag{1.91}$$

where $u_{gk}(x, \omega)$ and $\lambda_{gk}(\omega)$ are the homogeneous parts of the k-th order in the perturbations.

Substitution of (1.91) into (1.86) leads to the equations

$$M_0 u_{g0} = \lambda_{g0} r u_{g0}\,, \qquad U_{i0}[u_{g0}] = 0\,, \tag{1.92}$$

$$\begin{aligned} M_0 u_{gk} + M_1(\omega)\, u_{g,k-1} &= \sum_{s=0}^{k-1} \lambda_{gs}\big(r u_{g,k-s} + N_1(\omega)\, u_{g,k-s-1}\big) \\ &\quad + \lambda_{gk} r u_{g0}\,, \\ U_{i0}[u_{gk}] &= -U_{i1}[u_{g,k-1}]\,, \qquad k = 1, 2, \ldots \end{aligned} \tag{1.93}$$

From (1.92) we obtain $u_{g0}(x, \omega) \equiv w_g(x)$, $\lambda_{g0}(\omega) = \mu_g$; and from (1.93) for the successive calculation of $u_{gk}(x, \omega)$ and $\lambda_{gk}(\omega)$ the nonhomogeneous boundary value problem

$$\begin{aligned} M_0 u_{gk} - \mu_g r u_{gk} &= (\mu_g N_1 - M_1)\, u_{g,k-1} + \sum_{s=1}^{k-1} \lambda_{gs}(r u_{g,k-s} + N_1 u_{g,k-s-1}) \\ &\quad + \lambda_{gk} r w_g\,, \\ U_{i0}[u_{gk}] &= -U_{i1}[u_{g,k-1}]\,. \end{aligned}$$

In order to determine the k-th approximations to $u_g(x, \omega)$ and $\lambda_g(\omega)$, respectively, the nonhomogeneous problem

$$\begin{aligned} M_0 u_{gk} - \mu_g r u_{gk} &= \hat{q}_k + \lambda_{gk} r w_g\,, \\ U_{i0}[u_{gk}] &= h_{ik} \end{aligned} \tag{1.94}$$

has to be solved. Here the values

$$\begin{aligned} \hat{q}_k &\doteq (\mu_g N_1 - M_1)\, u_{g,k-1} + \sum_{s=1}^{k-1} \lambda_{gs}(r u_{g,k-s} + N_1 u_{g,k-s-1})\,, \\ h_{ik} &\doteq -U_{ik}[u_{g,k-1}] \end{aligned}$$

can be assumed to be known since they only depend on the already calculated terms u_{gs} and λ_{gs} of u_g and λ_g with $s < k$.

Assume that the problem (1.94) with the nonhomogeneous boundary conditions with the help of

$$u_{gk}(x, \omega) = v_{gk}(x, \omega) + z_{gk}(x, \omega)\,, \qquad z_{gk} \in \mathbf{C}^{2m}[0, 1] \quad \text{a.s.}$$

is transposed into a problem for v_{gk} with homogeneous boundary conditions

$$\begin{aligned} M_0 v_{gk} - \mu_g r v_{gk} &= q_k + \lambda_{gk} r w_g \,, \\ U_{i0}[v_{gk}] &= 0 \,. \end{aligned} \tag{1.95}$$

Here we have

$$q_k \doteq \hat{q}_k - M_0 z_{gk} + \mu_g r z_{gk}$$

and z_{gk} must satisfy the conditions

$$U_{i0}[z_{gk}] = -U_{i1}[u_{g,k-1}] \,.$$

Let $G(x, y)$ denote the Green's function corresponding to M_0 and the boundary conditions $U_{i0}[u] = 0$. Defining

$$Tv \doteq \int_0^1 G(x, y)\, v(y)\, dy \,,$$

we obtain from (1.95)

$$v_{gk} - \mu_g T r v_{gk} = T q_k + \lambda_{gk} T r w_g \,. \tag{1.96}$$

The compact self-adjoint operator on $\mathbf{L}_2(0, 1)$ defined by

$$\tilde{T} v \doteq \int_0^1 G(x, y) \sqrt{r(x)\, r(y)}\, v(y)\, dy$$

has eigenfunctions $\tilde{w}_l(x) \doteq \sqrt{r(x)}\, w_l(x)$ and eigenvalues $\mu_l^{-1} \doteq \tilde{\mu}_l$; that is $\tilde{T}\tilde{w}_l = \tilde{\mu}_l \tilde{w}_l$. This immediately follows by application of T to (1.89). From (1.96), with $\tilde{v}_{gk} \doteq \sqrt{r} v_{gk}$ we obtain

$$\tilde{v}_{gk} - \mu_g \tilde{T} \tilde{v}_{gk} = \tilde{p}_k$$

where

$$\tilde{p}_k \doteq \sqrt{r}\, T q_k + \lambda_{gk} \mu_g^{-1} \tilde{w}_g \,.$$

This equation has a solution if and only if $\tilde{p}_k$ is orthogonal in $\mathbf{L}_2(0, 1)$ to the eigenfunction $\tilde{w}_g$ associated with the eigenvalue μ_g^{-1} of the operator $\tilde{T}$, that is

$$0 = (\!(\tilde{p}_k, \tilde{w}_g)\!) = \mu_g^{-1}[(\!(q_k, w_g)\!) + \lambda_{gk}] \,.$$

$\lambda_{gk}(\omega)$ is to be determined in such a way that this condition holds:

$$\begin{aligned} \lambda_{gk}(\omega) = {} & (\!(\big(M_1(\omega) - \mu_g N_1(\omega)\big)\, u_{g,k-1}, w_g)\!) - \sum_{s=1}^{k-1} \lambda_{gs}[(\!(N_1(\omega)\, u_{g,k-s-1}, w_g)\!) \\ & + (\!(r u_{g,k-s}, w_g)\!)] + (\!((M_0 - \mu_g r)\, z_{gk}, w_g)\!) \,. \end{aligned} \tag{1.97}$$

With $\lambda_{gk}(\omega)$ determined as above, the solution of the equation

$$\tilde{v}_{gk} - \mu_g \tilde{T} \tilde{v}_{gk} = \tilde{p}_k$$

is given by

$$\tilde{v}_{gk}(x, \omega) = \tilde{p}_k(x, \omega) + \sum_{\substack{s=1 \\ g}}^{\infty} \frac{\mu_g}{\mu_s - \mu_g} (\!(\tilde{p}_k, \tilde{w}_s)\!)\, \tilde{w}_s(x) + c_k(\omega)\, \tilde{w}_g(x)$$

where $c_k(\omega)$ denotes an arbitrary random variable. Because

$$\tilde{T}u = \sum_{s=1}^{\infty} \mu_s^{-1} (\!(u, \tilde{w}_s)\!) \, \tilde{w}_s = \sqrt{r}\, T(\sqrt{r} u)$$

for any continuous function u, we obtain

$$\tilde{p}_k(x, \omega) = \sum_{s=1}^{\infty} (\!(\sqrt{r}\, T q_k, \tilde{w}_s)\!) \, \tilde{w}_s + \lambda_{gk} \mu_g^{-1} \tilde{w}_g \, .$$

In view of this we have

$$\tilde{v}_{gk}(x, \omega) = \sum_{\substack{s=1\\ s \neq g}}^{\infty} \frac{\mu_s}{\mu_s - \mu_g} (\!(\sqrt{r}\, T q_k, \tilde{w}_s)\!) \, \tilde{w}_s$$
$$+ [c_k(\omega) + \mu_g^{-1} \lambda_{gk} + (\!(\sqrt{r}\, T q_k, \tilde{w}_g)\!)] \, \tilde{w}_g \, .$$

From this the k-th homogeneous term associated with $u_g(x, \omega)$ (with $\mu_{sg} \doteq \mu_s - \mu_g$) results in

$$u_{gk}(x, \omega) = \sum_{\substack{s=1\\ s \neq g}}^{\infty} \frac{1}{\mu_{sg}} (\!(q_k, w_s)\!) \, w_s + [c_k(\omega) + \mu_g^{-1} \lambda_{gk} + \mu_g^{-1} (\!(q_k, w_g)\!)] \, w_g + z_{gk}$$
$$= \sum_{\substack{s=1\\ s \neq g}}^{\infty} \frac{1}{\mu_{sg}} (\!(q_k, w_s)\!) \, w_s + C_k(\omega) \, w_g + z_{gk}(x, \omega) \, .$$

One can satisfy the condition $(\!(u_g, u_g)\!)_r = 1$, as at the end of Section 1.3.1, with

$$(\!(u_{gk}, w_g)\!)_r = -\tfrac{1}{2} \sum_{p=1}^{k-1} (\!(u_{gp}, u_{g,k-p})\!)_r \, .$$

The term $(\!(u, v)\!)_r$ is defined by

$$(\!(u, v)\!)_r \doteq \int_0^1 u(x) \, v(x) \, r(x) \, \mathrm{d}x \, .$$

From this the constant $C_k(\omega)$ can be calculated by

$$C_k(\omega) = -(\!(z_{gk}, w_g)\!)_r - \tfrac{1}{2} \sum_{p=1}^{k-1} (\!(u_{gp}, u_{g,k-p})\!)_r$$

and therefore

$$u_{gk}(x, \omega) = \sum_{\substack{s=1\\ s \neq g}}^{\infty} \frac{1}{\mu_{sg}} (\!(q_k, w_s)\!) \, w_s - (\!(z_{gk}, w_g)\!)_r \, w_g + z_{gk}$$
$$- \frac{1}{2} \sum_{p=1}^{k-1} (\!(u_{gp}, u_{g,k-p})\!)_r \, w_g \qquad (1.98)$$

is obtained, where

$$(\!(q_k, w_s)\!) = -(\!((M_1 - \mu_g N_1) \, u_{g,k-1}, w_s)\!) - (\!((M_0 - \mu_g r) \, z_{gk}, w_s)\!)$$
$$+ \sum_{p=1}^{k-1} \lambda_{gp} (\!(r u_{g,k-p} + N_1 u_{g,k-p-1}, w_s)\!) \, .$$

Because of

$$\left|\sum_{\substack{s=1\\ s\neq g}}^{\infty} \frac{1}{\mu_{sg}} \langle\!\langle q_k, w_s\rangle\!\rangle\, w_s^{(p)}(x)\right| = \left|\sum_{\substack{s=1\\ s\neq g}}^{\infty} \frac{\mu_s}{\mu_{sg}} \langle\!\langle Tq_k, w_s\rangle\!\rangle_r\, w_s^{(p)}(x)\right|$$

$$\leqq \alpha \sum_{\substack{s=1\\ s\neq g}}^{\infty} |\langle\!\langle Tq_k, w_s\rangle\!\rangle_r\, w_s^{(p)}(x)|$$

with $|\mu_s/\mu_{sg}| \leqq \alpha$, the series $\sum_{\substack{s=1\\ s\neq g}}^{\infty} 1/\mu_{sg} \langle\!\langle q_k, w_s\rangle\!\rangle\, w_s^{(p)}(x)$ converges absolutely and uniformly in $0 \leqq x \leqq 1$ for $p = 0, 1, \dots, m-1$ a.s. for every admissible function (i.e. for every $2m$-times continuously differentiable function which satisfies the unperturbed boundary conditions $U_{i0}[u] = 0$). For $k = 1$ it follows from (1.97) and (1.98) with

$$\tilde{M}_1 \doteq M_1 - \mu_g N_1 \quad \text{and} \quad \tilde{M}_0 \doteq M_0 - \mu_g r$$

that

$$\lambda_{g1}(\omega) = \langle\!\langle \tilde{M}_1 w_g, w_g\rangle\!\rangle + \langle\!\langle \tilde{M}_0 z_{g1}, w_g\rangle\!\rangle\,,$$

$$u_{g1}(x, \omega) = -\sum_{\substack{s=1\\ s\neq g}}^{\infty} \frac{1}{\mu_{sg}} [\langle\!\langle \tilde{M}_1 w_g, w_s\rangle\!\rangle + \langle\!\langle \tilde{M}_0 z_{g1}, w_s\rangle\!\rangle]\, w_s(x)$$

$$+ z_{g1}(x, \omega) - \langle\!\langle z_{g1}, w_g\rangle\!\rangle_r\, w_g(x)\,.$$

For the operator adjoint to $\tilde{M}_1 = M_1 - \mu_g N_1$

$$\tilde{M}_1^* u = \sum_{s=0}^{2m} (-1)^s \left[\big(h_s(x, \omega) - \mu_g g_s(x, \omega)\big)\, u\right]^{(s)}$$

($h_s \equiv 0$ if $s > p$, $g_s \equiv 0$ if $s > q$) holds

$$\langle\!\langle \tilde{M}_1 u, v\rangle\!\rangle = \langle\!\langle u, \tilde{M}_1^* v\rangle\!\rangle + D(u, v)\,,$$

where

$$D(u, v) \doteq \sum_{s=0}^{2m} \sum_{t=1}^{s} (-1)^{t-1}\, u^{(s-t)}\, [(h_s - \mu_g g_s)\, v]^{(t-1)}\Big|_0^1\,.$$

Thus from (1.97) we obtain, for $k = 2$,

$$\lambda_{g2}(\omega) = \langle\!\langle u_{g1}, \tilde{M}_1^* w_g\rangle\!\rangle + D(u_{g1}, w_g) - \lambda_{g1}[\langle\!\langle N_1 w_g, w_g\rangle\!\rangle + \langle\!\langle u_{g1}, w_g\rangle\!\rangle_r]$$

$$+ \langle\!\langle \tilde{M}_0 z_{g2}, w_g\rangle\!\rangle$$

$$= \sum_{\substack{s=1\\ s\neq g}}^{\infty} \frac{1}{\mu_{sg}} \langle\!\langle q_1, w_s\rangle\!\rangle\, [\langle\!\langle \tilde{M}_1 w_s, w_g\rangle\!\rangle - D(w_s, w_g)]$$

$$- \langle\!\langle z_{g1}, w_g\rangle\!\rangle_r \langle\!\langle \tilde{M}_1 w_g, w_g\rangle\!\rangle$$

$$+ D\left(\sum_{\substack{s=1\\ s\neq g}}^{\infty} \frac{1}{\mu_{sg}} \langle\!\langle q_1, w_s\rangle\!\rangle\, w_s, w_g\right) + \langle\!\langle \tilde{M}_1 z_{g1}, w_g\rangle\!\rangle$$

$$- \lambda_{g1} \langle\!\langle N_1 w_g, w_g\rangle\!\rangle + \langle\!\langle \tilde{M}_0 z_{g2}, w_g\rangle\!\rangle$$

where

$$q_1 = -\tilde{M}_1 w_g - \tilde{M}_0 z_{g1} \,.$$

In the last equation we have used $\langle\!\langle u_{g1}, w_g \rangle\!\rangle_r = 0$. For $k = 2$, with

$$q_2 = -\tilde{M}_1 u_{g1} + \lambda_{g1}(r u_{g1} + N_1 w_g) - \tilde{M}_0 z_{g2} \,,$$

it follows from (1.98) that

$$\begin{aligned}
u_{g2}(x, \omega) &= \sum_{\substack{s=1\\ s\neq g}}^{\infty} \frac{1}{\mu_{sg}} [-\langle\!\langle u_{g1}, \tilde{M}_1^* w_s \rangle\!\rangle - D(u_{g1}, w_s) + \lambda_{g1}(\langle\!\langle u_{g1}, w_s \rangle\!\rangle_r \\
&\qquad + \langle\!\langle N_1 w_g, w_s \rangle\!\rangle) - \langle\!\langle \tilde{M}_0 z_{g2}, w_s \rangle\!\rangle]\, w_s(x) \\
&\quad + z_{g2}(x, \omega) - \langle\!\langle z_{g2}, w_g \rangle\!\rangle_r\, w_g(x) - \frac{1}{2} \langle\!\langle u_{g1}, u_{g1} \rangle\!\rangle_r\, w_g(x) \\
&= \sum_{\substack{s=1\\ s\neq g}}^{\infty} \frac{1}{\mu_{sg}} \Bigg[- \sum_{\substack{t=1\\ t\neq g}}^{\infty} \frac{1}{\mu_{tg}} \langle\!\langle q_1, w_t \rangle\!\rangle \{\langle\!\langle \tilde{M}_1 w_t, w_s \rangle\!\rangle - D(w_t, w_s)\} \\
&\qquad - \langle\!\langle \tilde{M}_1 z_{g1}, w_s \rangle\!\rangle + \langle\!\langle z_{g1}, w_g \rangle\!\rangle_r \langle\!\langle \tilde{M}_1 w_g, w_s \rangle\!\rangle \\
&\qquad - D\Bigg(\sum_{\substack{t=1\\ t\neq g}}^{\infty} \frac{1}{\mu_{tg}} \langle\!\langle q_1, w_t \rangle\!\rangle\, w_t, w_s \Bigg) \\
&\qquad + \lambda_{g1} \left\{ \frac{1}{\mu_{sg}} \langle\!\langle q_1, w_s \rangle\!\rangle + \langle\!\langle z_{g1}, w_s \rangle\!\rangle_r + \langle\!\langle N_1 w_g, w_s \rangle\!\rangle \right\} \\
&\qquad - \langle\!\langle \tilde{M}_0 z_{g2}, w_s \rangle\!\rangle \Bigg] w_s(x) + z_{g2} - \langle\!\langle z_{g2}, w_g \rangle\!\rangle_r\, w_g(x) \\
&\quad - \frac{1}{2} \Bigg[\sum_{\substack{s=1\\ s\neq g}}^{\infty} \frac{1}{\mu_{sg}^2} \langle\!\langle q_1, w_s \rangle\!\rangle^2 + 2 \sum_{\substack{s=1\\ s\neq g}}^{\infty} \frac{1}{\mu_{sg}} \langle\!\langle q_1, w_s \rangle\!\rangle \langle\!\langle z_{g1}, w_s \rangle\!\rangle_r \\
&\qquad + \langle\!\langle z_{g1}, z_{g1} \rangle\!\rangle_r - \langle\!\langle z_{g1}, w_g \rangle\!\rangle_r^2 \Bigg] w_g(x) \,.
\end{aligned}$$

If for the operator $\tilde{M}_1 = M_1 - \mu_g N_1$ with the boundary conditions $U_{i0}[u] = 0$ for all continuous functions $u(x)$ the following relation holds

$$D\Bigg(\sum_{\substack{s=1\\ s\neq g}}^{\infty} \frac{1}{\mu_{sg}} \langle\!\langle u, w_s \rangle\!\rangle\, w_s, w_t \Bigg) = \sum_{\substack{s=1\\ s\neq g}}^{\infty} \frac{1}{\mu_{sg}} \langle\!\langle u, w_s \rangle\!\rangle\, D(w_s, w_t) \tag{1.99}$$

then earlier calculations yield

$$\begin{aligned}
\lambda_{g2}(\omega) &= - \sum_{\substack{s=1\\ s\neq g}}^{\infty} \frac{1}{\mu_{sg}} [\langle\!\langle \tilde{M}_1 w_g, w_s \rangle\!\rangle + \langle\!\langle \tilde{M}_0 z_{g1}, w_s \rangle\!\rangle] \langle\!\langle \tilde{M}_1 w_s, w_g \rangle\!\rangle \\
&\quad + \langle\!\langle \tilde{M}_1 z_{g1}, w_g \rangle\!\rangle - \langle\!\langle z_{g1}, w_g \rangle\!\rangle_r \langle\!\langle \tilde{M}_1 w_g, w_g \rangle\!\rangle - \lambda_{g1} \langle\!\langle N_1 w_g, w_g \rangle\!\rangle \\
&\quad + \langle\!\langle \tilde{M}_0 z_{g2}, w_g \rangle\!\rangle
\end{aligned}$$

and

$$u_{g2}(x, \omega) = \sum_{\substack{s=1 \\ s \neq g}}^{\infty} \left\{ \sum_{\substack{t=1 \\ t \neq g}}^{\infty} \frac{1}{\mu_{sg}\mu_{tg}} [\langle\!\langle \tilde{M}_1 w_g, w_t \rangle\!\rangle + \langle\!\langle \tilde{M}_0 z_{g1}, w_t \rangle\!\rangle] \langle\!\langle \tilde{M}_1 w_t, w_s \rangle\!\rangle \right\} w_s(x)$$

$$+ \sum_{\substack{s=1 \\ s \neq g}}^{\infty} \frac{1}{\mu_{sg}} \Bigg[\langle\!\langle z_{g1}, w_g \rangle\!\rangle_r \langle\!\langle \tilde{M}_1 w_g, w_s \rangle\!\rangle - \langle\!\langle \tilde{M}_1 z_{g1}, w_s \rangle\!\rangle - \langle\!\langle \tilde{M}_0 z_{g2}, w_s \rangle\!\rangle$$

$$+ \lambda_{g1} \Bigg\{ -\frac{1}{\mu_{sg}} (\langle\!\langle \tilde{M}_1 w_g, w_s \rangle\!\rangle + \langle\!\langle \tilde{M}_0 z_{g1}, w_s \rangle\!\rangle)$$

$$+ \langle\!\langle z_{g1}, w_s \rangle\!\rangle_r + \langle\!\langle N_1 w_g, w_s \rangle\!\rangle \Bigg\} \Bigg] w_s(x)$$

$$- \frac{1}{2} \Bigg[2 \langle\!\langle z_{g2}, w_g \rangle\!\rangle_r + \sum_{\substack{s=1 \\ s \neq g}}^{\infty} \frac{1}{\mu_{sg}^2} [\langle\!\langle \tilde{M}_1 w_g, w_s \rangle\!\rangle + \langle\!\langle \tilde{M}_0 z_{g1}, w_s \rangle\!\rangle]^2$$

$$+ \langle\!\langle z_{g1}, z_{g1} \rangle\!\rangle_r - 2 \sum_{\substack{s=1 \\ s \neq g}}^{\infty} \frac{1}{\mu_{sg}} [\langle\!\langle \tilde{M}_1 w_g, w_s \rangle\!\rangle + \langle\!\langle \tilde{M}_0 z_{g1}, w_s \rangle\!\rangle]$$

$$\times \langle\!\langle z_{g1}, w_s \rangle\!\rangle_r - \langle\!\langle z_{g1}, w_g \rangle\!\rangle_r^2 \Bigg] w_g(x) + z_{g2}(x, \omega) \,.$$

Summarizing these results we have the following theorem.

Theorem 1.15. *For the eigenvalues and the eigenfunctions of the random eigenvalue problem* (1.86), *with perturbation operators* M_1, N_1 *and perturbation parts* U_{i1} *of the boundary conditions, the perturbation expansions are*

$$\lambda_g(\omega) = \mu_g + \lambda_{g1}(\omega) + \lambda_{g2}(\omega) + \dots \,,$$

$$u_g(x, \omega) = w_g(x) + u_{g1}(x, \omega) + u_{g2}(x, \omega) + \dots \,,$$

where

$$\lambda_{g1}(\omega) = b_{gg} + Z^0_{1gg} \,,$$

$$\lambda_{g2}(\omega) = - \sum_{\substack{s=1 \\ s \neq g}}^{\infty} \frac{1}{\mu_{sg}} [b_{gs} + Z^0_{1gs}] \, b_{sg} - Z_{1gg} b_{gg} + Z^1_{1gg} - \lambda_{g1} a_{gg} + Z^0_{2gg} \,,$$

$$u_{g1}(x, \omega) = - \sum_{\substack{s=1 \\ s \neq g}}^{\infty} \frac{1}{\mu_{sg}} [b_{gs} + Z^0_{1gs}] \, w_s(x) + z_{g1}(x, \omega) - Z_{1gg} w_g(x) \,,$$

$$u_{g2}(x, \omega) = \sum_{\substack{s=1 \\ s \neq g}}^{\infty} \left[\sum_{\substack{t=1 \\ t \neq g}}^{\infty} \frac{1}{\mu_{sg}\mu_{tg}} \{b_{gt} + Z^0_{1gt}\} \, b_{ts} \right] w_s(x) + \sum_{\substack{s=1 \\ s \neq g}}^{\infty} \frac{1}{\mu_{sg}}$$

$$\times \left[Z_{1gg} b_{gs} - Z^1_{1gs} - Z^9_{2gs} + \lambda_{g1} \left\{ Z_{1gs} + a_{gs} - \frac{1}{\mu_{sg}} (b_{gs} + Z^0_{1gs}) \right\} \right]$$

$$\times w_s(x) - \frac{1}{2} \Bigg[2 Z_{2gg} + \sum_{\substack{s=1 \\ s \neq g}}^{\infty} \frac{1}{\mu_{sg}^2} (b_{gs} + Z^0_{1gs})^2$$

$$- 2 \sum_{\substack{s=1 \\ s \neq g}}^{\infty} \frac{1}{\mu_{sg}} (b_{gs} + Z^0_{1gs}) \, Z_{1gs} + \langle\!\langle z_{g1}, z_{g1} \rangle\!\rangle_r - (Z_{1gg})^2 \Bigg] w_g(x)$$

$$+ z_{g2}(x, \omega) \,,$$

provided condition (1.99) *is satisfied. Here, with* $\tilde{M}_1 \doteq M_1 - \mu_g N_1$ *and* $\tilde{M}_0 \doteq M_0 - \mu_g r$, *the following expressions have been introduced:*

$$a_{ij}(\omega) \doteq \langle\!\langle N_1 w_i, w_j \rangle\!\rangle, \qquad Z^0_{ikj}(\omega) \doteq \langle\!\langle \tilde{M}_0 z_{ki}, w_j \rangle\!\rangle,$$
$$b_{ij}(\omega) \doteq \langle\!\langle \tilde{M}_1 w_i, w_j \rangle\!\rangle, \qquad Z^1_{ikj}(\omega) \doteq \langle\!\langle \tilde{M}_1 z_{ki}, w_j \rangle\!\rangle,$$
$$Z_{ikj}(\omega) \doteq \langle\!\langle r z_{ki}, w_j \rangle\!\rangle$$

where μ_l *denote the eigenvalues and* $w_l(x)$ *the eigenfunctions of the unperturbed eigenvalue problem* (1.89). *The* $z_{gk}(x, \omega)$ *are random processes with a.s.* $2m$ *times continuously differentiable trajectories satisfying* $U_{i0}[z_{gk}] = -U_{i1}[u_{gk-1}]$. *Moreover let be* $\langle z_{g1}(x)\rangle = 0$. *Also, we have put* $\mu_{sg} \doteq \mu_s - \mu_g$. *The infinite series in* u_{g1} *and* u_{g2} *and its* p *times differentiated series,* $p = 1, 2, \ldots, m-1$, *are a.s. uniformly and absolutely convergent. The series for* λ_{g2} *is also a.s. convergent.*

Remark 1.3. (a) *Condition* (1.99) *is satisfied if* $\tilde{M}_1$ *is a differential operator of order less than or equal to* m.

(b) *Condition* (1.99) *is satisfied, also, if* $\tilde{M}_1$ *is symmetric with respect to all admissible functions. This is clearly satisfied if the boundary conditions* $U_{i0}[u] = 0$ *contain the following ones:*

$$u(0) = u'(0) = \ldots = u^{(k)}(0) = u(1) = u'(1) = \ldots = u^{(k)}(1) = 0,$$

where $k \doteq [(s_0 - 1)/2]$ ($[x]$ *is the greatest integer not greater than* x). s_0 *denotes the greatest positive integer satisfying*

$$h_{s_0}(x, \omega) - \mu_g g_{s_0}(x, \omega) \not\equiv 0 \quad \text{a.s.}$$

Proof. (a) Since the series $\sum_{\substack{s=1\\ s\neq g}}^{\infty} 1/\mu_{sg} \langle\!\langle u, w_s \rangle\!\rangle\, w_s$ and the series obtained by differentiating p times, $p = 1, 2, \ldots, m-1$, are uniformly and absolutely convergent, we have

$$D\left(\sum_{\substack{s=1\\ s\neq g}}^{\infty} \frac{1}{\mu_{sg}} \langle\!\langle u, w_s \rangle\!\rangle\, w_s, w_t\right)$$
$$= \sum_{i=1}^{m} \sum_{j=1}^{i} (-1)^{j-1} \left(\sum_{\substack{s=1\\ s\neq g}}^{\infty} \frac{1}{\mu_{sg}} \langle\!\langle u, w_s \rangle\!\rangle\, w_s^{(i-j)}\right) [(h_i - \mu_g g_i)\, w_t]^{(j-1)}\big|_0^1$$
$$= \sum_{\substack{s=1\\ s\neq g}}^{\infty} \frac{1}{\mu_{sg}} \langle\!\langle u, w_s \rangle\!\rangle\, D(w_s, w_t)\,. \blacktriangleleft$$

(b) First, $D(w_s, w_t) = 0$ since $w_i(x)$ are admissible functions. The function

$$f(x) = \sum_{\substack{s=1\\ s\neq g}}^{\infty} \frac{1}{\mu_{sg}} \langle\!\langle u, w_s \rangle\!\rangle\, w_s(x) + d w_g(x)$$

(d is an arbitrary constant) is a solution of the boundary value problem

$$M_0 f - \mu_g r f = u - \langle\!\langle u, w_g \rangle\!\rangle\, r w_g\,, \qquad U_{i0}[f] = 0\,, \qquad i = 1, 2, \ldots, 2m\,,$$

and, because $U_{i0}[w_g] = 0$, the equation

$$U_{i0}\left[\sum_{\substack{s=1\\ s\neq g}}^{\infty} \frac{1}{\mu_{sg}} (\!(u, w_s)\!)\, w_s(x)\right] = 0$$

also holds for arbitrary continuous functions $u(x)$. The function $f(x)$ is admissible and thus the assertion follows in view of the symmetry of $\tilde{M}_1$ with $D(f, w_t) = 0$. If the boundary conditions $U_{i0}[u]$ contain those mentioned above, then the symmetry of the operator $\tilde{M}_1$ follows for all admissible functions from

$$D(u, v) = \sum_{s=0}^{2m} \sum_{t=1}^{s} (-1)^t u^{(s-t)} [(h_s - \mu_g g_s)\, v]^{(t-1)}|_0^1 = 0$$

since

$$\max_{\substack{1\leqq s\leqq s_0\\ 1\leqq t\leqq s}} (\min \{s - t, t - 1\}) = [\tfrac{1}{2}(s_0 - 1)]\,. \blacktriangleleft$$

We now calculate approximately the first both moments of the eigenvalues and eigenfunctions of problem (1.86) by means of Theorem 1.15. For this we assume that the unperturbed problem (1.89) coincides with the averaged problem of (1.86), i.e. we assume that $\langle h_s(x, \omega)\rangle = \langle g_t(x, \omega)\rangle = 0$ holds for $s = 0, 1, \dots, p$, $t = 0, 1, \dots, q$ and $\langle U_{i1}\rangle = 0$ for $i = 1, 2, \dots, 2m$. In case $\langle N_1\rangle\, u = 0$, and since M_1 is symmetric the above relation can always be generated since the random eigenvalue problem (1.86) can be written in the form

$$\left(M_0' + M_1'(\omega)\right) u = \lambda\left(ru + N_1(\omega)\right) u\,,$$

$$U_{i0}'[u] + U_{i1}'[u] = 0\,, \qquad i = 1, 2, \dots, 2m\,,$$

where

$$M_0' \doteq M_0 + \langle M_1\rangle\,, \qquad M_1' \doteq M_1 - \langle M_1\rangle\,,$$

$$U_{i0}' \doteq U_{i0} + \langle U_{i1}\rangle\,, \qquad U_{i1}' \doteq U_{i1} - \langle U_{i1}\rangle$$

if $\langle M_1\rangle \neq 0$ and $\langle U_{i1}\rangle \neq 0$. If we can exchange the evaluation of the mean value and the summation

$$\langle\lambda_g\rangle = \sum_{k=0}^{\infty} \langle\lambda_{gk}\rangle\,, \qquad \langle u_g\rangle = \sum_{k=0}^{\infty} \langle u_{gk}\rangle$$

and if the same is allowed for the summands λ_{gk} and u_{gk}, then we obtain for the expectations of the eigenvalues $\lambda_g(\omega)$ and the eigenfunctions $u_g(x, \omega)$ (up to second order in the perturbations):

$$\langle\lambda_g\rangle = \mu_g + d_g\,,$$

$$\langle u_g(x)\rangle = w_g(x) + D_g(x)\,,$$

where

$$d_g = -\sum_{\substack{s=1\\s\neq g}}^{\infty} \frac{1}{\mu_{sg}} (\langle b_{gs} b_{sg}\rangle + \langle Z^0_{1gs} b_{sg}\rangle) - \langle Z_{1gg} b_{gg}\rangle + \langle Z^1_{1gg}\rangle + \langle Z^0_{2gg}\rangle$$
$$- \langle (b_{gg} + Z^0_{1gg})\, a_{gg}\rangle\,,$$

$$D_g(x) = \sum_{\substack{s=1\\s\neq g}}^{\infty} \sum_{\substack{t=1\\t\neq g}}^{\infty} \frac{1}{\mu_{sg}\mu_{tg}} (\langle b_{gt} b_{ts}\rangle + \langle Z^0_{1gt} b_{ts}\rangle)\, w_s(x)$$
$$+ \sum_{\substack{s=1\\s\neq g}}^{\infty} \frac{1}{\mu_{sg}} \Bigg[\langle Z_{1gg} b_{gs}\rangle - \langle Z^1_{1gs}\rangle - \langle Z^0_{2gs}\rangle$$
$$+ \left\langle (b_{gg} + Z^0_{1gg}) \left(-\frac{1}{\mu_{sg}} (b_{gs} + Z^0_{1gs}) + Z_{1gs} + a_{gs} \right) \right\rangle \Bigg] w_s(x)$$
$$- \frac{1}{2} \Bigg[2\langle Z_{2gg}\rangle + \sum_{\substack{s=1\\s\neq g}}^{\infty} \frac{1}{\mu^2_{sg}} \langle (b_{gs} + Z^0_{1gs})^2\rangle - 2 \sum_{\substack{s=1\\s\neq g}}^{\infty} \frac{1}{\mu_{sg}} \langle (b_{gs} + Z^0_{1gs})\, Z_{1gs}\rangle$$
$$+ \langle \mathfrak{C} z_{g1}, z_{g1} \mathfrak{D}_r\rangle - \langle (Z_{1gg})^2\rangle \Bigg] w_g(x) + \langle z_{g2}(x)\rangle\,,$$

since

$$\langle b_{gs}\rangle = \langle a_{gs}\rangle = \langle Z^0_{1gs}\rangle = \langle Z_{1gg}\rangle = \langle z_{g1}\rangle = 0\,.$$

In Theorem 1.17 sufficient conditions will be given which permit the interchange of the series and the expectation.

The quantities d_g and $D_g(x)$ give the approximate solution of the averaging problem, i.e. they give the difference between the mean value of the random solution and the solution of the averaged problem. In the case of deterministic (unperturbed) boundary conditions, we obtain

$$d_g = -\sum_{\substack{s=1\\s\neq g}}^{\infty} \frac{1}{\mu_{sg}} \langle b_{gs} b_{sg}\rangle - \langle b_{gg} a_{gg}\rangle$$

and

$$D_g(x) = \sum_{\substack{s,t=1\\s,t\neq g}}^{\infty} \frac{1}{\mu_{sg}\mu_{tg}} \langle b_{gt} b_{ts}\rangle\, w_s(x) + \sum_{\substack{s=1\\s\neq g}}^{\infty} \frac{1}{\mu_{sg}} \langle b_{gg} a_{gs}\rangle\, w_s(x)$$
$$- \sum_{\substack{s=1\\s\neq g}}^{\infty} \frac{1}{\mu^2_{sg}} \langle b_{gg} b_{gs}\rangle\, w_s(x) - \frac{1}{2} \sum_{\substack{s=1\\s\neq g}}^{\infty} \frac{1}{\mu^2_{sg}} \langle b^2_{gs}\rangle\, w_g(x)\,.$$

d_g and $D_g(x)$ can be determined from the second moments of the random processes $h_s(x, \omega)$ and $g_t(x, \omega)$. In fact, using

$$R_{st}(x, y) \doteq \langle h_s(x)\, h_t(y)\rangle\,,$$
$$Q_{st}(x, y) \doteq \langle h_s(x)\, g_t(y)\rangle\,,$$
$$P_{st}(x, y) \doteq \langle g_s(x)\, g_t(y)\rangle$$

we have

$$\langle b_{ij}b_{kl}\rangle = \sum_{s,t=0}^{2m} \int_0^1\int_0^1 [R_{st}(x, y) - \mu_g(Q_{st}(x, y) + Q_{ts}(x, y)) + \mu_g^2 P_{st}(x, y)]\, w_i^{(s)}(x)\, w_j(x)\, w_k^{(t)}(y)\, w_l(y)\, dx\, dy$$

and

$$\langle b_{ij}a_{kl}\rangle = \sum_{s,t=0}^{2m} \int_0^1\int_0^1 [Q_{st}(x, y) - \mu_g P_{st}(x, y)]\, w_i^{(s)}(x)\, w_j(x) \times w_k^{(t)}(y)\, w_l(y)\, dx\, dy$$

$\big(h_s(x, \omega) \equiv 0$ if $s > p$, $g_t(x, \omega) \equiv 0$ if $t > q\big)$.

If we have expansions of the eigenvalues and the eigenfunctions up to terms of the n-th order, then the correlation relations are given up to terms of the $(n + 1)$-th order. This way the correlation relations can be written down up to terms of the third order from Theorem 1.15. Using $\lambda_{g1}(\omega)$ and $u_{g1}(x, \omega)$ from Theorem 1.15 we have, up to second order terms

$$\langle(\lambda_g - \langle\lambda_g\rangle)(\lambda_h - \langle\lambda_h\rangle)\rangle = \langle b_{gg}b_{hh}\rangle + \langle b_{gg}Z_{1hh}^0\rangle + \langle b_{hh}Z_{1gg}^0\rangle + \langle Z_{1gg}^0 Z_{1hh}^0\rangle\,,$$

$$\begin{aligned}\langle\big(u_g(x) - \langle u_g(x)\rangle\big)\big(u_h(y) - \langle u_h(y)\rangle\big)\rangle &= \sum_{\substack{s=1\\ s\neq g}}^{\infty}\sum_{\substack{t=1\\ t\neq h}}^{\infty} \frac{1}{\mu_{sg}\mu_{th}} \langle(b_{gs} + Z_{1gs}^0)(b_{ht} + Z_{1ht}^0)\rangle\, w_s(x)\, w_t(y)\\ &- \sum_{\substack{s=1\\ s\neq g}}^{\infty} \frac{1}{\mu_{sg}} \langle(b_{gs} + Z_{1gs}^0)\big(z_{h1}(y) - Z_{1hh}w_h(y)\big)\rangle\, w_s(x)\\ &- \sum_{\substack{t=1\\ t\neq h}}^{\infty} \frac{1}{\mu_{th}} \langle(b_{ht} + Z_{1ht}^0)\big(z_{g1}(x) - Z_{1gg}w_g(x)\big)\rangle\, w_t(y)\\ &+ \langle\big(z_{g1}(x) - Z_{1gg}w_g(x)\big)\big(z_{h1}(y) - Z_{1hh}w_h(y)\big)\rangle\,,\end{aligned}$$

$$\begin{aligned}\langle(\lambda_g - \langle\lambda_g\rangle)\big(u_h(x) - \langle u_h(x)\rangle\big)\rangle &= \langle(b_{gg} + Z_{1gg}^0)\big(z_{h1}(x) - Z_{1hh}w_h(x)\big)\rangle\\ &- \sum_{\substack{s=1\\ s\neq h}}^{\infty} \frac{1}{\mu_{sh}} \langle(b_{hs} + Z_{1hs}^0)(b_{gg} + Z_{1gg}^0)\rangle\, w_s(x)\,.\end{aligned}$$

In particular, it follows that the variances are given by

$$\operatorname{var}\lambda_g = \langle b_{gg}^2\rangle + 2\langle b_{gg}Z_{1gg}^0\rangle + \langle(Z_{1gg}^0)^2\rangle\,,$$

$$\begin{aligned}\operatorname{var} u_g(x) &= \sum_{\substack{s,t=1\\ s,t\neq g}}^{\infty} \frac{1}{\mu_{sg}\mu_{tg}} \langle(b_{gs} + Z_{1gs}^0)(b_{gt} + Z_{1gt}^0)\rangle\, w_s(x)\, w_t(x)\\ &- 2\sum_{\substack{s=1\\ s\neq g}}^{\infty} \frac{1}{\mu_{sg}} \langle(b_{gs} + Z_{1gs}^0)\big(z_{g1}(x) - Z_{1gg}w_g(x)\big)\rangle\, w_s(x)\\ &+ \langle\big(z_{g1}(x) - Z_{1gg}w_g(x)\big)^2\rangle\,.\end{aligned}$$

1.4.2. Self-adjoint eigenvalue problems with deterministic boundary conditions

For the special case of self-adjoint positive definite eigenvalue problems with deterministic boundary conditions, we now give approximations higher than the second as well as convergence estimates by means of the method of Ritz.

Consider the self-adjoint positive definite random eigenvalue problem

$$M(\omega)\, u = \lambda u\,,$$
$$U_{i0}[u] = 0\,, \qquad i = 1, 2, \dots, 2m\,, \tag{1.100}$$

where

$$M(\omega)\, u \doteq \sum_{k=0}^{m} (-1)^k \left[f_k(x, \omega)\, u^{(k)}\right]^{(k)}$$

and $f_m(x, \omega) \neq 0$ a.s., $\langle f_m(x)\rangle \neq 0$. We assume that a.s. all realizations of the processes $f_k(x, \omega)$, $k = 0, 1, \dots, m$, are sufficiently often differentiable. Then the averaged problem corresponding to (1.100), namely

$$\langle M\rangle\, w = \mu w\,, \qquad U_{i0}[w] = 0\,, \qquad i = 1, 2, \dots, 2m\,, \tag{1.101}$$

is self-adjoint and positive definite (cf. Theorem 1.4). Furthermore, let the imbedding operator from $\mathscr{H}_{\langle M\rangle}$ into $\mathscr{H} = \mathbf{L}_2(0, 1)$ be compact. Hence, the random operator will have a discrete spectrum a.s. (cf. Theorem 1.8.)

The measurability of the eigenvalues $\lambda_l(\omega)$ and eigenfunctions $u_l(x, \omega)$ (as elements of the space $\mathbf{L}_2(0, 1)$; we have to assume that the eigenvalues are simple) follows from the method of Ritz with a suitable system $\{\varphi_i\}$ of coordinate functions (cf. Section 2.2.1). The Ritz' eigenvalue problem for (1.100) is

$$\sum_{p=1}^{n} \left[(\!(M(\omega)\, \varphi_q, \varphi_p)\!) - \Lambda \delta_{qp}\right] u_p = 0\,, \qquad q = 1, 2, \dots, n\,. \tag{1.102}$$

Let $\Lambda_l \doteq {}^n\lambda_l(\omega)$ be the eigenvalues of this problem, and let $\left({}^nu_{lp}(\omega)\right)_{1\leqq p\leqq n}$ denote the associated eigenvectors. Defining the functions

$${}^nu_l(x, \omega) = \sum_{p=1}^{n} {}^nu_{lp}(\omega)\, \varphi_p(x)$$

we can deduce the relations

$$\lim_{n\to\infty} {}^n\lambda_l(\omega) = \lambda_l(\omega) \quad \text{a.s.}, \tag{1.103}$$

$$\lim_{n\to\infty} \|{}^nu_l(\cdot\,, \omega) - u_l(\cdot\,, \omega)\|_{\mathbf{L}_2(0,1)} = 0 \quad \text{a.s.} \tag{1.104}$$

((1.104) under the additional assumption that the eigenvalue $\lambda_l(\omega)$ is simple). We remark that according to the results of Section 1.2, the quantities ${}^n\lambda_l(\omega)$ and ${}^nu_{lp}(\omega)$, for $p = 1, 2, \dots, n$, are random variables.

The random differential operator $M(\omega)$ is now written in the form

$$M(\omega) = \langle M\rangle + \left(M(\omega) - \langle M\rangle\right) \doteq \langle M\rangle + M_1(\omega)\,,$$

where

$$\langle M \rangle u = \sum_{k=0}^{m} (-1)^k [\langle f_k \rangle u^{(k)}]^{(k)} ,$$

$$M_1(\omega) u = \sum_{k=0}^{m} (-1)^k [(f_k - \langle f_k \rangle) u^{(k)}]^{(k)} \doteq \sum_{k=0}^{m} (-1)^k [\tilde{f}_k u^{(k)}]^{(k)} ,$$

and $\langle M_1 \rangle = 0$.

In the following we will prove a theorem on the expansion of the eigenvalues and the eigenfunctions using the method of Ritz and the results from Section 1.3.

Theorem 1.16. *The eigenvalues and the eigenfunctions of the self-adjoint positive definite random eigenvalue problem*

$$\langle M \rangle u + M_1(\omega) u = \lambda u , \qquad U_{i0}[u] = 0 , \qquad i = 1,.2, \dots , 2m ,$$

have the expansions

$$\lambda_g(\omega) = \sum_{p=0}^{\infty} \lambda_{gp}(\omega) , \qquad u_g(x, \omega) = \sum_{p=0}^{\infty} u_{gp}(x, \omega)$$

where $\lambda_{gp}(\omega)$ and $u_{gp}(x, \omega)$ are the corresponding homogeneous terms of the p-th order in the perturbation quantities. $u_g(x, \omega)$ is normed by

$$⟪u_g(. , \omega), u_g(. , \omega)⟫ = 1 .$$

Let μ_l be the eigenvalues assumed to be simple and let $w_l(x)$ be the associated eigenfunctions of the averaged problem

$$\langle M \rangle w = \mu w , \quad U_{i0}[w] = 0 , \quad i = 1, 2, \dots , 2m .$$

If the random perturbation operator $M_1(\omega)$ satiesfies the condition

$$||M_1 u|| \leqq a(\omega) ||u|| + b(\omega) ||\langle M \rangle u|| \tag{1.105}$$

for all $u \in \mathrm{D}(\langle M \rangle)$ with a.s. non-negative constants $a(\omega)$ and $b(\omega)$, then the expansions written above converge when $C_g(\omega) < 1$. $C_g(\omega)$ is defined by

$$C_g(\omega) \doteq 16 [a(\omega)/d_g + b(\omega) (1 + \mu_g/d_g)] , \tag{1.106}$$

where $0 < d_g \leqq \min \{\mu_g - \mu_{g-1}, \mu_{g+1} - \mu_g\}$. More precisely, the estimates

$$|\lambda_g(\omega) - \sum_{p=0}^{s} \lambda_{gp}(\omega)| \leqq \tfrac{1}{4} d_g C_g^{s+1} \quad a.s., \qquad s = 0, 1, 2, \dots , \tag{1.107}$$

and

$$||u_g(. , \omega) - \sum_{p=0}^{s} u_{gp}(. , \omega)|| \leqq \tfrac{1}{2} C_g^{s+1} \quad a.s., \qquad s = 0, 1, 2, \dots , \tag{1.108}$$

are obtained. Setting

$$b_{ij}(\omega) \doteq ⟪M_1(\omega) w_i, w_j⟫ \quad \text{and} \quad \mu_{ij} \doteq \mu_i - \mu_j$$

the homogeneous terms $\lambda_{gp}(\omega)$ and

$$u_{gp}(x, \omega) = \sum_{i=1}^{\infty} {}^g y_{pi}(\omega) w_i(x)$$

(up to the fourth order) are given by

$$\lambda_{g0} = \mu_g ,$$

$$\lambda_{g1} = b_{gg} ,$$

$$\lambda_{g2} = - \sum_{\substack{i=1 \\ i \neq g}}^{\infty} \frac{b_{ig}^2}{\mu_{ig}} ,$$

$$\lambda_{g3} = \sum_{\substack{i,j=1 \\ i,j \neq g}}^{\infty} \frac{b_{gi} b_{ij} b_{jg}}{\mu_{ig} \mu_{jg}} - b_{gg} \sum_{\substack{i=1 \\ i \neq g}}^{\infty} \frac{b_{ig}^2}{\mu_{ig}^2} ,$$

$$\lambda_{g4} = - \sum_{\substack{i,j,k=1 \\ i,j,k \neq g}}^{\infty} \frac{b_{gi} b_{ij} b_{jk} b_{kg}}{\mu_{ig} \mu_{jg} \mu_{kg}} + \sum_{\substack{i,j=1 \\ i,j \neq g}}^{\infty} \frac{b_{ig}^2 b_{jg}^2}{\mu_{ig}^2 \mu_{jg}} + 2 b_{gg} \sum_{\substack{i,j=1 \\ i,j \neq g}}^{\infty} \frac{b_{gi} b_{ij} b_{jg}}{\mu_{ig}^2 \mu_{jg}} - b_{gg}^2 \sum_{\substack{i=1 \\ i \neq g}}^{\infty} \frac{b_{ig}^2}{\mu_{ig}^3} ,$$

and

$${}^g y_0 = {}^g z_0 , \qquad {}^g y_1 = {}^g z_1 ,$$

$${}^g y_p = {}^g z_p - \tfrac{1}{2} \sum_{q=2}^{p} {}^g z_{p-q} \sum_{r=1}^{q-1} \langle\!\langle {}^g y_r , {}^g y_{q-r} \rangle\!\rangle , \qquad p = 2, 3, \ldots ,$$

where the ${}^g z_p$ *are expressed in terms of the* $b_{ij}(\omega)$ *as follows* $({}^g z_p = ({}^g z_{p1}, {}^g z_{p2}, \ldots) = ({}^g z_{pi})_{1 \leqq i < \infty})$:

$${}^g z_{0i} = \delta_{gi} ,$$

$${}^g z_{1i} = - \frac{b_{ig}}{\mu_{ig}} \quad \text{for} \quad i \neq g , \qquad {}^g z_{1g} = 0 ,$$

$${}^g z_{2i} = \frac{1}{\mu_{ig}} \left[\sum_{\substack{j=1 \\ j \neq g}}^{\infty} \frac{b_{ij} b_{jg}}{\mu_{jg}} - \frac{b_{gg} b_{ig}}{\mu_{ig}} \right] \quad \text{for} \quad i \neq g , \qquad {}^g z_{2g} = 0 ,$$

$${}^g z_{3i} = \frac{1}{\mu_{ig}} \left[- \sum_{\substack{j,k=1 \\ j,k \neq g}}^{\infty} \frac{b_{ij} b_{jk} b_{kg}}{\mu_{jg} \mu_{kg}} + \sum_{\substack{j=1 \\ j \neq g}}^{\infty} \frac{b_{gj}^2 b_{ig}}{\mu_{jg} \mu_{ig}} - \frac{b_{gg}^2 b_{ig}}{\mu_{ig}^2} + \sum_{\substack{j=1 \\ j \neq g}}^{\infty} \left(\frac{1}{\mu_{ig}} + \frac{1}{\mu_{jg}} \right) \frac{b_{gg} b_{ij} b_{jg}}{\mu_{jg}} \right] \quad \text{for} \quad i \neq g , \qquad {}^g z_{3g} = 0 ,$$

$$\begin{aligned} {}^g z_{4i} = \frac{1}{\mu_{ig}} \Bigg[& \sum_{\substack{j,k,m=1 \\ j,k,m \neq g}}^{\infty} \frac{b_{ij} b_{jk} b_{km} b_{mg}}{\mu_{jg} \mu_{kg} \mu_{mg}} - \sum_{\substack{j,k=1 \\ j,k \neq g}}^{\infty} \left(\frac{1}{\mu_{jg}} + \frac{1}{\mu_{ig}} \right) \frac{b_{ij} b_{jg} b_{kg}^2}{\mu_{jg} \mu_{kg}} \\ & - b_{ig} \sum_{\substack{j,k=1 \\ j,k \neq g}}^{\infty} \frac{b_{gj} b_{jk} b_{kg}}{\mu_{ig} \mu_{jg} \mu_{kg}} - b_{gg} \sum_{\substack{j,k=1 \\ j,k \neq g}}^{\infty} \left(\frac{1}{\mu_{jg}} + \frac{1}{\mu_{kg}} + \frac{1}{\mu_{ig}} \right) \frac{b_{ij} b_{jk} b_{kg}}{\mu_{jg} \mu_{kg}} \\ & + b_{ig} b_{gg} \sum_{\substack{j=1 \\ j \neq g}}^{\infty} \left(\frac{1}{\mu_{jg}} + \frac{2}{\mu_{ig}} \right) \frac{b_{gj}^2}{\mu_{ig} \mu_{jg}} - \frac{b_{gg}^3 b_{ig}}{\mu_{ig}^3} \\ & + b_{gg}^2 \sum_{\substack{j=1 \\ j \neq g}}^{\infty} \left(\frac{1}{\mu_{jg}^2} + \frac{1}{\mu_{jg} \mu_{ig}} + \frac{1}{\mu_{ig}^2} \right) \frac{b_{ij} b_{jg}}{\mu_{jg}} \Bigg] \quad \text{for} \quad i \neq g , \qquad {}^g z_{4g} = 0 . \end{aligned}$$

Remark 1.4. *The functions*

$$u_{gp}(x, \omega) \doteq \sum_{i=1}^{\infty} {}^g z_{pi} w_i(x) \,, \qquad p = 0, 1, 2, \dots \,,$$

are the terms in the expansion of the eigenfunctions $u_g(x, \omega) = \sum_{p=0}^{\infty} u_{gp}(x, \omega)$ *corresponding to* $\lambda_g(\omega)$ *if the eigenfunction is normalized by*

$$\langle\!\langle u_g(.\,, \omega), w_g \rangle\!\rangle = 1 \,.$$

Proof. We choose the eigenfunctions $w_i(x)$ of the averaged problem as the coordinate functions for the Ritz method. Then the Ritz eigenvalue problem associated with the eigenvalue problem

$$\langle M \rangle\, u + M_1(\omega)\, u = \lambda u \,, \qquad U_{i0}[u] = 0 \,, \qquad i = 1, 2, \dots, 2m \,,$$

can be written as

$$\sum_{j=1}^{n} [a_{ij} + b_{ij}(\omega) - \Lambda \delta_{ij}]\, z_j = 0 \,, \qquad i = 1, 2, \dots, n \,,$$

using (1.102), where

$$a_{ij} \doteq \langle\!\langle \langle M \rangle\, w_i, w_j \rangle\!\rangle = \mu_j \delta_{ij} \,, \qquad b_{ij}(\omega) \doteq \langle\!\langle M_1(\omega)\, w_i, w_j \rangle\!\rangle \,.$$

The system $\{w_i(x)\}_{1 \leqq i < \infty}$ is complete in $\mathcal{H}_{\langle M \rangle}$.

We immediately obtain the expansions of the Ritz approximate solution

$${}^n\lambda_g(\omega) = \sum_{p=0}^{\infty} {}^n\lambda_{gp}(\omega) \,, \qquad {}^n u_g(x, \omega) = \sum_{p=0}^{\infty} {}^n u_{gp}(x, \omega)$$

from Theorem 1.10, where

$${}^n u_{gp}(x, \omega) = \sum_{i=1}^{n} {}^g y_{pi}(\omega)\, w_i(x) \,.$$

First we must prove the convergence (in a sense which is to be defined) of the Ritz approximate solution to the expansions given in this theorem. Thereby, the perturbation operator must satisfy a condition adapted to the Ritz method. From this convergence, the assertion of this theorem follows using the results of Rellich [4] and taking into consideration the uniqueness of the expansions of $\lambda_g(\omega)$ and $u_g(x, \omega)$.

Assume that the perturbation operator satisfies the condition

$$|\langle\!\langle M_1(\omega)\, u, u \rangle\!\rangle| \leqq \tilde{a}(\omega)\, ||u||^2 + \tilde{b}(\omega)\, \langle\!\langle \langle M \rangle\, u, u \rangle\!\rangle \,. \tag{1.109}$$

This condition is more suitable than the condition (1.105) for the Ritz method since for $u = \sum_{i=1}^{n} c_i w_i$ it follows that

$$\Big| \sum_{i,j=1}^{n} b_{ij}(\omega)\, c_i c_j \Big| \leqq \tilde{a}(\omega) \sum_{i=1}^{n} c_i^2 + \tilde{b}(\omega) \sum_{i=1}^{n} \mu_i c_i^2 \,.$$

This implies the convergence of the expansions of the eigenvalues and eigenfunctions

$${}^n\lambda_g(\omega) = \sum_{p=0}^{\infty} \varepsilon^p\, {}^n\lambda_{gp}(\omega) \quad \text{and} \quad {}^n u_g(x, \omega) = \sum_{p=0}^{\infty} \varepsilon^p\, {}^n u_{gp}(x, \omega)$$

from the conditions

$$\sum_{j=1}^{n} [\mu_i \delta + \varepsilon b_{ij} - \Lambda \delta_{ij}] \, y_j = 0 \,, \qquad i = 1, 2, \dots, n \,,$$

if $|\varepsilon D_g| < 1$ where D_g is defined by

$$D_g \doteq 16 d_g \, [\tilde{b}(\omega) + \big(1 + \tilde{a}(\omega) + \tilde{b}(\omega) \, \mu_g\big)/d_g]^2$$

(cf. RELLICH [4]). More precisely, the following estimates hold:

$$\begin{aligned} |{}^n\lambda_g(\omega) - \sum_{p=0}^{s} \varepsilon^p \, {}^n\lambda_{gp}(\omega)| &\leqq \frac{1}{4} d_g \, |\varepsilon D_g|^{s+1} \quad \text{a.s.}, \\ [{}^n u_g(. \,, \omega) - \sum_{p=0}^{s} \varepsilon^p \, {}^n u_{gp}(. \,, \omega)] &\leqq \frac{1}{2} \frac{|\varepsilon D_g|^{s+1}}{\sqrt{1 + \tilde{a} + \tilde{b} \mu_g}} \quad \text{a.s.}, \end{aligned} \tag{1.110}$$

where

$$[u] \doteq \big(1 + \tilde{a}(\omega)\big) \, ||u|| + \tilde{b}(\omega) \, ||u||_{\langle M \rangle} \,.$$

The condition (1.109) yields the convergence estimate (1.110) uniformly in n. ${}^n u_g(x, \omega)$ is normalized by $(\!({}^n u_g, {}^n u_g)\!) = 1$ from which

$$(\!({}^n u_{gp}, w_g)\!) = - \tfrac{1}{2} \sum_{q=1}^{p-1} (\!({}^n u_{gq}, {}^n u_{g, p-q})\!) \,, \qquad p = 1, 2, \dots \,,$$

follows. In view of the fact that the convergence in the space $\mathcal{H}_{\langle M \rangle}$ implies convergence with respect to the norm [.], one obtains from the Ritz method

$$\lim_{n \to \infty} {}^n\lambda_g = \lambda_g \quad \text{a.s.},$$

$$\lim_{n \to \infty} [{}^n u_g - u_g] = 0 \quad \text{a.s.}$$

(see also the proof of Theorem 1.9). To continue the proof we need the following lemma.

Lemma 1.4. *Let ${}^n v$ for $n = 1, 2, \dots$ and v be admissible functions belonging to $\langle M \rangle$ and $U_{i0}[.] = 0$, $i = 1, 2, \dots, 2m$, satisfying the condition*

$$\lim_{n \to \infty} [{}^n v - v] = 0 \quad \text{a.s.}$$

Suppose that the condition (1.109) *is satisfied for* $0 \leqq \tilde{b}(\omega) < 2$. *Then the following holds:*

$$\lim_{n \to \infty} [{}^n r - r] = 0 \,, \qquad \lim_{n \to \infty} [{}^n u - u] = 0 \,,$$

where

$${}^n r = \sum_{\substack{s=1 \\ s \neq g}}^{n} \frac{1}{\mu_{sg}} (\!({}^n v, w_s)\!) \, w_s \,, \qquad {}^n u = \sum_{\substack{s=1 \\ s \neq g}}^{n} \frac{1}{\mu_{sg}} (\!({}^n v, M_1 w_s)\!) \, w_s \,,$$

$$r = \sum_{\substack{s=1 \\ s \neq g}}^{\infty} \frac{1}{\mu_{sg}} (\!(v, w_s)\!) \, w_s \,, \qquad u = \sum_{\substack{s=1 \\ s \neq g}}^{\infty} \frac{1}{\mu_{sg}} (\!(v, M_1 w_s)\!) \, w_s \,.$$

Proof of Lemma 1.4 First, for those $\omega \in \Omega$ such that $\tilde{b}(\omega) = 0$, the convergence of $({}^n u)$ to u and $({}^n r)$ to r is to be proved in the norm of $\mathbf{L}_2$. Let $G(x, y)$ denote the Green's function belonging to $\langle M \rangle$ and the boundary conditions $U_{i0}[u] = 0$, $i = 1, 2, \dots, 2m$. We define

$$Tu \doteq \int_0^1 G(x, y)\, u(y)\, \mathrm{d}y$$

and obtain the estimates

$$0 \leqq \lim_{n\to\infty} ||{}^n r - r||^2 = \lim_{n\to\infty} \left[\sum_{\substack{s=1\\ s\neq g}}^{n} \frac{\mu_s^2}{\mu_{sg}^2} (\!(T({}^n v - v), w_s)\!)^2 + \sum_{s=n+1}^{\infty} \frac{\mu_s^2}{\mu_{sg}^2} (\!(Tv, w_s)\!)^2 \right]$$

$$\leqq \alpha^2 \lim_{n\to\infty} \Big[||T({}^n v - v)||^2 + \sum_{s=n+1}^{\infty} (\!(Tv, w_s)\!)^2 \Big] = 0$$

and

$$0 \leqq \lim_{n\to\infty} ||{}^n u - u||^2 = \lim_{n\to\infty} \left[\sum_{\substack{s=1\\ s\neq g}}^{n} \frac{\mu_s^2}{\mu_{sg}^2} (\!(TM_1({}^n v - v), w_s)\!)^2 + \sum_{s=n+1}^{\infty} \frac{\mu_s^2}{\mu_{sg}^2} (\!(TM_1 v, w_s)\!)^2 \right]$$

$$\leqq \alpha^2 \lim_{n\to\infty} \Big[||TM_1({}^n v - v)||^2 + \sum_{s=n+1}^{\infty} (\!(TM_1 v, w_s)\!)^2 \Big] = 0\,,$$

because

$$||TM_1({}^n v - v)||^2 \leqq \int_0^1\!\!\int_0^1 \big(M_1 G(x, y)\big)^2\, \mathrm{d}x\, \mathrm{d}y\, ||{}^n v - v||^2$$

(in the term $M_1 G(x, y)$, the operator M_1 acts on G as a function of y). We have assumed that α is a real number with $|\mu_s/\mu_{sg}| \leqq \alpha$, for $s = 1, 2, \dots$. For $\tilde{b}(\omega) > 0$ the norm $[.]$ is equivalent to the norm $||.||_{\langle M \rangle}$. We put

$${}^n r - r \doteq \bar{a}_n - \bar{b}_n\,,$$

where

$$\bar{a}_n \doteq \sum_{\substack{s=1\\ s\neq g}}^{n} \frac{1}{\mu_{sg}} (\!({}^n v - v, w_s)\!)\, w_s\,, \qquad \bar{b}_n \doteq \sum_{s=n+1}^{\infty} \frac{1}{\mu_{sg}} (\!(v, w_s)\!)\, w_s\,.$$

Since $(\!(\langle M \rangle \bar{a}_n, \bar{b}_n)\!) = (\!(\bar{a}_n, \langle M \rangle \bar{b}_n)\!) = 0$, it follows that

$$(\!(\langle M \rangle ({}^n r - r), {}^n r - r)\!) = (\!(\langle M \rangle \bar{a}_n, \bar{a}_n)\!) + (\!(\langle M \rangle \bar{b}_n, \bar{b}_n)\!)$$

$$= \sum_{\substack{s=1\\ s\neq g}}^{n} \frac{\mu_s}{\mu_{sg}^2} (\!({}^n v - v, w_s)\!)^2 + \sum_{s=n+1}^{\infty} \frac{\mu_s}{\mu_{sg}^2} (\!(v, w_s)\!)^2$$

$$\leqq \alpha^2 \Big[(\!(T({}^n v - v), {}^n v - v)\!) + \sum_{s=n+1}^{\infty} (\!(Tv, w_s)\!)(\!(v, w_s)\!) \Big]\,.$$

This implies that

$$\lim_{n\to\infty} [{}^n r - r] = 0 .$$

Analogously, we obtain

$${}^n u - u = a_n - b_n ,$$

where

$$a_n \doteq \sum_{\substack{s=1\\ s\neq g}}^{n} \frac{1}{\mu_{sg}} ⟪{}^n v - v, M_1 w_s⟫ w_s , \qquad b_n \doteq \sum_{s=n+1}^{\infty} \frac{1}{\mu_{sg}} ⟪v, M_1 w_s⟫ w_s .$$

Again, since $⟪\langle M\rangle a_n, b_n⟫ = ⟪a_n, \langle M\rangle b_n⟫ = 0$, we obtain the relation

$$⟪\langle M\rangle ({}^n u - u), {}^n u - u⟫ = ⟪\langle M\rangle a_n, a_n⟫ + ⟪\langle M\rangle b_n, b_n⟫ .$$

For $⟪\langle M\rangle b_n, b_n⟫$ the estimate

$$⟪\langle M\rangle b_n, b_n⟫ = \sum_{s=n+1}^{\infty} \frac{1}{\mu_{sg}} ⟪v, M_1 w_s⟫⟪w_s, \langle M\rangle b_n⟫$$

$$= \sum_{s=n+1}^{\infty} \frac{\mu_s}{\mu_{sg}^2} ⟪v, M_1 w_s⟫^2 \leqq \alpha^2 \sum_{s=n+1}^{\infty} ⟪TM_1 v, w_s⟫⟪M_1 v, w_s⟫$$

can be deduced. Since the series $\sum_{s=1}^{\infty} ⟪TM_1 v, w_s⟫⟪M_1 v, w_s⟫$ converges this implies

$$\lim_{n\to\infty} ⟪\langle M\rangle b_n, b_n⟫ = 0 .$$

For $⟪\langle M\rangle a_n, a_n⟫$ we obtain

$$⟪\langle M\rangle a_n, a_n⟫ = \sum_{\substack{s=1\\ s\neq g}}^{n} \frac{\mu_s}{\mu_{sg}^2} ⟪{}^n v - v, M_1 w_s⟫^2$$

$$\leqq \alpha^2 \sum_{s=1}^{\infty} ⟪TM_1({}^n v - v), w_s⟫⟪M_1({}^n v - v), w_s⟫$$

$$= \alpha^2 ⟪TM_1({}^n v - v), M_1({}^n v - v)⟫ .$$

Defining

$${}^n v - v \doteq v_n \quad \text{and} \quad TM_1 v_n \doteq c_n$$

we derive

$$⟪TM_1({}^n v - v), M_1({}^n v - v)⟫ = ⟪M_1 v_n, c_n⟫$$

$$= \tfrac{1}{4} \{⟪M_1(c_n + v_n), c_n + v_n⟫ - ⟪M_1(c_n - v_n), c_n - v_n⟫\}$$

$$= \tfrac{1}{4} \{\tilde{a}(\omega) (\|c_n + v_n\|^2 + \|c_n - v_n\|^2) + 2\tilde{b}(\omega) (⟪\langle M\rangle c_n, c_n⟫ + ⟪\langle M\rangle v_n, v_n⟫)\} .$$

Hence,

$$⟪\langle M\rangle c_n, c_n⟫ = ⟪M_1 v_n, TM_1 v_n⟫$$

and $0 \leqq \tilde{b}(\omega) < 2$ a.s. yield

$$0 \leqq \lim_{n\to\infty} \langle\!\langle T M_1 v_n, M_1 v_n \rangle\!\rangle$$

$$\leqq \frac{1}{4\left(1 - \frac{1}{2}\tilde{b}(\omega)\right)} \lim_{n\to\infty} \{\tilde{a}(\omega)\,(||c_n + v_n||^2 + ||c_n - v_n||^2)$$

$$+ 2\tilde{b}(\omega)\,\langle\!\langle \langle M \rangle\, v_n, v_n \rangle\!\rangle\} = 0;$$

and, therefore,

$$\lim_{n\to\infty} \langle\!\langle \langle M \rangle\, ({}^n u - u), {}^n u - u \rangle\!\rangle = 0\,. \blacktriangleleft$$

Using Lemma 1.4 we will now show that

$$\lim_{n\to\infty} [{}^n u_{gk} - u_{gk}] = 0 \quad \text{and} \quad \lim_{n\to\infty} {}^n\lambda_{gk} = \lambda_{gk} \quad \text{a.s.,} \tag{1.111}$$

where

$$u_{gk} = \sum_{\substack{s=1\\ s\neq g}}^{\infty} \frac{1}{\mu_{sg}} \langle\!\langle q_k, w_s \rangle\!\rangle\, w_s - \frac{1}{2} \sum_{r=1}^{k-1} \langle\!\langle u_{gr}, u_{g,k-r} \rangle\!\rangle\, w_g\,,$$

$$\lambda_{gk} = - \sum_{r=1}^{k-1} \lambda_{gr} \langle\!\langle u_{g,k-r}, w_g \rangle\!\rangle + \langle\!\langle M_1 u_{g,k-1}, w_g \rangle\!\rangle$$

for $k = 1, 2, \ldots$, and

$$q_k = -M_1 u_{g,k-1} + \sum_{r=1}^{k-1} \lambda_{gr} u_{g,k-r}$$

(cf. (1.97), (1.98)). From now on in the proof of this theorem we denote the solution of (1.24) by ${}^n\lambda_{gk}$, ${}^n y_{gk}$ (in this case the upper index n denotes the number of the coordinate functions of the Ritz method and the lower index g the number of the eigenvalue or the eigenvector, respectively; λ_{gk} in (1.24) here corresponds to $-{}^n\lambda_{gk}$) and have

$$\begin{aligned} {}^n y_{gki} &= -\frac{1}{\mu_{ig}} [(B\, {}^n y_{g,k-1})_i - \sum_{r=1}^{k} {}^n y_{gr}\, {}^n\lambda_{g,k-r,i}] \quad \text{for} \quad i \neq g\,, \\ {}^n y_{gkg} &= -\frac{1}{2} \sum_{s=1}^{k-1} \langle\!\langle {}^n y_{gs}, {}^n y_{g,k-s} \rangle\!\rangle\,, \quad \text{where} \quad {}^n y_{gk} \doteq ({}^n y_{gki})_{1 \leqq i \leqq n}\,, \end{aligned} \tag{1.112}$$

and

$${}^n\lambda_{gk} = \langle\!\langle B\, {}^n y_{g,k-1}, {}^n y_{g0} \rangle\!\rangle - \sum_{r=1}^{k-1} {}^n\lambda_{gr} \langle\!\langle {}^n y_{g,k-r}, {}^n y_{g0} \rangle\!\rangle\,.$$

Using the relations

$$\langle\!\langle {}^n u_{gp}, w_s \rangle\!\rangle = {}^n y_{gps}\,, \qquad \sum_{s=1}^{k-1} \langle\!\langle {}^n u_{gs}, {}^n u_{g,k-s} \rangle\!\rangle = \sum_{s=1}^{k-1} \langle\!\langle {}^n y_{gs}, {}^n y_{g,k-s} \rangle\!\rangle\,,$$

$$(B\, {}^n y_{gp})_i = \sum_{r=1}^{n} \langle\!\langle M_1 w_i, w_r \rangle\!\rangle\, {}^n y_{gpr} = \langle\!\langle M_1 w_i, {}^n u_{gp} \rangle\!\rangle$$

we obtain

$$ {}^n u_{gk} = - \sum_{\substack{s=1 \\ s \neq g}}^{n} \frac{1}{\mu_{sg}} \langle\!\langle {}^n u_{g,k-1}, M_1 w_s \rangle\!\rangle w_s + \sum_{r=1}^{k-1} {}^n\lambda_{gr} \sum_{\substack{s=1 \\ s \neq g}}^{n} \frac{1}{\mu_{sg}} \langle\!\langle {}^n u_{g,k-r}, w_s \rangle\!\rangle w_s $$

$$ - \frac{1}{2} \sum_{r=1}^{k-1} \langle\!\langle {}^n u_{gr}, {}^n u_{g,k-r} \rangle\!\rangle w_g \qquad (1.113) $$

if we multiply (1.112) by w_i and take the sum over i. Applying

$$ \langle\!\langle B\, {}^n y_{g,k-1}, {}^n y_{g0} \rangle\!\rangle = \langle\!\langle M_1\, {}^n u_{g,k-1}, w_g \rangle\!\rangle $$

it follows that

$$ {}^n\lambda_{gk} = \langle\!\langle M_1\, {}^n u_{g,k-1}, w_g \rangle\!\rangle - \sum_{r=1}^{k-1} {}^n\lambda_{gr} \langle\!\langle {}^n u_{g,k-r}, w_g \rangle\!\rangle \quad \text{for} \quad k = 1, 2, \ldots \qquad (1.114) $$

Because ${}^n u_{g0} = w_g = u_{g0}$, ${}^n\lambda_{g0} = \mu_g = \lambda_{g0}$, and using Lemma 1.4, the sequence

$$ {}^n u_{g1} = - \sum_{\substack{s=1 \\ s \neq g}}^{n} \frac{1}{\mu_{sg}} \langle\!\langle w_g, M_1 w_s \rangle\!\rangle w_s $$

converges in the norm [.] to

$$ u_{g1} = - \sum_{\substack{s=1 \\ s \neq g}}^{\infty} \frac{1}{\mu_{sg}} \langle\!\langle w_g, M_1 w_s \rangle\!\rangle w_s $$

almost surely as $n \to \infty$ and

$$ {}^n\lambda_{g1} = \langle\!\langle M_1 w_g, w_g \rangle\!\rangle $$

converges almost surely to

$$ \lambda_{g1} = \langle\!\langle M_1 w_g, w_g \rangle\!\rangle . $$

Suppose that the relations (1.111) have already been proved for $k = p - 1$. Then by means of Lemma 1.4, (1.113) implies the convergence of ${}^n u_{gp}$; and (1.114) implies the convergence of ${}^n\lambda_{gp}$ a.s. where the limits u_{gp}, λ_{gp}, respectively, satisfy the last two relations of (1.111) for $k = p$.

From condition (1.110) and the formulas (1.111) we now obtain

$$ [u_g(.\,,\omega) - \sum_{p=0}^{s} \varepsilon^p u_{gp}(.\,,\omega)] \leqq \frac{1}{2} \frac{1}{\sqrt{1 + \tilde{a} + \tilde{b}\mu_g}} (\varepsilon D_g)^{s+1} \quad \text{a.s.}, $$

$$ |\lambda_g(\omega) - \sum_{p=0}^{s} \varepsilon^p \lambda_{gp}(\omega)| \leqq \frac{1}{4} d_g (\varepsilon D_g)^{s+1} \quad \text{a.s.} $$

Since the convergence in the norm [.] implies the convergence in the $\mathbf{L}_2$-norm, we have

$$ \lim_{n \to \infty} {}^n y_{gpi} = \lim_{n \to \infty} \langle\!\langle {}^n u_{gp}, w_i \rangle\!\rangle = \langle\!\langle u_{gp}, w_i \rangle\!\rangle = {}^g y_{pi} . $$

Therefore, the ${}^n y_{gpi}$ evaluated from the Ritz' eigenvalue problem of order n according to Theorem 1.10 converge to ${}^g y_{pi}$ given in this theorem. The same is true for ${}^n\lambda_{gp}$ from Theorem 1.10 which converge to λ_{gp} given by this theorem.

The considerations so far have been based on condition (1.109). Finally, the propositions of this theorem follow from the results given above using the uniqueness of the perturbation expansions as well as Rellich's result that the expansions of the eigenvalues and the eigenfunctions converge for $C_g < 1$ if the condition

$$||M_1 u|| \leqq a||u|| + b||\langle M\rangle u||$$

is satisfied. Furthermore, the estimates (1.107), (1.108) hold. ◀

The following theorem yields the calculation of the mean values.

Theorem 1.17. *Using the notation and assumptions of Theorem* 1.16, *the following relations hold:*

$$\langle \lambda_g \rangle = \sum_{p=0}^{\infty} \langle \lambda_{gp} \rangle ,$$

$$\langle u_g(x) \rangle = \sum_{p=0}^{\infty} \langle u_{gp}(x) \rangle \quad (\textit{in mean-square}).$$

Especially, the estimates

$$|\langle \lambda_g \rangle - \sum_{p=0}^{s} \langle \lambda_{gp} \rangle| \leqq \tfrac{1}{4}\, d_g \langle C_g^{s+1} \rangle ,$$

$$||\langle u_g(.)\rangle - \sum_{p=0}^{s} \langle u_{gp}(.)\rangle|| \leqq \tfrac{1}{2} \sqrt{\langle C_g^{2s+2}\rangle}$$

can be deduced.

Proof. The inequality (1.107) yields

$$-\tfrac{1}{4}\, d_g C_g^{s+1} \leqq \lambda_g(\omega) - \sum_{p=0}^{s} \lambda_{gp}(\omega) \leqq \tfrac{1}{4}\, d_g C_g^{s+1}$$

and taking the expectation value the assertion follows for the eigenvalues. Let us remark that the relation

$$\lim_{k\to\infty} \langle C_g^k \rangle = 0$$

can be derived from $C_g^k \geqq C_g^{k+1}$ a.s. and $\lim_{k\to\infty} C_g^k = 0$ a.s. According to (1.108) we have for the eigenfunctions

$$\int_0^1 \Big(u_g(x, \omega) - \sum_{p=0}^{s} u_{gp}(x, \omega)\Big)^2 \,\mathrm{d}x \leqq \tfrac{1}{4}\, C_g^{2s+2} .$$

Using the inequality $\langle X^2 \rangle \geqq \langle X \rangle^2$ the assertion of this theorem for the eigenfunctions

$$\int_0^1 \Big(\langle u_g(x)\rangle - \sum_{p=0}^{s} \langle u_{gp}(x)\rangle\Big)^2 \,\mathrm{d}x \leqq \int_0^1 \Big\langle \Big(u_g(x) - \sum_{p=0}^{s} u_{gp}(x)\Big)^2 \Big\rangle \,\mathrm{d}x$$
$$\leqq \tfrac{1}{4} \langle C_g^{2s+2} \rangle$$

is obtained by taking expectations. ◀

The expansions given in Theorem 1.16 and Theorem 1.17 allow approximations of mean values up to the fourth order to be evaluated if the moments of the random coefficients are known up to fourth order. The correlation quantities are obtained up to the fifth order according to the remarks at the end of Section 1.4.1.

1.4.3. A characteristic special class of the general case

In this section we consider eigenvalue problems of the form

$$\begin{aligned} -u'' + a(x, \omega)\, u &= \lambda\big(1 + b(x, \omega)\big)\, u\,, \\ u(0) = u(1) + \eta(\omega)\, u'(1) &= 0\,. \end{aligned} \tag{1.115}$$

This is a characteristic example for the eigenvalue problem (1.86), since $M_1 \neq 0$, $N_1 \neq 0$, and the boundary conditions are stochastic. (1.115) has interesting applications. Thus for $b(x, \omega) \equiv \eta(\omega) = 0$ the one-dimensional Schrödinger equation is obtained with random potential $a(x, \omega)$. The investigation of eigenvibrations of an elastic string with random density which is fixed at the one end ($x = 0$) and elastically supported at the other end ($x = 1$) (the random variable $\eta(\omega)$ defines the mobility at $x = 1$) leads to (1.115) with $b(x, \omega) \equiv 0$.

Let $\langle a(x)\rangle \equiv \langle b(x)\rangle \equiv \langle \eta\rangle = 0$ so that the averaged or unperturbed problem

$$-w'' = \mu w\,, \qquad w(0) = w(1) = 0$$

admits the simple eigenvalues $\mu_g = (\pi g)^2$ with the associated eigenfunctions $w_g(x) = \sqrt{2}\sin(g\pi x)$, $g = 1, 2, \ldots$ All of the assumptions of Theorem 1.15 are satisfied (in particular, (1.99)), since $\tilde{M}_1 u = (a - \mu_g b)\, u$ is symmetric. In order to use the perturbation theory we assume that $|a(x, \omega)|$, $|b(x, \omega)|$, $|\eta(\omega)|$ are sufficiently small. Furthermore, let $a(x, \omega)$, $b(x, \omega)$ have a.s. continuous sample functions. For numerical calculations of the moments we assume the stationarity in the wide-sense of the vector process $\big(a(x, \omega), b(x, \omega)\big)$.

In the proof of Theorem 1.15 series of the form

$$\sum_{\substack{s=1 \\ s \neq g}}^{\infty} \frac{1}{\mu_{sg}} (\!(f, w_s)\!)\, w_s(x)$$

are given. The following lemma permits this series for the special μ_l and w_l of our problem to be computed explicitly.

Lemma 1.5. *Let $f(x, \omega)$ be a stochastic process for which almost all realizations are continuous. Put $w_g(x) = \sqrt{2}\sin(g\pi x)$ and $\mu_g = (g\pi)^2$. Then the equation*

$$\begin{aligned} &\sum_{\substack{s=1 \\ s \neq g}}^{\infty} \frac{1}{\mu_{sg}} (\!(f(\omega), w_s)\!)\, w_s(x) \\ &= c_1 w_g(x) + \frac{1}{g\pi} \int_0^x f(t, \omega) \sin\big(g\pi(t - x)\big)\, dt + \frac{c}{\sqrt{2}\, g\pi}\, x \cos(g\pi x) \end{aligned}$$

can be obtained where

$$c(\omega) = -\langle\!\langle f, w_g \rangle\!\rangle$$

and

$$c_1(\omega) = -\frac{c}{4\mu_g} + \frac{1}{\sqrt{2}\, g\pi} \langle\!\langle f(t, \omega), (1 - t) \cos (g\pi t) \rangle\!\rangle .$$

Proof. For

$$s(x, \omega) = \sum_{\substack{k=1\\ k \neq g}}^{\infty} \frac{1}{\mu_{kg}} \langle\!\langle f, w_k \rangle\!\rangle\, w_k(x) = \sum_{\substack{k=1\\ k \neq g}}^{\infty} \frac{1}{\mu_{kg}} \langle\!\langle v, w_k \rangle\!\rangle\, w_k(x)$$

with

$$v(x, \omega) \doteq f(x, \omega) + c w_g(x)$$

we have

$$s - \mu_g T s = T v$$

if $\langle\!\langle v, w_g \rangle\!\rangle = 0$ holds. Let T denote the integral operator with the Green's function $G(x, y)$ associated with $M_0 w = -w''$ and $w(0) = w(1) = 0$ as kernel. From $\langle\!\langle v, w_g \rangle\!\rangle = 0$ it follows $c = -\langle\!\langle f, w_g \rangle\!\rangle$. Thus $s(x, \omega)$ is a solution of the boundary value problem

$$u'' + \mu_g u = -v , \qquad u(0) = u(1) = 0$$

which admits the general solution

$$u(x, \omega) = \tilde{c}_1 w_g(x) + \frac{1}{\pi g} \int_0^x v(t, \omega) \sin \big(g\pi(t - x)\big)\, dt .$$

From $\langle\!\langle s, w_g \rangle\!\rangle = 0$ the constant $\tilde{c}_1$ can be found; therefore

$$\tilde{c}_1(\omega) = -\frac{1}{\pi g} \langle\!\langle \int_0^x v(t, \omega) \sin \big(g\pi(t - x)\big)\, dt, w_g(x) \rangle\!\rangle .$$

By an easy computation one verifies that

$$\int_0^x v(t, \omega) \sin \big(g\pi(t - x)\big)\, dt = \int_0^x f(t, \omega) \sin \big(g\pi(t - x)\big)\, dt - \frac{c}{2g\pi} w_g(x) + \frac{c}{\sqrt{2}} x \cos (g\pi x)$$

so that

$$\tilde{c}_1(\omega) = -\frac{1}{\pi g} \langle\!\langle \int_0^x f(t, \omega) \sin \big(g\pi(t - x)\big)\, dt, w_g(x) \rangle\!\rangle + \frac{3c}{4\mu_g}$$

$$= -\frac{1}{\pi g} \int_0^1 \int_0^x f(t, \omega) \sin \big(g\pi(t - x)\big)\, w_g(x)\, dx\, dt + \frac{3c}{4\mu_g}$$

$$= \frac{c}{4\mu_g} + \frac{1}{\sqrt{2}\, g\pi} \langle\!\langle f(t, \omega), (1 - t) \cos (g\pi t) \rangle\!\rangle . \quad \blacktriangleleft$$

1.4.3.1. Calculation of moments and asymptotic results

In order to evaluate the approximations for the eigenvalues and the eigenfunctions utilizing to Theorem 1.15 we have to make use of the stochastic processes $z_{gk}(x, \omega)$ which satisfy the condition

$$U_{i0}[z_{gk}] = -U_{i1}[u_{g,k-1}],$$

that is, in the case of (1.115)

$$z_{gk}(0) = 0, \qquad z_{gk}(1) = -\eta u'_{g,k-1}(1).$$

The processes $z_{gk}(x, \omega) \doteq -\eta u'_{g,k-1}(1)\, x$ satisfy these conditions. Using

$$\tilde{M}_1(\omega)\, u \doteq \tilde{a}(x, \omega)\, u \doteq \big(a(x, \omega) - \mu_g b(x, \omega)\big)\, u,$$

$$N_1(\omega)\, u \doteq b(x, \omega)\, u,$$

$$\tilde{M}_0 u \doteq -u'' - \mu_g u$$

we obtain the expansions of Theorem 1.15 where

$$b_{ij} \doteq \langle\!\langle \tilde{a} w_i, w_j \rangle\!\rangle \quad \text{and} \quad a_{ij} \doteq \langle\!\langle b w_i, w_j \rangle\!\rangle.$$

To calculate the first approximations we take into consideration that

$$z_{g1}(x, \omega) = -\eta(\omega)\, w'_g(1)\, x$$

and

$$\langle\!\langle z_{g1}, w_g \rangle\!\rangle = 2\eta.$$

Then we obtain, from Lemma 1.5 applied to $\sum_{\substack{s=1\\ s\neq g}}^{\infty} 1/\mu_{sg} \langle\!\langle z_{g1}, w_s \rangle\!\rangle\, w_s$ the equation

$$\sum_{\substack{s=1\\ s\neq g}}^{\infty} \frac{1}{\mu_{sg}} \langle\!\langle z_{g1}, w_s \rangle\!\rangle\, w_s = -\eta w'_g(1) \left[\frac{(-1)^g}{(g\pi)^2}\, x \cos(g\pi x) - \frac{x}{(g\pi)^2} - \frac{3}{2}\frac{(-1)^g}{(g\pi)^3} \sin(g\pi x)\right]. \tag{1.116}$$

This implies

$$\lambda_{g1}(\omega) = b_{gg} - 2\mu_g \eta, \tag{1.117}$$

$$u_{g1}(x, \omega) = -\sum_{\substack{s=1\\ s\neq g}}^{\infty} \frac{1}{\mu_{sg}}\, b_{gs} w_s(x) - \eta x w'_g(x) - \frac{1}{2}\eta w_g(x). \tag{1.118}$$

Theorem 1.15 yields for $\lambda_{g2}(\omega)$ and $u_{g2}(x, \omega)$

$$\lambda_{g2}(\omega) = -\sum_{\substack{s=1\\ s\neq g}}^{\infty} \frac{1}{\mu_{sg}} b_{gs}^2 + \mu_g \sum_{\substack{s=1\\ s\neq g}}^{\infty} \frac{1}{\mu_{gs}} \langle\langle z_{g1}, w_s\rangle\rangle b_{gs}$$
$$- \langle\langle z_{g1}, w_g\rangle\rangle b_{gg} + \langle\langle \tilde{a} z_{g1}, w_g\rangle\rangle - \lambda_{g1} a_{gg} - \mu_g \langle\langle z_{g2}, w_g\rangle\rangle ,$$

$$u_{g2}(x, \omega) = \sum_{\substack{s,t=1\\ s,t\neq g}}^{\infty} \frac{1}{\mu_{tg}\mu_{sg}} b_{gt} b_{ts} w_s(x) + z_{g2} - \langle\langle z_{g2}, w_g\rangle\rangle w_g$$
$$- \mu_g \sum_{\substack{s=1\\ s\neq g}}^{\infty} \frac{1}{\mu_{sg}} \left[\sum_{\substack{t=1\\ t\neq g}}^{\infty} \frac{1}{\mu_{tg}} \langle\langle z_{g1}, w_t\rangle\rangle b_{ts} \right] w_s(x) - \frac{1}{2} \langle\langle u_{g1}, u_{g1}\rangle\rangle w_g$$
$$+ \sum_{\substack{s=1\\ s\neq g}}^{\infty} \frac{1}{\mu_{sg}} \left[\langle\langle z_{g1}, w_g\rangle\rangle b_{gs} - \langle\langle \tilde{a} z_{g1}, w_s\rangle\rangle + \mu_g \langle\langle z_{g2}, w_s\rangle\rangle \right.$$
$$\left. + \lambda_{g1} \left\{ \frac{\mu_s}{\mu_{sg}} \langle\langle z_{g1}, w_s\rangle\rangle + a_{gs} - \frac{1}{\mu_{sg}} b_{gs} \right\} \right] w_s(x) .$$

Multiplying (1.116) by $\tilde{a}(x, \omega)\, w_t(x)$ and integrating over [0, 1] lead to

$$\mu_g \sum_{\substack{s=1\\ s\neq g}}^{\infty} \frac{1}{\mu_{sg}} \langle\langle z_{g1}, w_s\rangle\rangle b_{ts} = \frac{3}{2} \eta b_{gt} - \eta \langle\langle \tilde{a} w_t, x w_g'\rangle\rangle - \langle\langle \tilde{a} z_{g1}, w_t\rangle\rangle \tag{1.119}$$

and, therefore,

$$\lambda_{g2}(\omega) = -\sum_{\substack{s=1\\ s\neq g}}^{\infty} \frac{1}{\mu_{sg}} b_{gs}^2 - b_{gg} a_{gg}$$
$$- \eta \left[\frac{1}{2} b_{gg} + \langle\langle \tilde{a} w_g, x w_g'\rangle\rangle - 2\mu_g a_{gg} + w_g'(1)\, u_{g1}'(1) \right] . \tag{1.120}$$

By means of (1.119) we calculate

$$\mu_g \sum_{\substack{s=1\\ s\neq g}}^{\infty} \frac{1}{\mu_{sg}} \left[\sum_{\substack{t=1\\ t\neq g}}^{\infty} \frac{1}{\mu_{tg}} \langle\langle z_{g1}, w_t\rangle\rangle b_{ts} \right] w_s + \sum_{\substack{s=1\\ s\neq g}}^{\infty} \frac{1}{\mu_{sg}} \langle\langle \tilde{a} z_{g1}, w_s\rangle\rangle w_s$$
$$= \frac{3}{2} \eta \sum_{\substack{s=1\\ s\neq g}}^{\infty} \frac{1}{\mu_{sg}} b_{gs} w_s - \eta \sum_{\substack{s=1\\ s\neq g}}^{\infty} \frac{1}{\mu_{sg}} \langle\langle \tilde{a} x w_g', w_s\rangle\rangle w_s .$$

Finally, if we use

$$\sum_{\substack{s=1\\ s\neq g}}^{\infty} \frac{1}{\mu_{sg}} \langle\langle z_{g1}, w_g\rangle\rangle b_{gs} w_s = 2\eta \sum_{\substack{s=1\\ s\neq g}}^{\infty} \frac{1}{\mu_{sg}} b_{gs} w_s$$

the $u_{g2}(x, \omega)$ can be determined as

$$u_{g2}(x, \omega) = \sum_{\substack{s,t=1 \\ s,t \neq g}}^{\infty} \frac{1}{\mu_{sg}\mu_{tg}} b_{gt} b_{ts} w_s + \eta \sum_{\substack{s=1 \\ s \neq g}}^{\infty} \frac{1}{\mu_{sg}} \langle\!\langle \tilde{a}(xw_g' + \tfrac{1}{2} w_g), w_s \rangle\!\rangle w_s$$
$$+ \lambda_{g1} \sum_{\substack{s=1 \\ s \neq g}}^{\infty} \frac{1}{\mu_{sg}} \{\langle\!\langle u_{g1}, w_s \rangle\!\rangle + a_{gs}\} w_s$$
$$+ \left(xw_g' + \frac{1}{2} w_g\right) \frac{1}{\sqrt{2}\, g\pi} (-1)^{g+1} \eta u_{g1}'(1) - \frac{1}{2} \langle\!\langle u_{g1}, u_{g1} \rangle\!\rangle w_g . \tag{1.121}$$

All of the infinite series in (1.118), (1.120), and (1.121) can be evaluated by using Lemma 1.5.

First, we have

$$\sum_{\substack{s=1 \\ s \neq g}}^{\infty} \frac{1}{\mu_{sg}} b_{gs} w_s = \frac{b_{gg}}{2\mu_g} \left(\frac{1}{2} w_g - xw_g'\right) + \frac{1}{2\mu_g} \langle\!\langle \tilde{a} w_g, (1 - t) w_g' \rangle\!\rangle w_g$$
$$+ \frac{1}{g\pi} \int_0^x \tilde{a} w_g \sin\big(g\pi(t - x)\big)\, dt \tag{1.122}$$

and thus we obtain from (1.118)

$$u_{g1}(x, \omega) = -\frac{1}{g\pi} \int_0^x \tilde{a} w_g \sin\big(g\pi(t - x)\big)\, dt + \left(\frac{b_{gg}}{2\mu_g} - \eta\right) xw_g'$$
$$- \left[\frac{b_{gg}}{4\mu_g} + \frac{1}{2\mu_g} \langle\!\langle \tilde{a} w_g, (1 - t) w_g' \rangle\!\rangle + \frac{1}{2} \eta\right] w_g . \tag{1.123}$$

From (1.122) we deduce by scalar multiplication with $\tilde{a}(x, \omega)\, w_r(x)$

$$\sum_{\substack{s=1 \\ s \neq g}}^{\infty} \frac{1}{\mu_{sg}} b_{gs} b_{sr} = \left[\frac{1}{2} b_{gg} + \langle\!\langle \tilde{a} w_g, (1 - t) w_g' \rangle\!\rangle\right] \frac{b_{gr}}{2\mu_g} - \frac{b_{gg}}{2\mu_g} \langle\!\langle \tilde{a} w_r, xw_g' \rangle\!\rangle$$
$$+ \frac{1}{\pi g} \langle\!\langle \int_0^x \tilde{a}(t)\, w_g(t) \sin\big(g\pi(t - x)\big)\, dt, \tilde{a} w_r \rangle\!\rangle . \tag{1.124}$$

(1.123) yields

$$u_{g1}'(1, \omega) = (-1)^g \left[\langle\!\langle \tilde{a} w_g, t \cos(g\pi t) \rangle\!\rangle + \sqrt{2}\, b_{gg}/4g\pi - 3g\pi\eta/\sqrt{2}\right] .$$

In a similar way $\lambda_{g2}(\omega)$ can also be explicitly calculated from (1.120). We have

$$\lambda_{g2}(\omega) = -b_{gg}^2/4\mu_g - \langle\!\langle \tilde{a} w_g, (1 - 2t) w_g' \rangle\!\rangle b_{gg}/2\mu_g - b_{gg} a_{gg}$$
$$- \frac{1}{\pi g} \langle\!\langle \int_0^x \tilde{a} w_g \sin\big(g\pi(t - x)\big)\, dt, \tilde{a} w_g \rangle\!\rangle \tag{1.125}$$
$$- \eta[b_{gg} + 2\langle\!\langle \tilde{a} w_g', xw_g \rangle\!\rangle - 2\mu_g a_{gg} - 3\mu_g \eta] .$$

Similarly, there are not fundamental difficulties in calculating the infinite series in $u_{g2}(x, \omega)$ by using Lemma 1.5.

We now determine the mean values of the eigenvalues and the eigenfunctions up to second order terms in the random perturbations. Since $\langle \lambda_{g1} \rangle = \langle u_{g1} \rangle = 0$, $\langle \lambda_{g2} \rangle$ and $\langle u_{g2} \rangle$ are to be calculated. If we neglect terms of the third and higher orders then we obtain

$$\langle \lambda_g \rangle - \mu_g = \langle \lambda_{g2} \rangle , \qquad \langle u_g \rangle - w_g = \langle u_{g2} \rangle ;$$

and thus $\langle \lambda_{g2} \rangle$, $\langle u_{g2} \rangle$ are approximate solutions of the averaging problem.

The following lemma is used for the calculation of the means of u_{g2} and λ_{g2}.

Lemma 1.6. *Let a, b, c be arbitrary real numbers. Then the quantities*

$$B \doteq \int_0^1 \int_0^1 R(y - x)\, x \sin\big(g\pi(ax + by) + c\big)\, dx\, dy ,$$

$$C \doteq \int_0^1 \int_0^1 R(y - x) \sin\big(g\pi(ax + by) + c\big)\, dx\, dy$$

are given by

$$B = \begin{cases} -\dfrac{1}{(g\pi)^2 (a+b)^2} \displaystyle\int_0^1 \{R(u)\,[\sin\big(g\pi a u - g\pi(a+b) - c\big) \\ \qquad + \sin(g\pi b u + c)] \\ \qquad + R(-u)\,[\sin\big(g\pi b u - g\pi(a+b) - c\big) \\ \qquad + \sin(g\pi a u + c)]\}\, du \\ + \dfrac{1}{g\pi(a+b)} \displaystyle\int_0^1 \{R(u)\,(u-1)\cos\big(g\pi a u - g\pi(a+b) - c\big) \\ \qquad + R(-u)\,[u \cos(g\pi a u + c) \\ \qquad - \cos\big(g\pi b u - g\pi(a+b) - c\big)]\}\, du \\ \qquad\qquad\qquad \textit{for} \quad a + b \neq 0 , \\ \frac{1}{2} \displaystyle\int_0^1 \{-R(u)\,(1-u)^2 \sin(g\pi a u - c) \\ \qquad + R(-u)\,(1-u^2) \sin(g\pi a u + c)\}\, du \quad \textit{for} \quad a + b = 0; \end{cases}$$

$$C = \begin{cases} -\dfrac{1}{g\pi(a+b)} \displaystyle\int_0^1 \{R(u)\,[\cos\big(g\pi a u - g\pi(a+b) - c\big) \\ \qquad - \cos(g\pi b u + c)] \\ \qquad + R(-u)\,[\cos\big(g\pi b u - g\pi(a+b) - c\big) \\ \qquad - \cos(g\pi a u + c)]\}\, du \quad \text{for} \quad a + b \neq 0 , \\ \displaystyle\int_0^1 \{(u-1)\,R(u)\sin(g\pi a u - c) + (1-u)\,R(-u)\sin(g\pi a u + c)\}\, du \\ \qquad\qquad\qquad \textit{for} \quad a + b = 0 \end{cases}$$

where $R(t)$ denotes an arbitrary continuous function on $[-1, 1]$.

Proof. The assertion can be proved by substituting the variables

$$y = (v + u)/2\,, \qquad x = (v - u)/2$$

and then by integrating with respect to v. ◀

To calculate $\langle \lambda_{g2} \rangle$ and $\langle u_{g2} \rangle$ we also assume that $\big(a(x, \omega), b(x, \omega)\big)$ is stationary in the wide sense, where

$$\begin{aligned} \tilde{R}(y - x) &\doteq \langle \tilde{a}(y)\, \tilde{a}(x) \rangle \\ &= R_a(y - x) - \mu_g[R_{ab}(y - x) + R_{ab}(x - y)] + \mu_g^2 R_b(y - x)\,, \\ \tilde{Q}(y - x) &\doteq \langle \tilde{a}(y)\, b(x) \rangle = R_{ab}(y - x) - \mu_g R_b(y - x)\,. \end{aligned} \tag{1.126}$$

The functions $R_a(u)$ and $R_b(u)$ denote the correlation functions of the processes $a(x, \omega)$ and $b(x, \omega)$, respectively, and $R_{ab}(u)$ denotes the cross-correlation function of the processes $a(x, \omega)$ and $b(x, \omega)$.

If we take the expectation of (1.125), then this leads to

$$\begin{aligned} \langle \lambda_{g2} \rangle = &- \int_0^1\!\!\int_0^1 \left[\frac{1}{4\mu_g} \tilde{R}(y - x) + \tilde{Q}(y - x)\right] w_g^2(x)\, w_g^2(y)\, \mathrm{d}x\, \mathrm{d}y \\ &- \frac{1}{2\mu_g} \int_0^1\!\!\int_0^1 \tilde{R}(y - x)\, w_g^2(y)\, w_g(x)\, w_g'(x)\, (1 - 2x)\, \mathrm{d}x\, \mathrm{d}y \\ &- \frac{1}{\pi g} \int_0^1\!\!\int_0^x \tilde{R}(y - x)\, w_g(y) \sin\big(g\pi(y - x)\big)\, w_g(x)\, \mathrm{d}y\, \mathrm{d}x + 3\mu_g \langle \eta^2 \rangle \\ &- \int_0^1 \langle \eta \tilde{a}(x) \rangle\, [w_g^2(x) + 2x w_g(x)\, w_g'(x)]\, \mathrm{d}x \\ &+ 2\mu_g \int_0^1 \langle \eta b(x) \rangle\, w_g^2(x)\, \mathrm{d}x\,. \end{aligned} \tag{1.127}$$

By decomposing the products of the trigonometric functions under the integrals in (1.127) into linear factors, and using Lemma 1.6, we have

$$\begin{aligned} \langle b_{gg}^2 \rangle &= \int_0^1\!\!\int_0^1 \tilde{R}(y - x)\, w_g^2(x)\, w_g^2(y)\, \mathrm{d}x\, \mathrm{d}y \\ &= \langle\!\langle \tilde{R}(u), (1 - u)\big(2 + \cos(2g\pi u)\big) + \frac{3}{2g\pi} \sin(2g\pi u) \rangle\!\rangle\,, \end{aligned}$$

$$\langle b_{gg}a_{gg}\rangle = \int_0^1\int_0^1 \tilde{Q}(y-x)\, w_g^2(x)\, w_g^2(y)\, dx\, dy$$

$$= ⟪\tilde{Q}(u) + \tilde{Q}(-u), 1 - u + \frac{3}{4g\pi}\sin(2g\pi u) + \frac{1}{2}(1-u)\cos(2g\pi u)⟫,$$

$$\langle b_{gg}⟪\tilde{a}w_g, (1-2t)\, w_g'⟫\rangle = \int_0^1\int_0^1 \tilde{R}(y-x)\, w_g^2(y)\, w_g(x)\, w_g'(x)\, (1-2x)\, dx\, dy$$

$$= \langle b_{gg}⟪\tilde{a}w_g, -2tw_g'⟫\rangle$$

$$= ⟪\tilde{R}(u), \frac{3}{4g\pi}\sin(2g\pi u) + g\pi u(1-u)\sin(2g\pi u) + 1 + \frac{1}{2}(1-3u)\cos(2g\pi u)⟫,$$

$$\left\langle ⟪\int_0^x \tilde{a}(t)\, w_g(t)\sin\big(g\pi(t-x)\big)\, dt, \tilde{a}w_g⟫\right\rangle$$

$$= \int_0^1\int_0^x \tilde{R}(y-x)\, w_g(y)\, w_g(x)\sin\big(g\pi(y-x)\big)\, dy\, dx$$

$$= ⟪\tilde{R}(u), \frac{1}{2g\pi}\big(\cos(2g\pi u) - 1\big) - \frac{1}{2}(1-u)\sin(2g\pi u)⟫.$$

Finally, using (1.126) and (1.127) we have

$$\begin{aligned}
\langle\lambda_{g2}\rangle = {}& \frac{1}{\mu_g}⟪R_a(u), \frac{1}{2}(u-1) - \frac{3}{4g\pi}\sin(2g\pi u) + (u-1)\cos(2g\pi u) \\
& \qquad + \frac{1}{2}g\pi(u-1)^2\sin(2g\pi u)⟫ \\
& + \mu_g⟪R_b(u), \frac{2}{3}(1-u) + \frac{3}{4g\pi}\sin(2g\pi u) \\
& \qquad + \frac{1}{2}g\pi(u-1)^2\sin(2g\pi u)⟫ \\
& + ⟪R_{ab}(u) + R_{ab}(-u), \frac{1}{2}(u-1) - \frac{1}{2}(u-1)\cos(2g\pi u) \\
& \qquad - \frac{1}{2}g\pi(u-1)^2\sin(2g\pi u)⟫ \\
& + ⟪\langle a(u)\,\eta\rangle, \cos(2g\pi u) - 1 - 2\, g\pi u\sin(2g\pi u)⟫ \\
& + \mu_g⟪\langle b(u)\,\eta\rangle, 3 - 3\cos(2g\pi u) + 2g\pi u\sin(2g\pi u)⟫ + 3\mu_g\langle\eta\rangle^2 .
\end{aligned} \tag{1.128}$$

Analogously, the function $\langle u_{g2}(x)\rangle$ can be evaluated from (1.121). Defining

$$B_{gtrs} \doteq \langle b_{gt}b_{rs}\rangle = \int_0^1\int_0^1 \tilde{R}(y-x)\, w_g(y)\, w_t(y)\, w_r(x)\, w_s(x)\, \mathrm{d}x\, \mathrm{d}y\,,$$

$$C_{gtrs} \doteq \langle b_{gt}a_{rs}\rangle = \int_0^1\int_0^1 \tilde{Q}(y-x)\, w_g(y)\, w_t(y)\, w_r(x)\, w_s(x)\, \mathrm{d}x\, \mathrm{d}y$$

the expression

$$\begin{aligned}
\langle u_{g2}(x)\rangle = {} & \sum_{\substack{s=1\\ s\neq g}}^{\infty} \frac{1}{\mu_{sg}}\left[\sum_{\substack{t=1\\ t\neq g}}^{\infty}\frac{1}{\mu_{tg}} B_{gtts} - \frac{1}{\mu_{sg}} B_{gggs} + C_{gggs}\right] w_s \\
& - \frac{1}{2}\sum_{\substack{s=1\\ s\neq g}}^{\infty} \frac{1}{\mu_{sg}^2} B_{gsgs} w_g \\
& + \Big[-\Big(\!\Big(\frac{1}{8\mu_g}\langle\eta\tilde{a}\rangle\, w_g,\, w_g\Big)\!\Big) + \frac{1}{2}\Big(\!\Big(\langle\eta\tilde{a}\rangle\, w_g,\, t(t-1)\, w_g\Big)\!\Big) \\
& \quad - \Big(\!\Big(\langle\eta b\rangle\, w_g,\, (1-t)\, w_g'\Big)\!\Big) + \frac{1}{2\mu_g}\Big(\!\Big(\langle\eta\tilde{a}\rangle\, w_g',\, t(1-t)\, w_g'\Big)\!\Big) \\
& \quad + \frac{1}{4\mu_g}\Big(\!\Big(\langle\eta\tilde{a}\rangle\, w_g, (t-1)\, w_g'\Big)\!\Big) - \frac{1}{2}\Big(\!\Big(\langle\eta b\rangle\, w_g,\, w_g\Big)\!\Big)\Big] w_g(x) \\
& + \Big[\frac{1}{2\mu_g}\Big(\!\Big(\langle\eta\tilde{a}\rangle\, w_g,\, (1-3t)\, w_g'\Big)\!\Big) + \Big(\!\Big(\langle\eta b\rangle\, w_g,\, w_g\Big)\!\Big)\Big] x w_g'(x) \\
& + \frac{1}{2}\Big(\!\Big(\langle\eta\tilde{a}\rangle\, w_g,\, w_g\Big)\!\Big)\, x^2 w_g(x) \\
& - 2g\pi \int_0^x \langle\eta b\rangle\, w_g(t)\sin\big(g\pi(t-x)\big)\, \mathrm{d}t \\
& + \int_0^x \langle\eta\tilde{a}\rangle\Big[-\frac{1}{2} w_g(t)\Big\{\frac{1}{\pi g}\sin\big(g\pi(t-x)\big) + x\cos\big(g\pi(t-x)\big)\Big\} \\
& \qquad + \sqrt{2}\, t\sin\big(g\pi(2t-x)\big)\Big]\, \mathrm{d}t \\
& + \langle\eta^2\rangle\Big[\Big(\frac{3}{8} - \frac{1}{2}\mu_g x^2\Big) w_g(x) + \frac{3}{2} x w_g'(x)\Big]
\end{aligned} \tag{1.129}$$

can be derived.

Numerical results are given in the following section. For numerical computation, the expressions B_{gtrs} and C_{gtrs} are determined by simple scalar products which contain $R(u)$ and $Q(u)$. Using Lemma 1.6, we obtain for $s, t \neq g$

$$B_{gtts} = \begin{cases} (1 + (-1)^{g+s}) \left[A_{gst} + A_{gs,-t} + A_{sgt} + A_{sg,-t}\right] \\ \qquad \text{for} \quad t \neq \dfrac{g+s}{2}, \left|\dfrac{g-s}{2}\right|, \\ 2A_{gst} + 2A_{sgt} - 2\pi \left(\dfrac{g+s}{\mu_{gs}} - \dfrac{g-s}{\mu_{g-s,\,g+s}}\right) \tilde{V}_{(g-s)/2} + \tilde{W}_{(g-s)/2} \\ \qquad \text{for} \quad t = \dfrac{g+s}{2}, \\ 2A_{gs,\,(s-g)/2} + 2A_{sg,\,(g-s)/2} + 2\pi \left(\dfrac{g-s}{\mu_{gs}} - \dfrac{g+s}{\mu_{g+s,\,g-s}}\right) \tilde{V}_{(g+s)/2} - \tilde{W}_{(g+s)/2} \\ \qquad \text{for} \quad t = \left|\dfrac{g-s}{2}\right|; \end{cases} \tag{1.130}$$

and for $s \neq g$

$$B_{ggg s} = \begin{cases} (1 + (-1)^{g+s}) \dfrac{4g^2\pi}{\mu_{gs}} \left[\dfrac{1}{3g+s} \tilde{V}_{g+s} + \dfrac{1}{3g-s} \tilde{V}_{s-g} \right. \\ \qquad \left. - \dfrac{2s}{\mu_{3g,\,s}} \tilde{V}_{2g}\right] \quad \text{for} \quad s \neq 3g, \\ -\dfrac{1}{6g\pi} (\tilde{V}_{4g} + \tilde{V}_{2g}) - \tilde{W}_{2g} \quad \text{for} \quad s = 3g \end{cases} \tag{1.131}$$

where

$$A_{gst} \doteq \pi g \left(\frac{1}{\mu_{gs}} - \frac{1}{\mu_{g,\,s+2t}}\right) \tilde{V}_{s+t}$$

and

$$\tilde{V}_p \doteq \langle\!\langle \tilde{R}(u), \sin(p\pi u) \rangle\!\rangle ,$$

$$\tilde{W}_p \doteq \langle\!\langle \tilde{R}(u)\,(1-u), \cos(p\pi u) \rangle\!\rangle .$$

The values $C_{ggg s}$ can be calculated as

$$C_{ggg s} = \begin{cases} \dfrac{4g^2\pi}{\mu_{gs}} \left[\dfrac{1}{3g+s} \{(-1)^{g+s} V_{g+s} + V_{g+s}^-\} \right. \\ \qquad + \dfrac{1}{3g-s} \{(-1)^{g+s} V_{s-g} + V_{s-g}^-\} \\ \qquad \left. - \dfrac{2s}{\mu_{3g,\,s}} \{V_{2g} + (-1)^{g+s} V_{2g}^-\}\right] \quad \text{for} \quad s \neq 3g, \\ -\dfrac{1}{12\pi g} [V_{4g} + V_{4g}^- + V_{2g} + V_{2g}^-] - \dfrac{1}{2} W_{2g} - \dfrac{1}{2} W_{2g}^- \\ \qquad \text{for} \quad s = 3g \end{cases} \tag{1.132}$$

where

$$V_p \doteq ⟪\tilde{Q}(u), \sin(p\pi u)⟫, \qquad W_p \doteq ⟪\tilde{Q}(u)(1-u), \cos(p\pi u)⟫,$$

$$V_p^- \doteq ⟪\tilde{Q}(-u), \sin(p\pi u)⟫, \qquad W_p^- \doteq ⟪\tilde{Q}(-u)(1-u), \cos(p\pi u)⟫.$$

Now we take up the calculation of the correlation quantities for the eigenvalues and eigenfunctions. For the correlation function of the g-th eigenfunction, we have up to terms of the second order

$$K_g(x, y) \doteq \langle (u_g(x) - \langle u_g(x)\rangle)(u_g(y) - \langle u_g(y)\rangle)\rangle = \langle u_{g1}(x)\, u_{g1}(y)\rangle + \dots;$$

and for the variance of the g-th eigenvalue

$$\sigma_g^2 \doteq \langle(\lambda_g - \langle\lambda_g\rangle)^2\rangle = \langle\lambda_{g1}^2\rangle + \dots$$

(1.117) yields

$$\langle\lambda_{g1}^2\rangle = \langle b_{gg}^2\rangle - 4\mu_g\langle\eta b_{gg}\rangle + 4\mu_g\langle\eta^2\rangle .$$

Using the expression for $\langle b_{gg}^2\rangle$ we get

$$\langle\lambda_{g1}^2\rangle = ⟪\tilde{R}(u), (1-u)\big(2 + \cos(2g\pi u)\big) + \frac{3}{2g\pi}\sin(2g\pi u)⟫ - 4\mu_g⟪\langle\eta\tilde{a}\rangle, 1 - \cos(2g\pi u)⟫ + 4\mu_g^2\langle\eta^2\rangle . \tag{1.133}$$

(1.123) leads to

$$\begin{aligned}\langle u_{g1}(x)\, u_{g1}(y)\rangle = {} & \frac{1}{\mu_g}\int_0^x\int_0^y \langle\tilde{a}(t)\,\tilde{a}(s)\rangle\, w_g(t)\, w_g(s)\sin\big(g\pi(t-x)\big) \\ & \times \sin\big(g\pi(s-y)\big)\, dt\, ds \\ & + \langle\xi^2\rangle\, w_g(x)\, w_g(y) + \langle\zeta^2\rangle\, xy w_g'(x)\, w_g'(y) \\ & - \langle\xi\zeta\rangle[x w_g'(x)\, w_g(y) + y w_g'(y)\, w_g(x)] \\ & + \frac{1}{\pi g}\int_0^x \langle\tilde{a}\xi\rangle\, w_g(t)\sin\big(g\pi(t-x)\big)\, dt\, w_g(y) \\ & + \frac{1}{\pi g}\int_0^y \langle\tilde{a}\xi\rangle\, w_g(t)\sin\big(g\pi(t-y)\big)\, dt\, w_g(x) \\ & - \frac{1}{\pi g}\int_0^x \langle\tilde{a}\zeta\rangle\, w_g(t)\sin\big(g\pi(t-x)\big)\, dt\, y w_g'(y) \\ & - \frac{1}{\pi g}\int_0^y \langle\tilde{a}\zeta\rangle\, w_g(t)\sin\big(g\pi(t-y)\big)\, dt\, x w_g'(x) .\end{aligned} \tag{1.134}$$

Here we have put, for the sake of brevity,

$$\xi(\omega) \doteq \frac{b_{gg}}{4\mu_g} + \frac{c_{gg}}{2\mu_g} + \frac{1}{2}\eta\,, \qquad \zeta(\omega) \doteq \frac{b_{gg}}{2\mu_g} - \eta\,,$$

$$c_{gg} \doteq \langle\!\langle \tilde{a} w_g, (1-t)\, w_g' \rangle\!\rangle .$$

From (1.134) we derive the variance of $u_g(x, \omega)$ up to terms of the second order:

$$\begin{aligned}
\operatorname{var} u_g(x) &= \frac{\langle b_{gg}^2\rangle}{4\mu_g^2}\left[\frac{1}{2}w_g - xw_g'\right]^2 + \frac{\langle c_{gg}^2\rangle}{4\mu_g^2} w_g^2 \\
&+ \frac{\langle b_{gg}c_{gg}\rangle}{2\mu_g^2} w_g \left[\frac{1}{2}w_g - xw_g'\right] \\
&+ \frac{1}{\mu_g}\int_0^x\int_0^x \langle \tilde{a}(t)\,\tilde{a}(s)\rangle\, w_g(t)\, w_g(s) \sin\big(g\pi(t-x)\big) \\
&\qquad \times \sin\big(g\pi(s-x)\big)\, dt\, ds \\
&+ \frac{1}{\mu_g g\pi}\int_0^x \langle b_{gg}\tilde{a}(t)\rangle \sin\big(g\pi(t-x)\big)\, w_g(t)\, dt \left[\frac{1}{2}w_g - xw_g'\right] \\
&+ \frac{1}{\mu_g g\pi}\int_0^x \langle c_{gg}\tilde{a}(t)\rangle \sin\big(g\pi(t-x)\big)\, w_g(t)\, dt\, w_g \\
&+ \frac{1}{\pi g}\int_0^x \langle \eta\tilde{a}(t)\rangle \sin\big(g\pi(t-x)\big)\, w_g(t)\, dt\, [w_g + 2xw_g'] \\
&+ \frac{1}{\mu_g}\langle b_{gg}\eta\rangle \left[\frac{1}{4}w_g^2 - x^2 w_g'^2\right] + \langle\eta^2\rangle \left[\frac{1}{2}w_g + xw_g'\right]^2 \\
&+ \frac{1}{\mu_g}\langle c_{gg}\eta\rangle\, w_g \left[\frac{1}{2}w_g + xw_g'\right].
\end{aligned} \tag{1.135}$$

To compute numerically the variance of $u_g(x, \omega)$ it is advantageous to write it as a series in B_{ijst}. Thus, we get up to terms of the second order

$$\begin{aligned}
\operatorname{var} u_g(x) &= \sum_{\substack{s,t=1\\ s,t\neq g}}^{\infty} \frac{1}{\mu_{sg}\mu_{tg}} B_{sggt} w_s w_t + \left(xw_g' + \frac{1}{2}w_g\right)^2 \langle\eta^2\rangle \\
&+ 2\left(xw_g' + \frac{1}{2}w_g\right) \sum_{\substack{s=1\\ s\neq g}}^{\infty} \frac{1}{\mu_{sg}} \langle b_{gs}\eta\rangle\, w_g
\end{aligned}$$

where B_{sggt} for $s \neq t$ is given by (1.130) and for $s = t$, $s \neq g$ by

$$\begin{aligned}
B_{sggs} = \tilde{W}_{s-g} + \tilde{W}_{s+g} &- \frac{1}{\pi}\left(\frac{1}{s-g} + \frac{1}{g} - \frac{1}{s}\right)\tilde{V}_{s-g} \\
&- \frac{1}{\pi}\left(\frac{1}{s+g} - \frac{1}{g} - \frac{1}{s}\right)\tilde{V}_{s+g}\,.
\end{aligned}$$

At the end of this section we will give asymptotic results ($g \to \infty$) concerning the eigenvalues of (1.115) as well as their mean values and variances. Let $(a(x, \omega), b(x, \omega))$ be a wide-sense stationary vector process as given above. We assume that almost all realizations of $a(x, \omega)$ are continuous, and that those of $b(x, \omega)$ are continuously differentiable. For this it is sufficient that the correlation function $R_a(u)$ of $a(x, \omega)$ is twice and the correlation function of $b(x, \omega)$ is four times continuously differentiable at $u = 0$.

If almost all realizations of a stochastic process $f(x, \omega)$ are piecewise continuous then

$$\lim_{g\to\infty} \int_0^1 f(x, \omega) \cos(2g\pi x + c)\, dx = 0 \quad \text{a.s.} \tag{1.136}$$

holds for any real number c. If we assume that almost all realizations of $f(x, \omega)$ are piecewise continuously differentiable then we obtain, by integration by parts and in view of (1.136)

$$\lim_{g\to\infty} g\pi \int_0^1 f(x, \omega) \sin(2g\pi x + c)\, dx = \tfrac{1}{2} \big(f(0, \omega) - f(1, \omega)\big) \cos c\,. \tag{1.137}$$

After these remarks we can easily deduce the following relations:

$$\lim_{g\to\infty} \frac{1}{\mu_g} b_{gg} = -\int_0^1 b(t, \omega)\, dt \quad \text{a.s.},$$

$$\lim_{g\to\infty} a_{gg} = \int_0^1 b(t, \omega)\, dt \quad \text{a.s.},$$

$$\lim_{g\to\infty} \frac{1}{\mu_g} \langle\!\langle \tilde{a} w_g, (1 - 2t)\, w_g' \rangle\!\rangle = -\frac{1}{2} \big(b(0, \omega) + b(1, \omega)\big) \quad \text{a.s.},$$

$$\lim_{g\to\infty} \frac{1}{g\pi\mu_g} \Big\langle\!\!\Big\langle \int_0^x \tilde{a}(t, \omega)\, w_g(t) \sin\big(g\pi(t - x)\big)\, dt, \tilde{a} w_g \Big\rangle\!\!\Big\rangle$$

$$= -\frac{1}{4} \int_0^1 b^2(t, \omega)\, dt - \frac{1}{4} \int_0^1 b(t, \omega)\, [b(1, \omega) + b(0, \omega)]\, dt \quad \text{a.s.}$$

Using (1.117) and (1.125) it follows that

$$\lim_{g\to\infty} \frac{1}{\mu_g} \lambda_{g1}(\omega) = -\int_0^1 b(t, \omega)\, dt - 2\eta(\omega) \quad \text{a.s.},$$

$$\lim_{g\to\infty} \frac{1}{\mu_g} \lambda_{g2}(\omega) = \frac{3}{4} \left(\int_0^1 b(t, \omega)\, dt\right)^2 + \frac{1}{4} \int_0^1 b^2(t, \omega)\, dt$$

$$+ \eta \left[3 \int_0^1 b(t, \omega)\, dt - b(1, \omega) + 3\eta\right] \quad \text{a.s.}$$

Defining

$$\lambda_g^{\mathrm{II}} \doteq \mu_g + \lambda_{g1} + \lambda_{g2}$$

the expressions

$$\lim_{g\to\infty} \frac{1}{\mu_g} \lambda_g^{\mathrm{II}}(\omega) = 1 - \left[\int_0^1 b(t,\omega)\,\mathrm{d}t + 2\eta\right]$$
$$+ \left[\frac{3}{4}\left(\int_0^1 b(t,\omega)\,\mathrm{d}t\right)^2 + \frac{1}{4}\int_0^1 b^2(t,\omega)\,\mathrm{d}t\right.$$
$$\left. + \eta\left(3\int_0^1 b(t,\omega)\,\mathrm{d}t - b(1,\omega) + 3\eta\right)\right] \qquad (1.138)$$

can be calculated. This yields the asymptotic result for the mean of the eigenvalues up to terms of second order

$$\lim_{g\to\infty} \frac{1}{\mu_g} \langle\lambda_g^{\mathrm{II}}\rangle = 1 + \frac{3}{2} ⟪R_b(u), t - u⟫ + \frac{1}{4} R_b(0)$$
$$+ 3⟪\langle b(u)\,\eta\rangle, 1⟫ - \langle b(1)\,\eta\rangle + 3\langle\eta^2\rangle\,. \qquad (1.139)$$

By analogous computations, (1.133) leads to

$$\lim_{g\to\infty} \frac{1}{\mu_g^2} \operatorname{var} \lambda_g = ⟪R_b(u), 2(1-u)⟫ + 4⟪\langle\eta b(u)\rangle, 1⟫ + 4\langle\eta^2\rangle\,. \qquad (1.140)$$

This result can also be found in case $\eta = 0$ (see Boyce [1, 4]).

To evaluate the correlation between the eigenvalues λ_g and λ_h for large g we must consider the behaviour of $\langle {}^g b_{gg}\, {}^h b_{hh}\rangle$ as $g \to \infty$ since

$$\langle(\lambda_g - \langle\lambda_g\rangle)(\lambda_h - \langle\lambda_h\rangle)\rangle = \langle {}^g b_{gg}\, {}^h b_{hh}\rangle + \text{terms of third and higher order}$$

where

$${}^g b_{gg} = ⟪(a - \mu_g b)\, w_g, w_g⟫\,, \qquad {}^h b_{hh} = ⟪(a - \mu_h b)\, w_h, w_h⟫\,.$$

We obtain up to terms of the second order

$$\lim_{g\to\infty} \frac{1}{\mu_g} \langle(\lambda_g - \langle\lambda_g\rangle)(\lambda_h - \langle\lambda_h\rangle)\rangle$$
$$= \int_0^1\int_0^1 (\mu_h R_b(x-y) - R_{ab}(x-y))\, w_h^2(y)\,\mathrm{d}x\,\mathrm{d}y\,.$$

1.4.3.2. Numerical results

In the following we deal with the computation of the differences $\langle\lambda_g\rangle - \mu_g$, $\langle u_g\rangle - w_g$ (averaging problem), as well as $\operatorname{var} \lambda_g$ and $\operatorname{var} u_g$, for the eigenvalue problem (1.115) in each case up to second order terms in the random perturbations. The correlation quantities of the random parameters involved in problem (1.115) are assumed to be known.

Case 1. $a(x, \omega) \equiv b(x, \omega) \equiv 0$

From (1.117), (1.123), (1.125), (1.121) we immediately obtain up to terms of second order in η

$$\lambda_g(\omega) = \mu_g - 2\mu_g\eta + 3\mu_g\eta^2 ,$$
$$u_g(x, \omega) = w_g(x) - \left(\tfrac{1}{2} w_g(x) + x w_g'(x)\right) \eta + \tfrac{1}{2} \left(\left(\tfrac{3}{4} - \mu_g x^2\right) w_g(x) + 3x w_g'(x)\right) \eta^2 .$$

Hence, it follows that

$$\langle\lambda_g\rangle - \mu_g = 3\mu_g\langle\eta^2\rangle ,$$
$$\langle u_g\rangle - w_g = \tfrac{1}{2} \left(\left(\tfrac{3}{4} - \mu_g x^2\right) w_g + 3x w_g'\right) \langle\eta^2\rangle ,$$
$$\operatorname{var} \lambda_g = 4\mu_g^2 \operatorname{var} \eta ,$$
$$\operatorname{var} u_g = \left(\tfrac{1}{2} w_g + x w_g'\right)^2 \operatorname{var} \eta .$$

In this simple special case the relation between the random parameter η and the eigenvalues

$$\eta = \varphi(\lambda) = -\frac{1}{\sqrt{\lambda}} \tan \sqrt{\lambda}$$

can be written down explicitly. If $f(x)$, $g(x)$ denote the density functions of η or λ, respectively, then

$$g(x) = f(\varphi(x)) \, |\varphi'(x)|$$

is satisfied where the corresponding branch of the tangent function defines the eigenvalue (cf. Haines [1, 2]).

Case 2. $b(x, \omega) \equiv 0$

This case contains for $\eta = 0$ the one-dimensional stationary Schrödinger equation with random potential

$$M(\omega) u \doteq -u'' + a(x, \omega) u = \lambda u , \qquad u(0, \omega) = u(1, \omega) = 0 . \quad (1.141)$$

First, we make some remarks on the convergence of the expansions of the eigenvalues and the eigenfunctions in this case. If the stochastic process satisfies a.s. the inequality $a(x, \omega) \geqq -\mu_1 = -\pi^2$ then since

$$\langle\!\langle M(\omega) u, u\rangle\!\rangle \geqq \mu_1 \|u\|^2 + \langle\!\langle a(x, \omega) u, u\rangle\!\rangle > 0 \quad \text{a.s.}$$

the stochastic eigenvalue problem is positive definite. If we further assume that

$$|a(x, \omega)| \leqq S(\omega) \quad \text{a.s.}$$

we can apply Theorem 1.16 since

$$\|M_1 u\| = \left[\int_0^1 a^2(x, \omega) \, u^2(x) \, dx\right]^{1/2} \leqq S(\omega) \, \|u\| .$$

This theorem yields

$$\lambda_g(\omega) = \sum_{p=0}^{\infty} \lambda_{gp}(\omega) \quad \text{a.s.}, \qquad u_g(x, \omega) = \sum_{p=0}^{\infty} u_{gp}(x, \omega) \quad \text{a.s. in } \mathbf{L}_2,$$

assuming $S(\omega)$ satisfies the inequality

$$16S(\omega)/d_g < 1$$

where

$$0 < d_g = \min \{\mu_g - \mu_{g-1}, \mu_{g+1} - \mu_g\}.$$

Furthermore, it holds in this case that

$$|\lambda_g(\omega) - \sum_{p=0}^{s} \lambda_{gp}(\omega)| \leqq \tfrac{1}{4} d_g(16S(\omega)/d_g)^{s+1},$$

$$||u_g(x, \omega) - \sum_{p=0}^{s} u_{gp}(x, \omega)|| \leqq \tfrac{1}{2} (16S(\omega)/d_g)^{s+1}.$$

The mean values satisfy the estimates

$$|\langle\lambda_g\rangle - \sum_{p=0}^{s} \langle\lambda_{gp}\rangle| \leqq 4(16/d_g)^s \langle S^{s+1}\rangle,$$

$$||\langle u_g\rangle - \sum_{p=0}^{s} \langle u_{gp}\rangle|| \leqq \tfrac{1}{2}(16/d_g)^{s+1} \sqrt{\langle S^{2s+2}\rangle}$$

(cf. Theorem 1.17).

Now, we calculate the expectations and the variances of the eigenvalues of problem (1.141) using various correlation functions $R_a(u)$. From (1.128) and (1.133) we find up to second order terms

$$\langle\lambda_g\rangle - \mu_g = \frac{1}{\mu_g} \langle\!\langle R_a(u), \frac{1}{2}(u-1) - \frac{3}{4g\pi} \sin(2g\pi u) + (u-1)\cos(2g\pi u) + \frac{1}{2} g\pi(u-1)^2 \sin(2g\pi u)\rangle\!\rangle,$$

$$\operatorname{var} \lambda_g = \langle\!\langle R_a(u), (1-u)\big(2 + \cos(2g\pi u)\big) + \frac{3}{2g\pi} \sin(2g\pi u)\rangle\!\rangle.$$

Figs. 1.1a to 1.1c show $\langle\lambda_g\rangle - \mu_g$ for the correlation functions

$$R_a(u) = \sigma^2 e^{-\alpha|u|}, \qquad R_a(u) = \sigma^2 e^{-\alpha u^2} \cos\beta u, \quad \text{and} \quad R_a(u) = \sigma^2 \cos\beta u,$$

respectively ($\alpha, \beta > 0$). Figs. 1.2a to 1.2c illustrate the variances of the λ_g for these same correlation functions. The curves with broken lines are the asymptotic expressions as $g \to \infty$.

In connection with Fig. 1.1a we add some remarks on the bowing effect in mixed crystal theory (cf. e.g. Schulze, Unger [1]). If we restrict ourselves to the simple case of a one-dimensional crystal lattice, then the measured energy levels are just the means $\langle\lambda_g\rangle$ of the eigenvalues for a problem of the type (1.141) with a suitable potential $a(x, \omega)$. In order to compute the energy levels

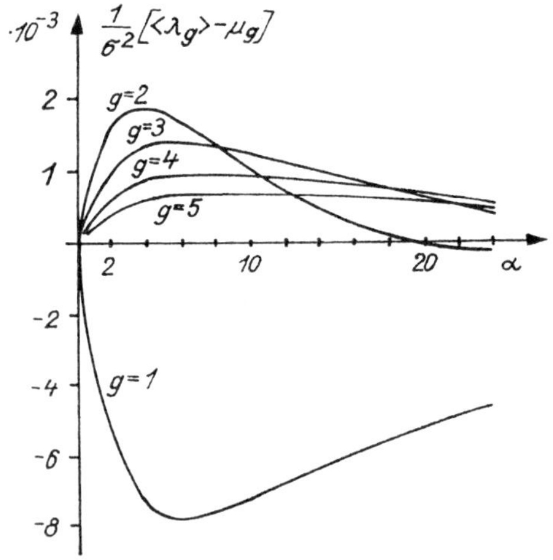

Fig. 1.1 a. The mean value $\langle\lambda_g\rangle$ for $R_a(u) = \sigma^2 e^{-\alpha|u|}$

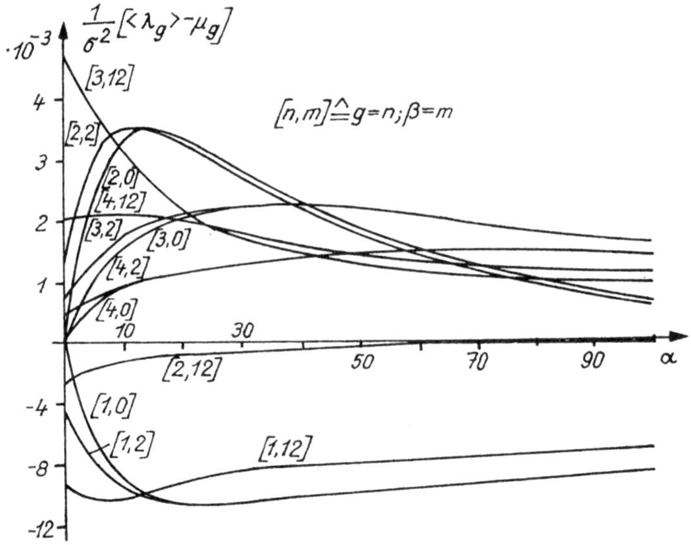

Fig. 1.1 b. The mean value $\langle\lambda_g\rangle$ for $R_a(u) = \sigma^2 e^{-\alpha u^2} \cos(\beta u)$

the so-called pseudopotential method will be used which consists in solving the Schrödinger equation with the averaged potential $\langle a(x)\rangle$. The energy levels computed by the pseudopotential method are just the eigenvalues μ_g of the averaged problem. The bowing effect is characterized by the fact that the measured energy levels differ from the computed ones, e.g. $\langle\lambda_3\rangle - \langle\lambda_2\rangle < \mu_3 - \mu_2$. The existence of this effect is immediately clear from results on stochastic eigenvalue problems. Fig. 1.1 a now shows that for stationary $a(x, \omega)$ with correlation

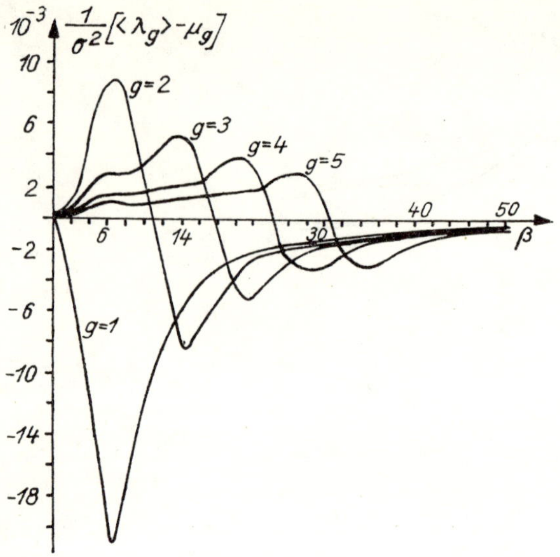

Fig. 1.1c. The mean value $\langle\lambda_g\rangle$ for $R_a(u) = \sigma^2 \cos(\beta u)$

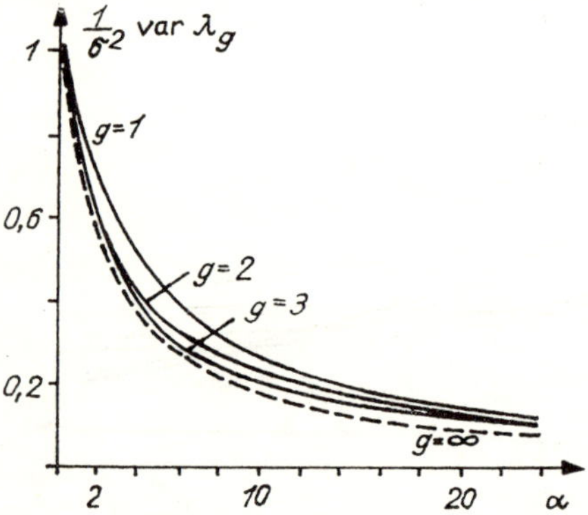

Fig. 1.2a. The variance var λ_g for $R_a(u) = \sigma^2\, e^{-\alpha|u|}$

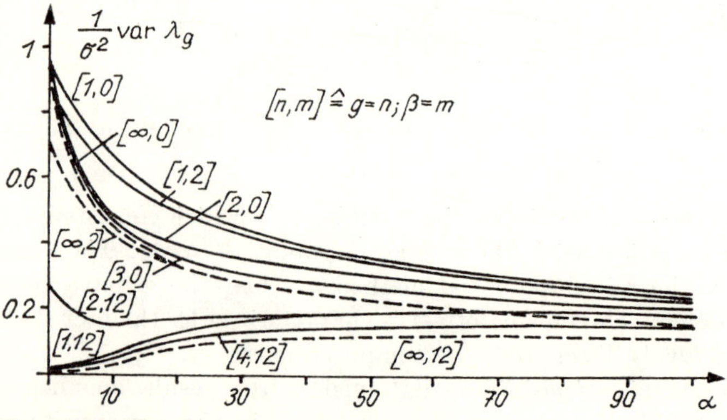

Fig. 1.2b. The variance var λ_g for $R_a(u) = \sigma^2\, e^{-\alpha u^2} \cos(\beta u)$

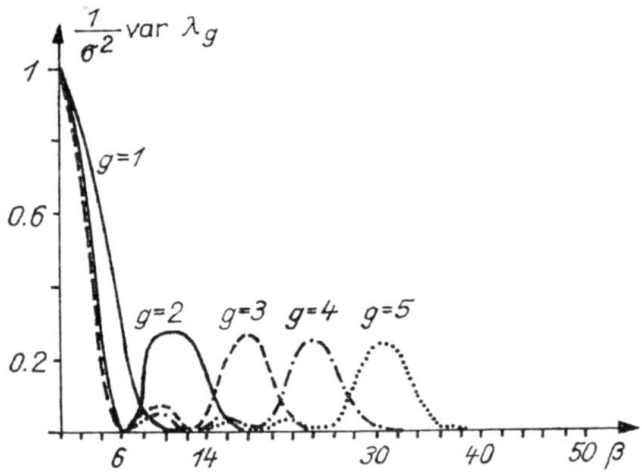

Fig. 1.2c. The variance var λ_g for $R_a(u) = \sigma^2 \cos(\beta u)$

function $R_a(u) = \sigma^2 e^{-\alpha|u|}$ the direction of the bowing depends on the parameter α. Thus, for example, $\langle\lambda_3\rangle - \langle\lambda_2\rangle < \mu_3 - \mu_2$ for $0 < \alpha < 7$ and $\langle\lambda_3\rangle - \langle\lambda_2\rangle > \mu_3 - \mu_2$ for $\alpha > 8$. The parameter α is a measure of the "distant-action", i.e. from the observed direction of the bowing effects it might be possible to conclude how the stochastic effects act in the crystal.

For $\eta \neq 0$ the terms containing η have to be added to the computed terms. We have

$$\langle\lambda_g\rangle - \mu_g = \frac{1}{\mu_g} \langle\!\langle R_a(u), \frac{1}{2}(u-1) - \frac{3}{4g\pi}\sin(2g\pi u) + (u-1)\cos(2g\pi u) + \frac{1}{2} g\pi(u-1)^2 \sin(2g\pi u)\rangle\!\rangle$$

$$+ \langle\!\langle\langle a(u)\,\eta\rangle, \cos(2g\pi u) - 1 - 2g\pi u \sin(2g\pi u)\rangle\!\rangle + 3\mu_g\langle\eta^2\rangle$$

and

$$\operatorname{var}\lambda_g = \langle\!\langle R_a(u), (1-u)\big(2+\cos(2g\pi u)\big) + \frac{3}{2g\pi}\sin(2g\pi u)\rangle\!\rangle$$

$$-4\mu_g\langle\!\langle\langle a(u)\,\eta\rangle, 1 - \cos(2g\pi u)\rangle\!\rangle + 4\mu_g^2\langle\eta^2\rangle\,.$$

The correlation of $\langle a(u)\,\eta\rangle$ is therefore needed from $R_a(u)$. This correlation is often given by $a(x, \omega)$ itself, for instance, if $\eta = a(1, \omega)$ or $\eta = a(0, \omega)$. Many static problems (buckling of a bar, and others) or vibration problems lead to eigenvalue problems of a type in which the stochastic parameters in the boundary conditions are given by stochastic processes involved in the differential operator.

We see from the asymptotic results of the preceding section that the terms containing η dominate for large g.

Now, we consider the numerical computation of the mean values and variances of the eigenfunctions of problem (1.141) using various assumptions on

$R_a(u)$. One obtains up to terms of the second order (cf. (1.118) and (1.129))

$$\langle u_g(x)\rangle - w_g(x) = \sum_{\substack{s=1\\ s\neq g}}^{\infty} \frac{1}{\mu_{sg}} \left[\sum_{\substack{t=1\\ t\neq g}}^{\infty} \frac{1}{\mu_{tg}} B_{gtts} - \frac{1}{\mu_{sg}} B_{gggs} \right] w_s$$

$$- \frac{1}{2} \sum_{\substack{s=1\\ s\neq g}}^{\infty} \frac{1}{\mu_{sg}^2} B_{gsgs} w_g \,, \tag{1.142}$$

and

$$\operatorname{var} u_g(x) = \sum_{\substack{s,t=1\\ s,t\neq g}}^{\infty} \frac{1}{\mu_{sg}\mu_{tg}} B_{sggt} w_s w_t \,, \tag{1.143}$$

with the expressions computed in (1.130) and (1.131) under the assumption that the process $a(x, \omega)$ is stationary in the wide sense. These expressions ((1.130) and (1.131)) reduce the double integral

$$B_{ijst} = \int_0^1 \int_0^1 \langle a(x)\, a(y)\rangle\, w_i(x)\, w_j(x)\, w_s(y)\, w_t(y)\, dx\, dy$$

to one-dimensional integrals which are numerically advantageous. From (1.130) and (1.131) it follows that

$$\sum_{\substack{t=1\\ t\neq g}}^{\infty} \frac{1}{\mu_{tg}} B_{gtts} = 0 \,, \qquad B_{gggs} = 0 \,,$$

if $g + s$ is odd. Using the estimation $|R(u)| \leqq R(0)$, which is satisfied for arbitrary correlation functions of a wide-sense stationary process $a(x, \omega)$, we estimate

$$|B_{ijst}| = \left| \int_0^1 \int_0^1 R_a(x - y)\, w_i(x)\, w_j(x)\, w_s(y)\, w_t(y)\, dx\, dy \right| \leqq 4R_a(0) \,.$$

Because of the boundedness of the B_{ijst} using (1.142) approximate expressions for $\langle u_g\rangle$ are obtained by only taking into consideration such summands with $s = g \pm 2$, $s = g \pm 4, \ldots,$ $s = g \pm 2k$ and $s > 0$, since $|\mu_{sg}| = \pi^2 |s^2 - g^2|$ quickly increases with $|s - g|$. The terms with $s = g \pm 1$, $s = g \pm 3$, ... are zero since in this case $s + g$ is odd. Thus

$$\langle u_g(x)\rangle - w_g(x) \approx \sum_{\substack{r=1\\ s=g\pm 2r\\ s>0}}^{k} \frac{1}{\mu_{sg}} \left[\sum_{\substack{t=1\\ t\neq g}}^{\infty} \frac{1}{\mu_{tg}} B_{gtts} - \frac{1}{\mu_{sg}} B_{gggs} \right] w_s$$

$$- \frac{1}{2} \sum_{\substack{s=g-2k\\ s>0}}^{s=g+2k} \frac{1}{\mu_{sg}^2} B_{gsgs} w_g \,.$$

The summand with $s = g - (2k + 2)$, assuming s is greater than zero, or the summand with $s = g + 2k + 2$ yield a rough estimate of the error, we have

$$\left| \frac{1}{\mu_{g\mp(2k+2),\,g}} \left[\sum_{\substack{t=1\\t\neq g}}^{\infty} \frac{1}{\mu_{tg}} B_{gtt,\,g\mp(2k+2)} - \frac{1}{\mu_{g\mp(2k+2),\,g}} B_{ggg,\,g\mp(2k+2)} \right] w_{g\mp(2k+2)} \right|$$

$$\leqq \frac{\sqrt{2}\,4R_a(0)}{|\mu_{g\mp(2k+2),\,g}|} \left[\sum_{\substack{t=1\\t\neq g}}^{\infty} \frac{1}{|\mu_{tg}|} + \frac{1}{|\mu_{g\mp(2k+2),\,g}|} \right].$$

Similar considerations for the variance of $u_g(x, \omega)$ lead to

$$\operatorname{var} u_g \approx \sum_{\substack{p=1\\s=g\pm 2p>0}}^{k} \sum_{\substack{q=1\\t=g\pm 2q>0}}^{k} \frac{1}{\mu_{sg}\mu_{tg}} B_{sggt} w_s w_t\,.$$

Figs. 1.3a to 1.3d show $\langle u_g\rangle - w_g$ for various g and different correlation functions. Figs. 1.4a to 1.4d illustrate the corresponding variances of u_g in each case for these correlation functions as in Fig. 1.3.

If we know not only the mean value and the correlation function of the stochastic process $a(x, \omega)$ but also the third and fourth moments, then we can easily compute the approximations up to the fourth order according to Theorem 1.16.

For example, assume that $a(x, \omega)$ is a stationary Gaussian process. Then we can make similar considerations as being used in Section 1.3.3.2. If $|a(x, \omega)| \leqq S(\omega)$

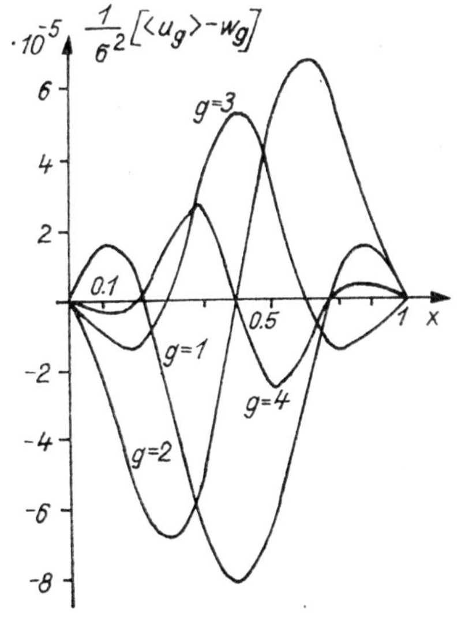

Fig. 1.3a. The mean value function $\langle u_g(x)\rangle$ for $R_a(u) = \sigma^2\,e^{-u^2}$

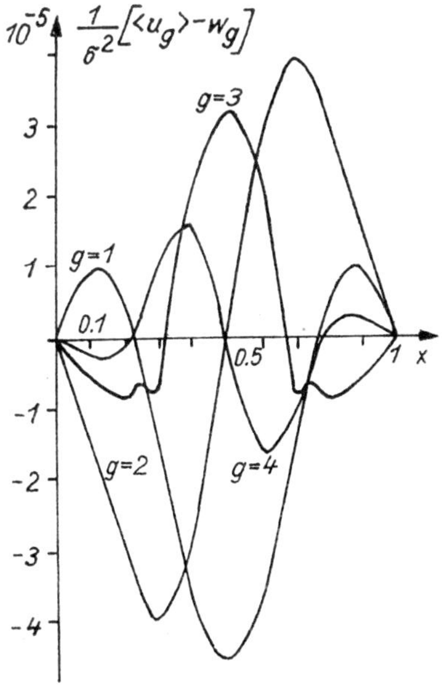

Fig. 1.3b. The mean value function $\langle u_g(x)\rangle$ for $R_a(u) = \sigma^2 \cos u$

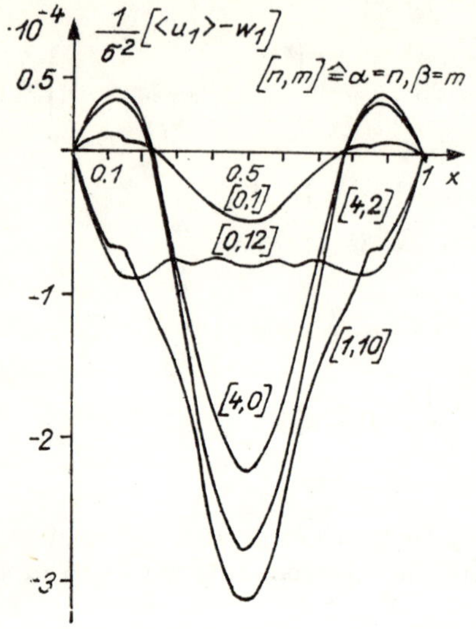

Fig. 1.3c. The mean value function $\langle u_1(x)\rangle$ for $R_a(u) = \sigma^2 \mathrm{e}^{-\alpha u^2} \cos(\beta u)$

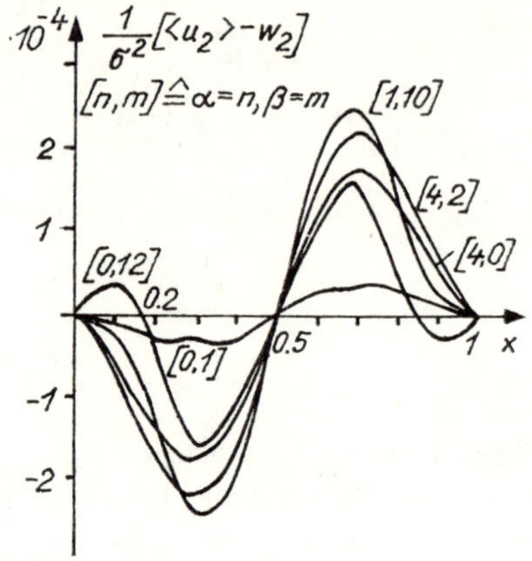

Fig. 1.3d. The mean value function $\langle u_2(x)\rangle$ for $R_a(u) = \sigma^2 \mathrm{e}^{-\alpha u^2} \cos(\beta u)$

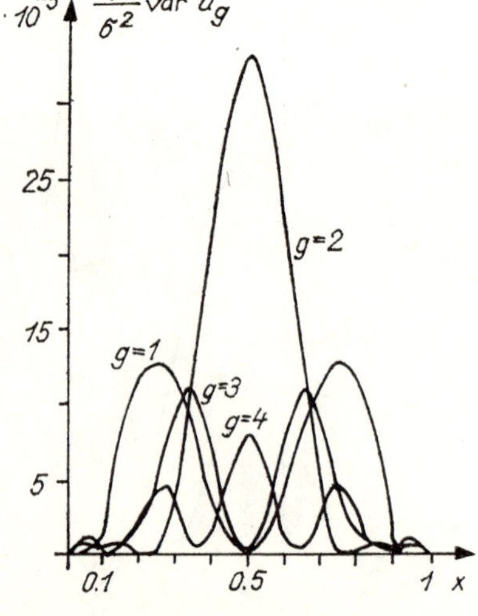

Fig. 1.4a. The variance var u_g for $R_a(u) = \sigma^2 \mathrm{e}^{-u^2}$

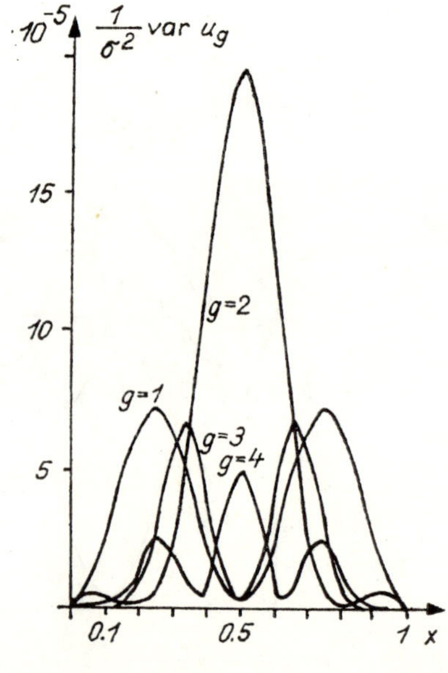

Fig. 1.4b. The variance var u_g for $R_a(u) = \sigma^2 \cos u$

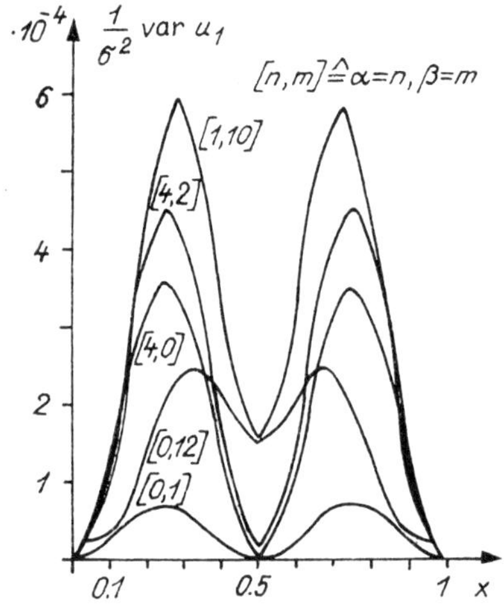

Fig. 1.4c. The variance var u_1 for $R_a(u) = \sigma^2 \mathrm{e}^{-\alpha u^2} \cos(\beta u)$

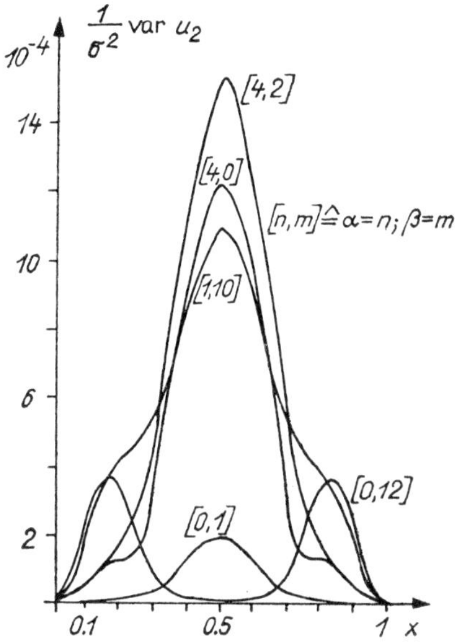

Fig. 1.4d. The variance var u_2 for $R_a(u) = \sigma^2 \mathrm{e}^{-\alpha u^2} \cos(\beta u)$

with $16S(\omega)/d_g < 1$ for $\omega \in A$, where $\mathsf{P}(A) = 1 - \varepsilon$, holds, the results of Theorem 1.16 remain true for all $\omega \in A$, and

$$\int_A \lambda_g(\omega)\, \mathsf{P}(\mathrm{d}\omega) = \sum_{p=0}^{\infty} \int_A \lambda_{gp}(\omega)\, \mathsf{P}(\mathrm{d}\omega)$$

is obtained. As in Section 1.3.3.2 the formula

$$\langle \lambda_g \rangle = \mu_g + \langle \lambda_{g1} \rangle + \langle \lambda_{g2} \rangle + \langle \lambda_{g3} \rangle + \langle \lambda_{g4} \rangle + \zeta$$

follows where ζ denotes an error which consists of two summands. The first summand comes from terminating the series after the fourth order terms. The second summand comes from substituting the integrals over A by integrals over Ω. If ε is sufficiently small then the second summand is essentially smaller than the first. For Gaussian processes $a(x, \omega)$ with $\langle a(x) \rangle = 0$, we have $\langle \lambda_{g1} \rangle = 0$; and since $\langle a(x_1)\, a(x_2)\, a(x_3) \rangle = 0$ we also have $\langle \lambda_{g3} \rangle = 0$. For the fourth moments of $a(x, \omega)$ it is well-known that

$$\begin{aligned} \langle a(x_1)\, a(x_2)\, a(x_3)\, a(x_4) \rangle = {} & \langle a(x_1)\, a(x_2) \rangle \langle a(x_3)\, a(x_4) \rangle \\ & + \langle a(x_1)\, a(x_3) \rangle \langle a(x_2)\, a(x_4) \rangle \\ & + \langle a(x_1)\, a(x_4) \rangle \langle a(x_2)\, a(x_3) \rangle\,. \end{aligned}$$

This implies for $b_{ij} = \langle\!\langle a w_i, w_j \rangle\!\rangle$ the expression

$$\langle b_{i_1 j_1} b_{i_2 j_2} b_{i_3 j_3} b_{i_4 j_4} \rangle = B_{i_1 j_1 i_2 j_2} B_{i_3 j_3 i_4 j_4} + B_{i_1 j_1 i_3 j_3} B_{i_2 j_2 i_4 j_4} + B_{i_1 j_1 i_4 j_4} B_{i_2 j_2 i_3 j_3}\,,$$

where $B_{ijst} = \langle b_{ij} b_{st} \rangle$.

With regard to this, we finally get from Theorem 1.16

$$\langle\lambda_{g2}\rangle = -\sum_{\substack{i=1\\ i\neq g}}^{\infty} \frac{1}{\mu_{ig}} B_{igig},$$

$$\langle\lambda_{g4}\rangle = -\sum_{\substack{i,j,k=1\\ i,j,k\neq g}}^{\infty} \frac{1}{\mu_{ig}\mu_{jg}\mu_{kg}} [B_{giij}B_{jkkg} + B_{gijk}B_{ijkg} + B_{gikg}B_{ijjk}]$$

$$+ \sum_{\substack{i,j=1\\ i,j\neq g}}^{\infty} \frac{1}{\mu_{ig}^2\mu_{jg}} [B_{gigi}B_{gjgj} + 2B_{gjgj}^2 + 2B_{gggi}B_{ijgj} + 2B_{ggij}B_{gigj} + 2B_{gggj}B_{giij}]$$

$$+ \sum_{\substack{i=1\\ i\neq g}}^{\infty} \frac{1}{\mu_{ig}^3} [B_{gggg}B_{gigi} + 2B_{gggi}B_{gggi}] .$$

These formulas can be numerically analyzed if the correlation function $\langle a(x)\, a(y)\rangle$ of the stationary Gaussian process is given. Provided the process is given by an explicit formula, or by its marginal distributions, we can likewise compute moments higher than the second and we can obtain higher approximations for $\langle\lambda_g\rangle$ and $\langle u_g\rangle$.

We consider for example the Rice noise (cf. Rice [1])

$$a(x, \omega) = \gamma \cos(\beta x + \alpha) ,$$

where α is a uniformly distributed real random variable on $[0, 2\pi]$; and β and γ are deterministic real constants. Then we have

$$\langle a(x)\rangle = 0 , \qquad \langle a(x_1)\, a(x_2)\, a(x_3)\rangle = 0 ,$$

$$\langle a(x_1)\, a(x_2)\rangle = \tfrac{1}{2}\gamma^2 \cos\big(\beta(x_1 - x_2)\big) ,$$

$$\langle a(x_1)\, a(x_2)\, a(x_3)\, a(x_4)\rangle = \tfrac{1}{8}\gamma^4[\cos\big(\beta(x_1 - x_2 - x_3 - x_4)\big) + \cos\big(\beta(x_1 - x_2 + x_3 - x_4)\big) + \cos\big(\beta(x_1 + x_2 - x_3 - x_4)\big)] .$$

From this $\langle b_{i_1j_1}b_{i_2j_2}\rangle$ and $\langle b_{i_1j_1}b_{i_2j_2}b_{i_3j_3}b_{i_4j_4}\rangle$ can be evaluated; $\langle b_{ij}\rangle$ and $\langle b_{i_1j_1}b_{i_2j_2}b_{i_3j_3}\rangle$ are equal to zero.

Case 3. $a(x, \omega) \equiv 0$

As in Case 2 the quantities $\langle\lambda_g\rangle - \mu_g$, $\langle u_g\rangle - w_g$, $\operatorname{var}\lambda_g$, and $\operatorname{var} u_g$ are investigated for dependence on the correlation function $R_b(u)$ of the stationary process $b(x, \omega)$. Therefore, we put $\eta = 0$. In case $\eta \neq 0$ the expressions in (1.128) and (1.129) which contain the random variable η can be calculated easily if $\langle b(x)\, \eta\rangle$ and $\langle\eta^2\rangle$ are known. In the investigated case it appears that the results for $\operatorname{var}\lambda_g$ and $\operatorname{var} u_g$ are equal to those of Case 2 multiplied by μ_g^2 if we put $R_b(u) = R_a(u)$. Thus Fig. 1.2 shows $\mu_g^{-2} \operatorname{var}\lambda_g$, especially,

in Fig. 1.2a this variance for $R_b(u) = \sigma^2 \exp(-\alpha\, |u|)$,
in Fig. 1.2b this variance for $R_b(u) = \sigma^2 \exp(-\alpha u^2) \cos(\beta u)$, and
in Fig. 1.2c this variance for $R_b(u) = \sigma^2 \cos(\beta u)$.

Fig. 1.4 shows μ_g^{-2} var $u_g(x)$ for various correlation functions, i.e.

Fig. 1.4a for $R_b(u) = \sigma^2 \exp(-u^2)$,
Fig. 1.4b for $R_b(u) = \sigma^2 \cos u$,
Fig. 1.4c with $g = 1$ } for $R_b(u) = \sigma^2 \exp(-\alpha u^2) \cos(\beta u)$ and different
Fig. 1.4d with $g = 2$ } parameters α, β.

For $\langle\lambda_g\rangle - \mu_g$, $\langle u_g\rangle - w_g$, respectively, (1.128) and (1.129) yield

$$\frac{1}{\mu_g}(\langle\lambda_g\rangle - \mu_g) = \langle\!\langle R_b(u), \frac{3}{2}(1-u) + \frac{3}{4g\pi}\sin(2g\pi u) + \frac{1}{2}g\pi(u-1)^2\sin(2g\pi u)\rangle\!\rangle ,$$

$$\frac{1}{\mu_g^2}(\langle u_g\rangle - w_g) = \sum_{\substack{s=1\\s\neq g}}^{\infty}\frac{1}{\mu_{sg}}\left(\sum_{\substack{t=1\\t\neq g}}^{\infty}\frac{1}{\mu_{tg}}B_{gtts} - \frac{1}{\mu_{sg}}B_{ggg s}\right)w_s - \frac{1}{\mu_g}\sum_{\substack{s=1\\s\neq g}}^{\infty}\frac{1}{\mu_{sg}}B_{gggs}w_s - \frac{1}{2}\sum_{\substack{s=1\\s\neq g}}^{\infty}\frac{1}{\mu_{sg}^2}B_{gsgs}w_g$$

where B_{ijst} is defined by

$$B_{ijst} \doteq \int_0^1\int_0^1 R_b(y-x)\, w_i(x)\, w_j(x)\, w_s(y)\, w_t(y)\,\mathrm{d}x\,\mathrm{d}y .$$

Furthermore, from (1.139) the asymptotic formula

$$\lim_{g\to\infty}\frac{1}{\mu_g}(\langle\lambda_g^{\mathrm{II}}\rangle - \mu_g) = \frac{3}{2}\langle\!\langle R_b(u), 1-u\rangle\!\rangle + \frac{1}{4}R_b(0)$$

can be derived. Figs. 1.5a to 1.5c show $\langle\lambda_g\rangle - \mu_g$ for various correlation functions $R_b(u)$. Figs. 1.6a and 1.6b illustrate $\langle u_g\rangle - w_g$ for $R_b(u) = \sigma^2\exp(-u^2)$, $R_b(u) = \sigma^2\cos u$, respectively. In Fig. 1.6c the difference $\langle u_2\rangle - w_2$ is plotted for $R_b(u) = \sigma^2\exp(-\alpha u^2)\cos(\beta u)$ and various pairs of parameters α, β.

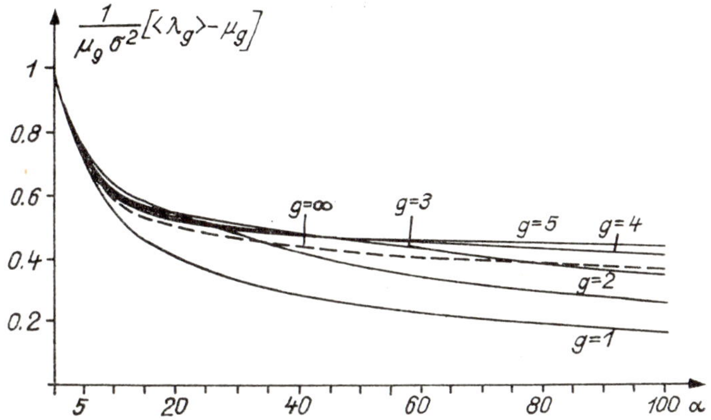

Fig. 1.5.a. The mean value $\langle\lambda_g\rangle$ for $R_b(u) = \sigma^2 e^{-\alpha u^2}$

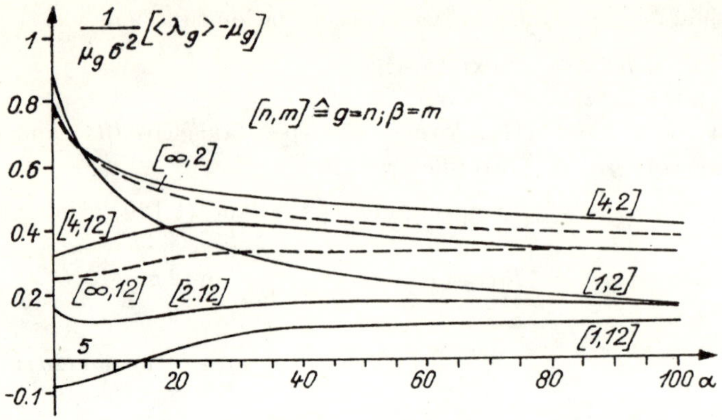

Fig. 1.5b. The mean value $\langle\lambda_g\rangle$ for $R_b(u) = \sigma^2 e^{-\alpha u^2} \cos(\beta u)$

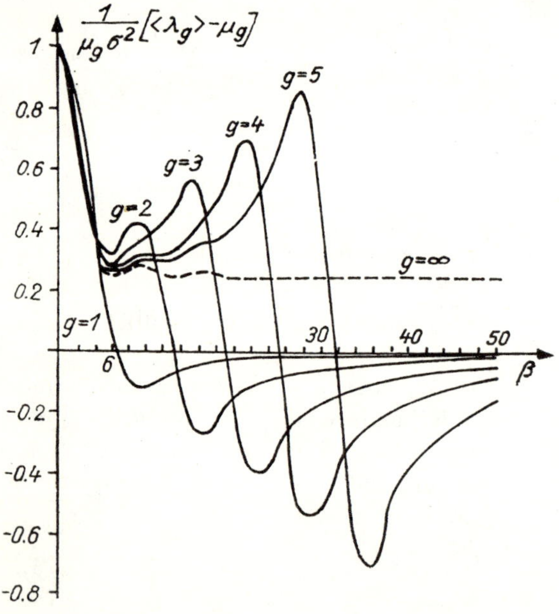

Fig. 1.5c. The mean value $\langle\lambda_g\rangle$ for $R_b(u) = \sigma^2 \cos(\beta u)$

1.4.3.3. „White noise“ as coefficient processes

We now consider the case where $a(x, \omega)$, $b(x, \omega)$ and $\eta(\omega)$ are assumed to be uncorrelated, and $a(x, \omega)$, $b(x, \omega)$ to be white noise processes, i.e. let $a(x, \omega)$, $b(x, \omega)$ be generalized stationary processes in the wide sense where $\langle a(x)\rangle = \langle b(x)\rangle = 0$ and the "correlation functions" are given by

$$\begin{aligned} \langle a(x)\, a(y)\rangle &= S_1 \delta(x-y)/2\,, \\ \langle b(x)\, b(y)\rangle &= S_2 \delta(x-y)/2\,. \end{aligned} \tag{1.144}$$

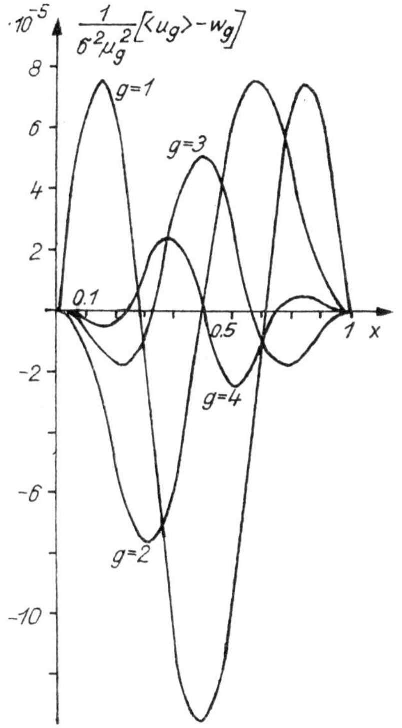

Fig. 1.6a. The mean value function $\langle u_g x)\rangle$ for $R_b(u) = \sigma^2 \, e^{-u^2}$

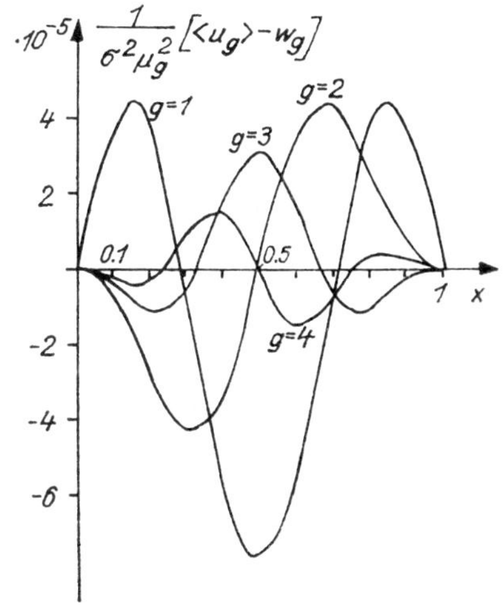

Fig. 1.6b. The mean value function $\langle u_g x)\rangle$ for $R_b(u) = \sigma^2 \cos u$

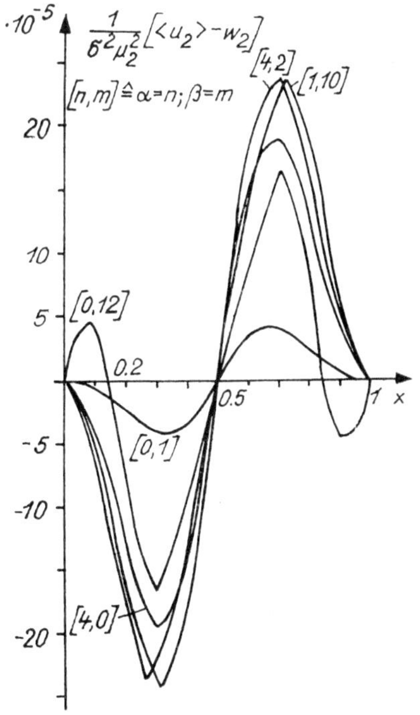

Fig. 1.6c. The mean value function $\langle u_2(x)\rangle$ for $R_b(u) = \sigma^2 \, e^{-\alpha u^2} \cos(\beta u)$

Stationary processes with correlation functions of the type (1.144) have completely discontinuous realizations, so that the eigenvalue problem (1.115) with such coefficient processes presents some difficulties. On the other hand, we get expressions which can easily be computed if we substitute in the formula for $\langle\lambda_{g2}\rangle$, $\langle u_{g2}\rangle$ and var u_g the white noise processes (the white noise processes we substitute in the terms B_{ijkl}). We will now interpret these expressions mathematically. Indeed, this is possible by means of weakly correlated processes defined in Chapter 2.

With the B_{ijkl} and C_{ijkl} given in Section 1.4.3.1 and under the assumption that $a(x, \omega)$, $b(x, \omega)$, $\eta(\omega)$ are uncorrelated we obtain from (1.120) and (1.129)

$$\langle\lambda_{g2}\rangle = -\sum_{\substack{s=1\\ s\neq g}}^{\infty} \frac{1}{\mu_{sg}} B_{gsgs} - C_{gggg} + 3\mu_g\langle\eta^2\rangle\,, \tag{1.145}$$

$$\begin{aligned}\langle u_{g2}(x)\rangle = &\sum_{\substack{s=1\\ s\neq g}}^{\infty} \frac{1}{\mu_{sg}}\left[\sum_{\substack{t=1\\ t\neq g}}^{\infty} \frac{1}{\mu_{tg}} B_{gtts} - \frac{1}{\mu_{sg}} B_{gggs} + C_{gggs}\right] w_s(x)\\ &- \frac{1}{2}\sum_{\substack{s=1\\ s\neq g}}^{\infty} \frac{1}{\mu_{sg}^2} B_{gsgs} w_g(x)\\ &+ \langle\eta^2\rangle\left[\frac{3}{8} w_g(x) + \frac{3}{2} x w_g'(x) - \frac{1}{2}\mu_g x^2 w_g(x)\right].\end{aligned} \tag{1.146}$$

Here we have for uncorrelated processes a and b

$$B_{ijkl} = B^a_{ijkl} + \mu_l^2 B^b_{ijkl}\,, \qquad C_{ijkl} = -\mu_l B^b_{ijkl}\,,$$

where

$$B^a_{ijkl} \doteq \langle\langle\!\langle a w_i, w_j\rangle\!\rangle\langle\!\langle a w_k, w_l\rangle\!\rangle\rangle\,,$$

$$B^b_{ijkl} \doteq \langle\langle\!\langle b w_i, w_j\rangle\!\rangle\langle\!\langle b w_k, w_l\rangle\!\rangle\rangle\,.$$

Let $a(x, \omega)$, $b(x, \omega)$ be weakly correlated processes with correlation length ε (see Chapter 2), with correlation functions

$$\sigma_1^2\varrho_{1\varepsilon}(t)\,, \qquad \sigma_2^2\varrho_{2\varepsilon}(t)\,,$$

respectively, where

$$\lim_{\varepsilon\to 0} \varepsilon^{-r} \int_{-\varepsilon}^{\varepsilon} \varrho_{1\varepsilon}(t)\,\mathrm{d}t = s_1 > 0\,,$$

$$\lim_{\varepsilon\to 0} \varepsilon^{-r} \int_{-\varepsilon}^{\varepsilon} \varrho_{2\varepsilon}(t)\,\mathrm{d}t = s_2 > 0\,.$$

Then it follows that

$$B^a_{ijkl} = \sigma_1^2 s_1 \varepsilon^r \int_0^1 w_i(x)\, w_j(x)\, w_k(x)\, w_l(x)\,\mathrm{d}x + O(\varepsilon^{r+1})\,,$$

$$B^b_{ijkl} = \sigma_2^2 s_2 \varepsilon^r \int_0^1 w_i(x)\, w_j(x)\, w_k(x)\, w_l(x)\,\mathrm{d}x + O(\varepsilon^{r+1})\,.$$

If we use assumption (1.144) in the formulas for B^a_{ijkl} and B^b_{ijkl}, respectively, then we obtain

$$\begin{aligned} B^a_{ijkl} &= \tfrac{1}{2} S_1 \int_0^1 w_i(x)\, w_j(x)\, w_k(x)\, w_l(x)\, \mathrm{d}x\,, \\ B^b_{ijkl} &= \tfrac{1}{2} S_2 \int_0^1 w_i(x)\, w_j(x)\, w_k(x)\, w_l(x)\, \mathrm{d}x\,. \end{aligned} \tag{1.147}$$

Therefore, the B^a_{ijkl} resulting from substituting the white noise process with $S_1 = 2\sigma_1^2 s_1$ are just the coefficients of the term with the lowest ε-power in the expansion of B^a_{ijkl} as a power series in ε (and analogously for b). Thus, the use of white noise processes according to formulas (1.147) is valid if this connection with weakly correlated processes is taken into consideration. Since weakly correlated processes can be assumed to be sufficiently often differentiable, the random eigenvalue problem (1.115) with weakly correlated processes $a(x, \omega)$ and $b(x, \omega)$ is a well-defined problem.

We now consider formulas (1.147). Because of

$$B_{lsls} = \begin{cases} \tfrac{1}{2} S_1 + \tfrac{1}{2} \mu_l^2 S_2 & \text{for} \quad l \neq s\,, \\ \tfrac{3}{4} S_1 + \tfrac{3}{4} \mu_l^2 S_2 & \text{for} \quad l = s\,, \end{cases}$$

$$C_{llll} = -\tfrac{3}{4} \mu_l S_2\,,$$

and

$$\sum_{\substack{s=1 \\ s \neq l}}^{\infty} \frac{1}{\mu_{sl}} = \frac{3}{4\mu_l}\,,$$

(1.145) leads to

$$\langle \lambda_{g2} \rangle = \frac{3}{8} \left(-\frac{1}{\mu_g} S_1 + \mu_g S_2 \right) + 3\mu_g \langle \eta^2 \rangle$$

and

$$\langle \lambda_g \rangle = \mu_g + \frac{3}{8} \left(-\frac{1}{\mu_g} S_1 + \mu_g S_2 \right) + 3\mu_g \langle \eta^2 \rangle\,.$$

The variance can be calculated by

$$\operatorname{var} \lambda_g \doteq \sigma_g^2 = \tfrac{3}{4} (S_1 + \mu_g^2 S_2) + 4\mu_g^2 \langle \eta^2 \rangle\,.$$

Furthermore, we obtain

$$B_{ltts} = \begin{cases} -\tfrac{1}{4} (S_1 + \mu_l^2 S_2) & \text{for } 2t - l + s = 0\,,\ 2t + l - s = 0\,,\ l \neq s\,, \\ \tfrac{1}{4} (S_1 + \mu_l^2 S_2) & \text{for } 2t - l - s = 0\,,\ l \neq s\,, \\ \tfrac{1}{2} (S_1 + \mu_l^2 S_2) & \text{for } l = s\,,\ t \neq l\,, \\ 0 & \text{otherwise}; \end{cases}$$

$$B_{llls} = \begin{cases} -\tfrac{1}{4} (S_1 + \mu_l^2 S_2) & \text{for} \quad s = 3l\,, \\ \tfrac{3}{4} (S_1 + \mu_l^2 S_2) & \text{for} \quad s = l\,, \\ 0 & \text{otherwise}; \end{cases}$$

$$C_{llls} = \begin{cases} \tfrac{1}{4} \mu_l S_2 & \text{for} \quad s = 3l\,, \\ -\tfrac{3}{4} \mu_l S_2 & \text{for} \quad s = l\,, \\ 0 & \text{otherwise} \end{cases}$$

from which for $\eta = 0$ we have

$$\langle u_{g2}(x)\rangle = \frac{1}{4}(S_1 + \mu_g^2 S_2)\left[\sum_{\substack{t=1\\ t\neq g,\, g/2}}^{\infty} \frac{w_{2t-g}(x)}{\mu_{tg}\mu_{g-2t,\,g}} - \sum_{\substack{t=1\\ t\neq g}}^{\infty} \frac{w_{2t+g}(x)}{\mu_{tg}\mu_{2t+g,\,g}} + \frac{w_{3g}(x)}{\mu_{3g,\,g}^2} + \left(\frac{11}{16\mu_g^2} - \frac{1}{12\mu_g}\right) w_g(x)\right] + \frac{1}{4}\mu_g S_2 \frac{w_{3g}(x)}{\mu_{3g,\,g}^2},$$

and

$$\operatorname{var} u_g(x) = \frac{1}{4}(S_1 + \mu_g^2 S_2)\left[2\sum_{\substack{t=1\\ t\neq g}}^{\infty} \frac{w_t^2(x)}{\mu_{tg}^2} - 2\sum_{\substack{t=1\\ t\neq g}}^{\infty} \frac{w_t(x)\, w_{2g+t}(x)}{\mu_{tg}\mu_{2g+t,\,g}} + \sum_{\substack{t=1\\ t\neq g}}^{2g-1} \frac{w_t(x)\, w_{2g-t}(x)}{\mu_{tg}\mu_{2g-t,\,g}}\right].$$

VAN DER LINDE [1] has discussed the variance of $u_g(x, \omega)$ by means of this formula. He investigated the variance relative to the extreme values. We obtain the results of van der Linde and additional results if we derive closed expressions for $\langle u_{g2}\rangle$ and var u_g by computations similar to those which led to (1.135). These calculations yield

$$\langle u_{g2}(x)\rangle = \frac{1}{32\mu_g^2} S_g' \left[\left(\frac{23}{4} + 2\mu_g\left(x^2 - x - \frac{1}{6}\right)\right) w_g - \frac{7}{8} w_{3g} + \left(\frac{1}{2} - x\right)\left(w_g' - \frac{1}{3} w_{3g}'\right)\right] + \frac{1}{32} S_2 w_{3g}, \tag{1.148}$$

$$\operatorname{var} u_g(x) = \frac{1}{8\mu_g} S_g' \left[-3x(x-1) + \left(\frac{1}{6} + 2x(x-1)\right) w_g^2 + \frac{1}{\mu_g}\left(x - \frac{1}{2}\right)(3 - w_g^2)\, w_g w_g' - \frac{5}{8\mu_g}\left(\frac{5}{2} - w_g^2\right) w_g^2\right], \tag{1.149}$$

where $S_g' \doteq S_1 + \mu_g^2 S_2$.

At first, we discuss the variance of the eigenfunctions. As critical points, i.e. as zeros of the derivative

$$(\operatorname{var} u_g(x))' = \frac{S_g'}{4\mu_g} w_g w_g' \left[\frac{1}{6} - \frac{1}{16\mu_g} + 2x(x-1) + \frac{3}{4\mu_g} w_g^2 - \frac{2}{\mu_g}\left(x - \frac{1}{2}\right) w_g w_g'\right],$$

we get the values

$$x_i = i/g \quad \text{for} \quad i = 1, 2, \ldots, g-1, \quad \text{from} \quad w_g(x) = 0,$$

$$y_j = (2j-1)/2g \quad \text{for } j = 1, 2, \ldots, g, \quad \text{from} \quad w_g'(x) = 0,$$

and

$$z_k \quad \text{from} \quad \frac{1}{6} - \frac{1}{16\mu_g} + 2x(x-1) + \frac{3}{4\mu_g} w_g^2 - \frac{2}{\mu_g}\left(x - \frac{1}{2}\right) w_g w_g' = 0 \, .$$

We have

$$\left(\operatorname{var} u_g(x)\right)''_{x=x_i} = \frac{1}{2} S_g' \left(\frac{1}{6} - \frac{1}{16\mu_g} + 2x_i(x_i - 1)\right),$$

$$\left(\operatorname{var} u_g(x)\right)''_{x=y_j} = -\frac{1}{2} S_g' \left(\frac{1}{6} + \frac{23}{16\mu_g} + 2y_j(y_j - 1)\right).$$

Therefore, the critical points x_i are relative maxima of the variance if

$$1/6 - 1/16\mu_g < 2x_i(1 - x_i) \, ,$$

otherwise relative minima; and the y_j are relative maxima if

$$1/6 + 23/16\mu_g > 2y_j(1 - y_j) \, ,$$

otherwise relative minima. This means that the variance possesses a relative maximum for $x_i = i/g$ if

$$g/2 - a_g < i < g/2 + a_g \, , \qquad a_g \doteq (1/32\pi^2 + g^2/6)^{1/2} \, ,$$

i.e. for

$$i = [g/2 - a_g] + 1, \ldots, [g/2 + a_g] - 1 \, .$$

For

$$i = 1, 2, \ldots, [g/2 - a_g], [g/2 + a_g], \ldots, g - 1 \, ,$$

x_i denotes the place of a relative minimum. For $y_j = (2j - 1)/2g$ the variance has a relative minimum if

$$(g+1)/2 - b_g < j < j < (g+1)/2 + b_g \, , \qquad b_g \doteq (g^2/6 - 23/32\pi^2)^{1/2}$$

holds, hence for

$$j = [(g+1)/2 - b_g] + 1, \ldots, [(g+1)/2 + b_g] - 1 \, .$$

Setting

$$j = 1, 2, \ldots, [(g+1)/2 - b_g], [(g+1)/2 + b_g], \ldots, g$$

the variance has a relative maximum at y_j.

Thus, all the x_i are relative maxima for $g = 2, 3, \ldots, 10$. For $g = 11$, x_1 and x_{10} are relative minima and $x_2, \ldots, x_9$ are relative maxima. On the other hand, all the y_j are relative minima for $g = 1, 2, \ldots, 5$. For $g = 6$, y_1 and y_6 are relative maxima and y_2, y_3, y_4, y_5 are relative minima. The relative extreme values of the variance at x_i, y_j are

$$\operatorname{var} u_g(x_i) = \frac{3}{8\mu_g} S_g' x_i(1 - x_i) \, ,$$

$$\operatorname{var} u_g(y_j) = \frac{1}{8\mu_g} S_g' \left[y_j(y_j - 1) + \frac{1}{3} - \frac{5}{8\mu_g} \right].$$

It is very difficult to evaluate explicitly the critical points z_k from the corresponding transcendental equation.

From (1.149) the relation

$$\lim_{g\to\infty}\left[\frac{1}{\mu_g}\operatorname{var}u_g(x)-\frac{1}{8}S_2\left\{3x(1-x)+\left(\frac{1}{6}+2x^2-2x\right)w_g^2(x)\right\}\right]=0$$

is easily obtained, so that one can put for large g

$$\operatorname{var}u_g(x)\approx\tfrac{1}{8}\,\mu_g S_2\,[3x(1-x)+(\tfrac{1}{6}+2x^2-2x)\,w_g^2(x)]\doteq G_g(x)\,.$$

This approximation is very good for $g=4$, as shown in Fig. 1.8 in which $G_4(x)$ is indicated with broken lines. Fig. 1.7 shows the variance of the eigenfunctions for $g=1,2,3$ with $a(x,\omega)=\eta(\omega)=0$.

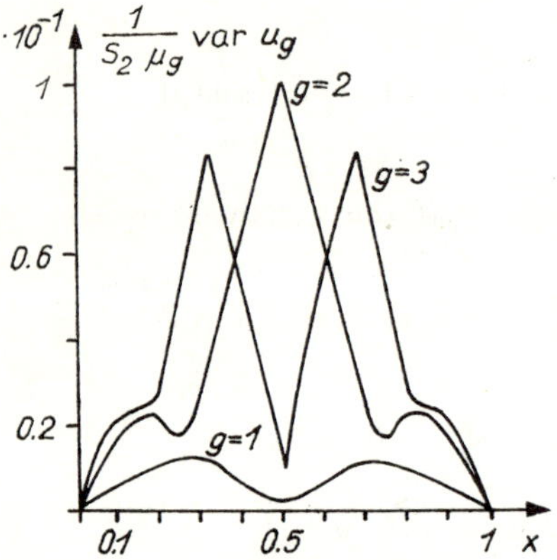

Fig. 1.7. The variance var u_g for $R_b(t)=\frac{1}{2}S_2\delta(t)$ and $a=\eta=0$

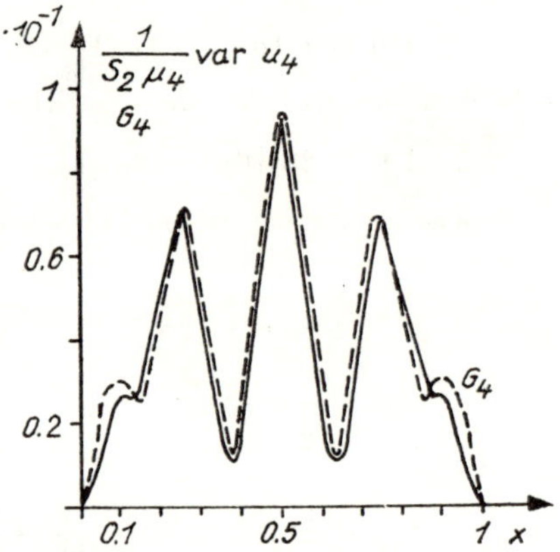

Fig. 1.8. The variance var u_4 and the asymptotic expression G_4 for $R_b(t)=\frac{1}{2}S_2\delta(t)$ and $a=\eta=0$

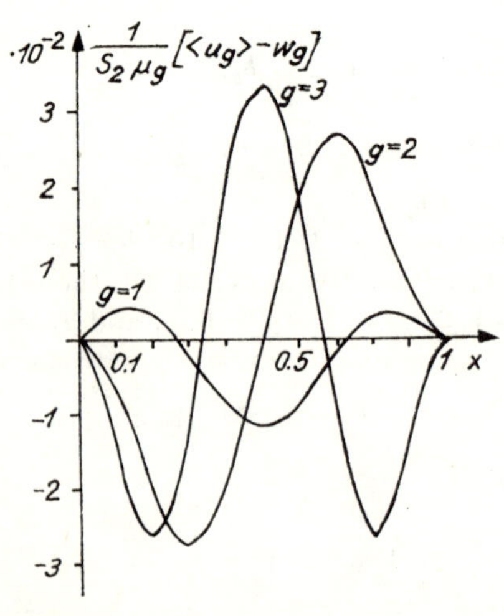

Fig. 1.9. The mean value function $\langle u_g\rangle$ for $R_b(t)=\frac{1}{2}S_2\delta(t)$ and $a=\eta=0$

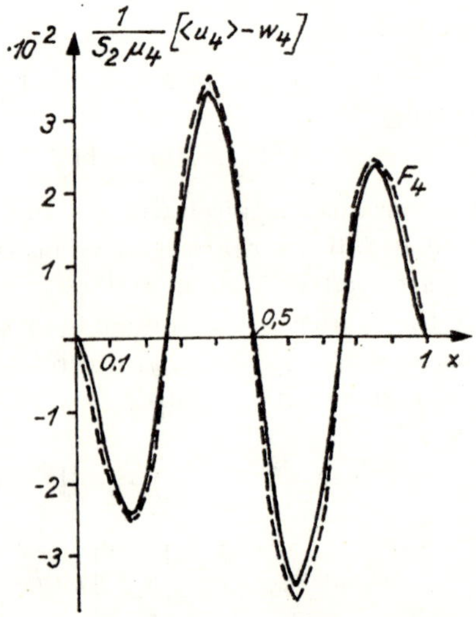

Fig. 1.10. The mean value function $\langle u_4\rangle$ and the asymptotic expression F_4 for $R_b(t)=\frac{1}{2}S_2\delta(t)$ and $a=\eta=0$

In the discussion of $\langle u_g(x)\rangle - w_g(x)$ we restrict ourselves to the case $a(x, \omega) = \eta(\omega) = 0$, since the results in the general case differ only slightly from this case. From (1.148) the expression

$$\langle u_{g2}(x)\rangle = \tfrac{1}{32} S_2 \left[\tfrac{49}{8} + 2\mu_g(x^2 - x - \tfrac{1}{6}) - \tfrac{1}{4} w_g^2 - 2(x - \tfrac{1}{2})\, w_g w_g'\right] w_g \tag{1.150}$$

follows. Fig. 1.9 shows $\langle u_g\rangle - w_g = \langle u_{g2}\rangle$ for $g = 1, 2, 3$. Assuming $g = 2m + 1$ the point $x = 1/2$ is for m odd the location of a maximum, and for m even the location of a minimum. For even g the function $\langle u_{g2}(x)\rangle$ has a zero at $x = 1/2$. For even g the function $\langle u_{g2}(x)\rangle$ is antisymmetric with respect to the point $x = 1/2$, and symmetric with respect to $x = 1/2$ for odd g. From the equation

$$\left(\langle u_{g2}(x)\rangle\right)'_{x=y_j} = \frac{1}{2\sqrt{2}} S_2\mu_g(-1)^{j+1}\left(y_j - \frac{1}{2}\right),$$

where $y_j = (2j - 1)/2g$ it follows that $\langle u_{g2}(x)\rangle$ has at least one critical point, i.e. a zero of the derivative, between y_j and y_{j+1} for $j = 1, 2, \dots, g/2 - 1$, $(g - 3)/2$. Furthermore, the points $x_i = i/g$ for $i = 1, 2, \dots, g - 1$ are zeros as well as points of inflexion for $\langle u_{g2}(x)\rangle$. Hence, relative extreme values of $\langle u_{g2}(x)\rangle$ exist between x_{i-1} and x_i for $i = 1, 2, \dots, g$. (1.150) implies

$$\lim_{g\to\infty}\left[\frac{1}{\mu_g}\langle u_{g2}(x)\rangle - \frac{1}{16} S_2\left(x^2 - x - \frac{1}{6}\right) w_g(x)\right] = 0\,;$$

and, therefore,

$$\langle u_{g2}(x)\rangle \approx \tfrac{1}{16} S_2(x^2 - x - \tfrac{1}{6})\, w_g(x) \doteq F_g(x)$$

is a good approximation for large g. Fig. 1.10 shows that $F_g(x)$ for $g = 4$ is very useful. From the equation

$$F_g'(x) = \tfrac{1}{16} S_2\mu_g\left((2x - 1)\, w_g(x) + (x^2 - x - \tfrac{1}{6})\, w_g'(x)\right) = 0$$

we can calculate the critical points of $\langle u_{g2}(x)\rangle$ for large g. These points will be approximate the zeros of $(x^2 - x - 1/6)\, w_g'(x)$, hence

$$y_j = \frac{2j - 1}{2g} \quad \text{for} \quad j = 1, 2, \dots, g\,.$$

Using

$$F_g''(x) = \tfrac{1}{16} S_2\mu_g\left(2w_g + 2(2x - 1)\, w_g' + (x^2 - x - \tfrac{1}{6})\, w_g''\right)$$

the function $F_g(x)$ has a relative maximum at the point y_j for even j, and a relative minimum for odd j since

$$y_j^2 - y_j - \tfrac{1}{6} < 0\,.$$

1.4.4. Application to random vibrations

In the following we consider small transversal vibrations of elastic beams. The differential equations is

$$(EIu'')'' - mAL^4\xi^2 u = 0\,, \tag{1.151}$$

where $u(x)$ denotes the transversal displacement and E the Young's modulous. I is the moment of inertia of the cross-section about an axis through the centroid perpendicular to the plane of vibration. m is the mass per unit length, A the cross-sectional area, L the length, and ξ the frequency of the transversal vibration.

Let E, m, and L be deterministic constants and $A(x, \omega)$ and $I(x, \omega)$ be stochastic processes. A somewhat simpler formulation results if we assume that only the size, and not the shape, of the cross-section varies with position. Then we can write

$$A(x, \omega) = k_1 \bar{r}^2(x, \omega), \qquad I(x, \omega) = k_2 \bar{r}^4(x, \omega)$$

where k_1 and k_2 are constants, and $\bar{r}(x, \omega)$ denotes some typical length in the cross-section. If we assume $\bar{r}(x, \omega)$ to be stationary in the wide sense and put

$$\bar{r}(x, \omega) = r_0\big(1 + r(x, \omega)\big)$$

where $r(x, \omega)$ is a stationary random process with $\langle r(x) \rangle = 0$, then it follows from (1.151) that

$$\big((1 + r(x, \omega))^4 \, u''(x)\big)'' - \lambda\big(1 + r(x, \omega)\big)^2 \, u(x) = 0 \,, \tag{1.152}$$

where

$$\lambda = k_1 m L^4 \xi^2 (k_2 E r_0^2)^{-1}$$

(cf. Boyce, Goodwin [1]). We can put $L = 1$ without loss of generality, and assume $r(x, \omega)$ to be sufficiently small. As boundary conditions we put

$$\begin{aligned} u(0) &= 0 \,, \qquad u''(0) - \delta_1 u'(0) = 0 \,, \\ u(1) &= 0 \,, \qquad u''(1) - \delta_2 u'(1) = 0 \,. \end{aligned} \tag{1.153}$$

As $\delta_1, \delta_2 \to 0$ the ends become hinged, and as $\delta_1, \delta_2 \to \infty$ they become clamped.

The differential equation (1.152) can be written in the form

$$M_0 u + M_{11} u + M_{12} u = \lambda u \,,$$

where the symmetric positive definite operator M_0 is given by

$$M_0 \, u = u'''' \,,$$

and the perturbation operators by

$$M_{11} u = 2 r u'''' + 8 r' u''' + 4 r'' u'' \,,$$

$$M_{12} u = r^2 u'''' + 8 r' r u''' + (12 r'^2 + 4 r r'') \, u'' \,.$$

M_{11} and M_{12} satisfy for all admissible functions u, v corresponding to M_0 and the boundary conditions (1.153), the following conditions

$$\langle\!\langle M_{11} u, v \rangle\!\rangle = \langle\!\langle u, M_{11}^* v \rangle\!\rangle + D_1(u, v) \,,$$

$$\langle\!\langle M_{12} u, v \rangle\!\rangle = \langle\!\langle u, M_{12}^* v \rangle\!\rangle + D_2(u, v)$$

where

$$D_1(u, v) \doteq -4 r' u' v' |_0^1$$

and

$$D_2(u, v) \doteq -4 r' r u' v' \, |_0^1 \,.$$

Assume that the unperturbed eigenvalue problem $M_0 w = \mu w$ with boundary conditions (1.153) has the simple eigenvalues μ_l with associated eigenfunctions $w_l(x)$. Any admissible function can be expanded as a uniformly convergent series associated with the eigenfunctions $w_l(x)$; that is

$$u(x) = \sum_{s=1}^{\infty} \langle\!\langle u, w_s \rangle\!\rangle w_s(x) .$$

This series can be once differentiated term by term. Hence, the series

$$\sum_{\substack{s=1 \\ s \neq g}}^{\infty} \frac{\mu_s}{\mu_{sg}} \langle\!\langle Tu, w_s \rangle\!\rangle w_s$$

can also be differentiated term by term. Here T is defined by

$$Tu \doteq \int_0^1 G(x, y)\, u(y)\, dy$$

where $G(x, y)$ denotes the Green's function belonging to M_0 and the given boundary conditions. From this it follows that condition (1.99) of Theorem 1.15 is satisfied, and this theorem can be used.

Let

$$R(x - y) \doteq \langle r(x)\, r(y) \rangle$$

denote the correlation function of the (stationary in the wide sense) process $r(x, \omega)$. If $R(u)$ is $(2m + 2)$ times differentiable at $u = 0$ then $r(x, \omega)$ is $(m + 1)$ times differentiable in the mean-square and m times differentiable a.s., and these derivatives of order k coincide for $k = 0, 1, \ldots, m$. For $p, q \leqq m$ the relation

$$(-1)^q R^{(p+q)}(x - y) = \langle r^{(p)}(x)\, r^{(q)}(y) \rangle$$

holds.

Hence, we obtain from Theorem 1.15

$$\begin{aligned} \langle \lambda_{g1} \rangle &= \langle\!\langle \langle M_{11} \rangle w_g, w_g \rangle\!\rangle + \langle\!\langle \langle M_{12} \rangle w_g, w_g \rangle\!\rangle \\ &= \mu_g R(0) - 8R''(0) \langle\!\langle w_g'', w_g \rangle\!\rangle \end{aligned}$$

and

$$\langle u_{g1}(x) \rangle = 8R''(0) \sum_{\substack{s=1 \\ s \neq g}}^{\infty} \frac{1}{\mu_{sg}} \langle\!\langle w_g'', w_s \rangle\!\rangle w_s(x)$$

since

$$\langle M_{11} \rangle u = 0 ,$$

$$\langle M_{12} \rangle u = R(0)\, u'''' - 8R''(0)\, u''$$

$\big(R'(0) = 0$ since $R(u) = R(-u)\big)$. We have to perform calculations up to second order terms in r and its derivatives. Hence, we must take into account only M_{11} in the second order terms given in Theorem 1.15. With

$$\langle b_{gt} b_{ts} \rangle = \langle \langle\!\langle M_{11} w_g, w_t \rangle\!\rangle \langle\!\langle M_{11} w_t, w_s \rangle\!\rangle \rangle + \text{terms of higher order}$$

we have all second order terms in Theorem 1.15. Thus

$$\begin{aligned}
&\langle \llbracket M_{11} w_g, w_t \rrbracket \llbracket M_{11} w_t, w_s \rrbracket \rangle \\
&= \int_0^1 \int_0^1 w_t(x)\, w_s(y)\, \{4\mu_t \mu_g w_g(x)\, w_t(y)\, R(x-y) \\
&\qquad + 16 w_g''(x)\, w_t''(y)\, R''''(x-y) \\
&\qquad + 16[\mu_t w_g'''(x)\, w_t(y) - \mu_g w_g(x)\, w_t'''(y)]\, R'(x-y) \\
&\qquad + 8[\mu_t w_t(y)\, w_g''(x) + \mu_g w_g(x)\, w_t''(x) - 8 w_t'''(y)\, w_g'''(x)] \\
&\qquad\qquad \times R''(x-y) \\
&\qquad + 32[-w_g''(x)\, w_t'''(y) + w_g'''(x)\, w_t''(y)]\, R'''(x-y)\} \\
&\qquad\qquad \times \mathrm{d}x\, \mathrm{d}y\,.
\end{aligned}$$

In order to reduce further computations we treat the case $\delta_1 = \delta_2 = 0$, i.e. the case of a simple supported beam. We will restrict ourselves to the computation of the eigenvalues.

Under the condition of $\delta_1 = \delta_2 = 0$ we have $w_l(x) = \sqrt{2} \sin(\nu_l x)$, $\mu_l = \nu_l^4$ with $\nu_l = l\pi$. From Lemma 1.6 we obtain for even $a + b$

$$\int_0^1 \int_0^1 Q(x-y) \cos\big(\pi(ax+by)\big)\, \mathrm{d}x\, \mathrm{d}y$$

$$= \begin{cases} -\dfrac{2}{\pi(a+b)} \displaystyle\int_0^1 Q(u)\, [\sin(\pi a u) + \sin(\pi b u)]\, \mathrm{d}u & \text{for} \quad a + b \neq 0\,, \\ 2 \displaystyle\int_0^1 Q(u)\, (1-u) \cos(\pi a u)\, \mathrm{d}u & \text{for} \quad a + b = 0; \end{cases}$$

$$\int_0^1 \int_0^1 P(x-y) \sin\big(\pi(ax+by)\big)\, \mathrm{d}x\, \mathrm{d}y$$

$$= \begin{cases} -\dfrac{2}{\pi(a+b)} \displaystyle\int_0^1 P(u)\, [\cos(\pi a u) - \cos(\pi b u)]\, \mathrm{d}u & \text{for} \quad a + b \neq 0\,, \\ 2 \displaystyle\int_0^1 P(u)\, (u-1) \sin(\pi a u)\, \mathrm{d}u & \text{for} \quad a + b = 0 \end{cases}$$

$\big(Q(u) = Q(-u),\ P(u) = -P(-u)\big)$, and from this by some additional computations

$$\begin{aligned}
&\langle \llbracket M_{11} w_g, w_t \rrbracket \llbracket M_{11} w_t, w_g \rrbracket \rangle \\
&= \sigma^2 [4\mu_g \mu_t X_{tg} - 8(\mu_t \nu_g^2 + \mu_g \nu_t^2)\, X_{tg}^{(2)} + 16 \nu_g^2 \nu_t^2 X_{tg}^{(4)} + 64 \nu_t^3 \nu_g^3 Z_{tg}^{(2)} \\
&\qquad - 16 \mu_t \nu_g^3 Y_{tg}^{(1)} - 16 \mu_g \nu_t^3 Y_{gt}^{(1)} + 32 \nu_g^3 \nu_t^2\, Y_{tg}^{(3)} + 32 \nu_t^3 \nu_g^2 Y_{gt}^{(3)}] \doteq \sigma^2 D_{gttg}\,.
\end{aligned} \tag{1.154}$$

Here we have used for brevity

$$X_{tg}^{(k)} \doteq P_{2,t-g}^{(k)} + P_{2,t+g}^{(k)} + \left(\frac{1}{\nu_t} - \frac{1}{\nu_g} - \frac{1}{\nu_t - \nu_g}\right) Q_{1,t-g}^{(k)} + \left(\frac{1}{\nu_t} + \frac{1}{\nu_g} - \frac{1}{\nu_t + \nu_g}\right) Q_{1,t+g}^{(k)} ,$$

$$Y_{tg}^{(k)} \doteq P_{1,t+g}^{(k)} - P_{1,t-g}^{(k)} + \frac{1}{\nu_t} (Q_{2,t-g}^{(k)} - Q_{2,t+g}^{(k)}) ,$$

$$Z_{tg}^{(k)} \doteq P_{2,t-g}^{(k)} - P_{2,t+g}^{(k)} + \frac{1}{\nu_t - \nu_g} Q_{1,t-g}^{(k)} - \frac{1}{\nu_t + \nu_g} Q_{1,t+g}^{(k)}$$

where

$$Q_{1p}^{(k)} \doteq \int_0^1 \varrho^{(k)}(t) \sin(p\pi t)\, dt ,$$

$$Q_{2p}^{(k)} \doteq \int_0^1 \varrho^{(k)}(t) \cos(p\pi t)\, dt ,$$

$$P_{1p}^{(k)} \doteq \int_0^1 \varrho^{(k)}(t)\, (1 - t) \sin(p\pi t)\, dt ,$$

$$P_{2p}^{(k)} \doteq \int_0^1 \varrho^{(k)}(t)\, (1 - t) \cos(p\pi t)\, dt .$$

For $k = 0$ the upper index is omitted, and we have put $R(t) \doteq \sigma^2 \varrho(t)$. As correlation function we shall use $R(u) = \sigma^2 \exp(-\alpha u^2)$. For the numerical computation of D_{gttg} it is advantageous to express $P_{1p}^{(k)}$, $P_{2p}^{(k)}$, $Q_{1p}^{(k)}$, $Q_{2p}^{(k)}$ by Q_{1p} and Q_{2p} since in this way the number of numerical integrations is considerably reduced. Consider

$$I_{n1} = \frac{1}{2\alpha} \left((n - 1)\, I_{n-2,1} + \nu_p I_{n-1,2}\right) ,$$

$$I_{n2} = \frac{1}{2\alpha} \left(-\beta_p + (n - 1)\, I_{n-2,2} - \nu_p I_{n-1,1}\right)$$

for $n \geqq 2$ and

$$I_{11} = \frac{1}{2\alpha} \nu_p Q_{2p} ,$$

$$I_{12} = \frac{1}{2\alpha} (1 - \beta_p - \nu_p Q_{1p}) ,$$

where I_{n1} and I_{n2} are defined by

$$I_{n1} \doteq \int_0^1 u^n \exp(-\alpha u^2) \sin(p\pi u)\, du ,$$

$$I_{n2} \doteq \int_0^1 u^n \exp(-\alpha u^2) \cos(p\pi u)\, du ,$$

and

$$\beta_p \doteq (-1)^p\, e^{-\alpha} .$$

We obtain

$$Q_{1p}^{(2)} = \nu_p(1-\beta_p) - \nu_p^2 Q_{1p}\,,$$

$$Q_{2p}^{(1)} = \beta_p - 1 + \nu_p Q_{1p}\,,$$

$$Q_{1p}^{(4)} = -\nu_p(1-\beta_p)\,(2\alpha + \nu_p^2) - (2\alpha)^2\,\nu_p\beta_p + \nu_p^4 Q_{1p}\,,$$

$$Q_{2p}^{(3)} = (2\alpha)^2\,\beta_p + (1-\beta_p)\,(2\alpha+\nu_p^2) - \nu_p^3 Q_{1p}\,,$$

$$P_{1p}^{(1)} = \frac{1}{2\alpha}\,\nu_p(1-\beta_p) - \nu_p Q_{2p} + \left(1 - \frac{1}{2\alpha}\,\nu_p^2\right) Q_{1p}\,,$$

$$P_{1p}^{(3)} = (1-\beta_p)\,\nu_p\left(2 - \frac{1}{2\alpha}\,\nu_p^2\right) + \nu_p^3 Q_{2p} + \nu_p^2\left(\frac{1}{2\alpha}\,\nu_p^2 - 3\right) Q_{1p}\,,$$

$$P_{2p} = \frac{1}{2\alpha}\,(\beta_p - 1) + Q_{2p} + \frac{1}{2\alpha}\,\nu_p Q_{1p}\,,$$

$$P_{2p}^{(2)} = (1-\beta_p)\left(\frac{1}{2\alpha}\,\nu_p^2 - 1\right) + \nu_p\left(2 - \frac{1}{2\alpha}\,\nu_p^2\right) Q_{1p} - \nu_p^2 Q_{2p}\,,$$

$$P_{2p}^{(4)} = (2\alpha)^2\,\beta_p + (1-\beta_p)\left(2\alpha + 3\nu_p^2 - \frac{1}{2\alpha}\,\nu_p^4\right) + \nu_p^4 Q_{2p} + \nu_p^3\left(\frac{1}{2\alpha}\,\nu_p^2 - 4\right) Q_{1p}\,.$$

From this

$$\langle\lambda_g\rangle - \mu_g = \sigma^2\left[\mu_g - 16\alpha\nu_g^2 - \sum_{\substack{t=1\\ t\neq g}}^{\infty} \frac{D_{gttg}}{\mu_{tg}}\right]$$

can be computed by relatively few numerical integrations, and it is easy to see that this series converges as $\sum_{t=1}^{\infty} 1/t^2$.

One obtains, with

$$(\!(M_{11}w_g, w_g)\!) = 2\mu_g \int_0^1 r(x,\omega)\, w_g^2(x)\,\mathrm{d}x\,,$$

the variance up to second order terms

$$\begin{aligned}\operatorname{var}\lambda_g &= \langle(\!(M_{11}w_g, w_g)\!)^2\rangle \\ &= 4\mu_g^2 \int_0^1\!\!\int_0^1 \langle r(x)\, r(y)\rangle\, w_g^2(x)\, w_g^2(y)\,\mathrm{d}x\,\mathrm{d}y \\ &= 4\mu_g^2\sigma^2(\!(\varrho(u), (1-u)\,\big(2+\cos(2g\pi u)\big) + \frac{3}{2g\pi}\sin(2g\pi u))\!)\end{aligned}$$

by using (1.133). The numerical results for various correlation functions $\varrho(u)$ can be obtained from Figs. 1.2a to 1.2c by multiplying the values given there by $4\mu_g^2$. We have the results of Table 1.6 for $g = 1$.

Table 1.6

	$\alpha = 1$	$\alpha = 4$
$\frac{1}{\sigma^2 \mu_1} (\langle \lambda_1 \rangle - \mu_1)$	1.3259	−13.9019
$\frac{1}{\sigma^2 \mu_1^2} \operatorname{var} \lambda_1$	1.1311	2.6858

Table 1.7

		$g = 1$	$g = 2$	$g = 3$	$g = 4$	$g = 5$
$\frac{1}{\sigma^2 \mu_g} (\langle \lambda_g \rangle - \mu_g)$	$\alpha=1$	0.2230	0.2621	0.2977	0.3138	0.3222
	$\alpha=2$	− 5.4007	−1.4338	−0.8297	−0.6258	−0.5318
	$\alpha=3$	−25.7700	−4.8839	−2.3096	−1.5492	−1.2256
$\frac{1}{\sigma^2 \mu_g^2} \operatorname{var} \lambda_g$	$\alpha=1$	3.2886	3.0317	2.9826	2.9648	2.9568
	$\alpha=2$	2.3540	1.9825	1.8959	1.8641	1.8487
	$\alpha=3$	1.7899	1.4724	1.3751	1.3358	1.3163

For the correlation function $R(u) = \sigma^2 \exp(-\alpha |u|)$ we get quite similar results, although the realizations of $r(x, \omega)$ with such a correlation function are not in general continuously differentiable. Table 1.7 gives some of these values. We remark that the computations are easier than in the case of $R(u) = \sigma^2 \exp(-\alpha u^2)$. If we use the formulas given for asymptotic calculations in Section 1.4.3.1 we obtain for our special case ($\delta_1 = \delta_2 = 0$) up to second order terms

$$\lim_{g \to \infty} \frac{1}{\mu_g} (\langle \lambda_g \rangle - \mu_g) = 5 \int_0^1 (1 - u)\, R(u)\, du - \frac{3}{2} R(0)\,.$$

For the correlation function $R(u) = \sigma^2 \exp(-3|u|/2)$ one obtains Table 1.8. For comparison, the exact computed values of second order and the asymptotic values for the correlation function $R(u) = \sigma^2 \exp(-3u^2/2)$ are also given. Corresponding to the considerations in Section 1.4.3.1 we have up to second

Table 1.8

$\frac{1}{\sigma^2} (\langle \lambda_g \rangle - \mu_g)$		$g = 5$	$g = 6$	$g = 10$	$g = 20$
$R = \sigma^2 \exp(-\frac{3}{2} \lvert u \rvert)$	exact	4087.5	9918.6	93686.0	1628299.0
	asymptotic	6511.5	13502.3	104184.4	1666951.3
deviation		59.3%	36.1%	11.0%	2.4%
$R = \sigma^2 \exp(-\frac{3}{2} u^2)$	asymptotic	31778.9	65896.8	508463.1	8135410.0

Table 1.9

	$g = 1$	$g = 3$	$g = 5$	$g = 6$	$g = 10$	$g = 20$
$\frac{1}{\sigma^2 \mu_g^2} \operatorname{var} \lambda_g$	3.00479	2.62259	2.59002	2.58410	2.57580	2.57230
Deviation of the exact value from the asymptotic value	14.8%	2.0%	0.7%	0.5%	0.2%	0.0%

order terms

$$\lim_{g \to \infty} \frac{1}{\mu_g^2} \operatorname{var} \lambda_g = 8 \int_0^1 (1 - u)\, R(u)\, du\,,$$

in particular, we obtain for $R(u) = \sigma^2 \exp(-3|u|/2)$ the asymptotic value

$$\lim_{g \to \infty} \frac{1}{\mu_g^2} \operatorname{var} \lambda_g = 2.57113\sigma^2\,.$$

Comparisons to the exact variance up to second order are given in Table 1.9. Using the method given in Section 1.4.1 to compute the mean values and correlation quantities of eigenvalues and eigenfunctions such cases can also be dealt with in which, e.g., the beam is elastically supported. There, the boundary conditions contain $r(x, \omega)$, and the correlation between the random variables in the differential equation and those in the boundary conditions must be considered. Without basic difficulties, numerical results can also be obtained in these cases.

2. Limit theorems

2.1. Limit theorems for functionals of weakly correlated fields

2.1.1. Weakly correlated fields

We consider random fields $f(x, \omega)$, $g(x, \omega)$, ... , where $x \in \mathcal{D}$, and $\mathcal{D}$ is an arbitrary subset of R^m. $|.|$ denotes the norm on R^m. The mean of a random field is denoted by $\langle f \rangle$.

Definition 2.1. Let $\{x_i, i \in I\}$, $I = \{1, 2, \dots, k\}$ be a finite set of points from R^m and $\varepsilon > 0$ an arbitrary real number. A subset

$$\{x_i, i \in \bar{I} \subset I\}, \qquad \bar{I} = \{i_1, i_2, \dots, i_l\}$$

is said to be *ε-adjoining* if a permutation $\begin{pmatrix} i_1 i_2 \dots i_l \\ j_1 j_2 \dots j_l \end{pmatrix}$ of $\bar{I}$ exists with

$$|x_{j_s} - x_{j_{s+1}}| \leqq \varepsilon, \qquad s = 1, 2, \dots, l-1.$$

A subset of one point is always called ε-adjoining. Furthermore, the subset $\{x_i, i \in \bar{I} \subset I\}$ is said to be *maximum ε-adjoining* relative to $\{x_i, i \in I\}$ if it is ε-adjoining but the subset $\{x_i, i \in \bar{I} \subset I\} \cup \{x_r\}$ is not ε-adjoining for every $x_r \in \{x_i, i \in I \setminus \bar{I}\}$.

We can now easily give the following result for the decomposition of a finite set into maximum ε-adjoining subsets.

Lemma 2.1. *Every finite set $\{x_i, i \in I\}$ of points from R^m decomposes uniquely into mutually exclusive maximum ε-adjoining subsets.*

Proof. We define the relation "$\sim$" in the set $\{x_i, i \in I\}$: $x \sim y$ if a ε-adjoining subset $\{x_i, i \in \bar{I} \subset I\}$ exists with $x, y \in \{x_i, i \in \bar{I}\}$. It easily follows that "$\sim$" is an equivalence relation and the equivalence classes are the maximum ε-adjoining subsets of $\{x_i, i \in I\}$. ◀

Let us now give the definition of a weakly correlated field.

Definition 2.2. A random field $f_\varepsilon(x, \omega)$, $x \in \mathcal{D} \subset R^m$, with $\langle f_\varepsilon(x) \rangle = 0$, is called *weakly correlated* with correlation length ε if the relation

$$\langle \prod_{i \in I} f_\varepsilon(x_i) \rangle = \prod_{j=1}^{p} \langle \prod_{i \in I_j} f_\varepsilon(x_i) \rangle$$

is satisfied for all k - th moments, $k = 2, 3, \ldots$, where $I = \{1, 2, \ldots, k\}$ and

$$\{x_i,\, i \in I_1\},\ \{x_i,\, i \in I_2\},\ \ldots,\ \{x_i,\, i \in I_p\}$$

with $\bigcup_{j=1}^{p} I_j = I$ denotes the decomposition of $\{x_i,\, i \in I\}$ in the maximum ε-adjoining subsets.

Let $f_\varepsilon(x, \omega)$ be a weakly correlated random field with correlation length ε. Then we have for the correlation function

$$\langle f_\varepsilon(x_1)\, f_\varepsilon(x_2)\rangle = \begin{cases} R_\varepsilon(x_1, x_2) & \text{for} \quad |x_1 - x_2| \leqq \varepsilon\,, \\ 0 & \text{for} \quad |x_1 - x_2| > \varepsilon\,, \end{cases}$$

i.e. the correlation function is zero if $|x_1 - x_2| > \varepsilon$.

We can prove the following theorem concerning the independence of the random field $f_\varepsilon(x, \omega)$ at the points x_1, x_2 where the distance between x_1 and x_2 is greater than ε.

Theorem 2.1. *Let $f_\varepsilon(x, \omega)$ be a weakly correlated field, and let ε denote the correlation length. We assume that $\sum_{k=0}^{\infty} 1/k!\, \langle |f_\varepsilon(x)|^k\rangle < \infty$ for every $x \in \mathcal{D}$. The set $\{x_i,\, i \in I\}$ with $x_i \in \mathcal{D}$ possesses the decomposition*

$$\{x_i,\, i \in I_1\},\ \{x_i,\, i \in I_2\},\ \ldots,\ \{x_i,\, i \in I_p\}$$

in the maximum ε-adjoining subsets. Then the random variables

$$f_\varepsilon(x_i, \omega),\ i \in I_r \quad \text{and} \quad f_\varepsilon(x_i, \omega),\ i \in I_s$$

are independent for $r \neq s$, $r, s \in \{1, 2, \ldots, p\}$.

In particular, the random variables $f_\varepsilon(x, \omega)$ and $f_\varepsilon(y, \omega)$ are independent if $|x - y| > \varepsilon$.

Proof. Denote

$$I_r = \{i_{r1}, \ldots, i_{rk_r}\}\,, \qquad r = 1, 2, \ldots, p\,,$$

and let $\xi_{rt} \doteq f_\varepsilon(x_{i_{rt}})$ where $r = 1, 2, \ldots, p$; $t = 1, 2, \ldots, k_r$ and s_{rt} are real numbers. Because of the property of weak correlation, we have

$$\begin{aligned}\langle \exp\,(i \sum_{r=1}^{p} \sum_{t=1}^{k_r} \xi_{rt} s_{rt})\rangle &= \left\langle \sum_{l=0}^{\infty} \frac{i^l}{l!} (\sum_{r=1}^{p} \sum_{t=1}^{k_r} \xi_{rt} s_{rt})^l \right\rangle \\ &= \sum_{l=0}^{\infty} \frac{i^l}{l!} \sum_{l_1 + \ldots + l_p = l} \frac{l!}{l_1! \ldots l_p!} \prod_{j=1}^{p} \langle (\sum_{t=1}^{k_j} \xi_{jt} s_{jt})^{l_j}\rangle\,.\end{aligned}$$

Furthermore, we have

$$\begin{aligned}\langle \exp\,(i \sum_{r=1}^{p} \sum_{t=1}^{k_r} \xi_{rt} s_{rt})\rangle &= \prod_{r=1}^{p} \left\langle \sum_{l_r=0}^{\infty} \frac{i^{l_r}}{l_r!} (\sum_{t=1}^{k_r} \xi_{rt} s_{rt})^{l_r} \right\rangle \\ &= \prod_{r=1}^{p} \langle \exp\,(i \sum_{t=1}^{k_r} \xi_{rt} s_{rt})\rangle\end{aligned}$$

and therefore the independence of the above random variables is proved. ◀

We need the following lemma for the proof of the existence of weakly correlated fields.

Lemma 2.2. *The function*

$$R_{2a}(z) \doteq (f_a * f_a)(z), \quad \text{with} \quad z = x_2 - x_1,\ x_1, x_2 \in \mathbb{R}^m,$$

is a correlation function of a wide-sense homogeneous random field possessing the property

$$R_{2a}(z) = 0, \quad \text{for} \quad |z| > 2a,$$

where

$$f_a(x) = f_a(|x|) = \begin{cases} \sigma \exp\left(\dfrac{1}{|x|^2 - a^2}\right) & \text{for} \quad |x| < a, \\ 0 & \text{for} \quad |x| \geqq a \end{cases} \qquad (\sigma > 0)$$

and $*$ stands for the operation of the convolution.

Proof. We first remark that $f_a(x) \in \mathbf{C}_0^\infty(\mathbb{R}^m)$ and

$$(f_a * f_a)(z) = \int_{\mathbb{R}^m} f_a(|y|) f_a(|z - y|)\, \mathrm{d}y$$

$$= \sigma^2 \int_{\mathcal{K}_a(0) \cap \mathcal{K}_a(z)} \exp\left(\frac{1}{|y|^2 - a^2}\right) \exp\left(\frac{1}{|z - y|^2 - a^2}\right) \mathrm{d}y$$

where $\mathcal{K}_r(x)$ is the sphere with centre x and radius r. The relation $(f_a * f_a)(z) = 0$ for $|z| > 2a$ is clear. The Fourier transform $\mathfrak{F}[R_{2a}]$ is non-negative, i.e.

$$\mathfrak{F}[R_{2a}] = \mathfrak{F}[f_a * f_a] = (\mathfrak{F}[f_a])^2 \geqq 0,$$

and $R_{2a}(z)$ has the properties of a correlation function of a wide-sense homogeneous random field. ◀

Theorem 2.2. *For any real number $\varepsilon > 0$ a weakly correlated Gaussian field exists with correlation length ε. A Gaussian field $u(x, \omega)$ satisfying $\langle u(x) \rangle = 0$ and $\langle u(x_1)\, u(x_2) \rangle = 0$ for $|x_1 - x_2| > \varepsilon$ is a weakly correlated field with correlation length ε', $\varepsilon' \leqq \varepsilon$.*

Proof. We consider a centered Gaussian field $u(x, \omega)$ whose correlation function $\langle u(x_1)\, u(x_2) \rangle$ is zero for $|x_1 - x_2| > \varepsilon$. The existence of such a correlation function follows from Lemma 2.2 with $a = \varepsilon/2$. We will prove that this random field is weakly correlated.

The k-th moments of a Gaussian field have the structure

$$\langle u_1 u_2 \ldots u_k \rangle$$

$$= \begin{cases} \sum\limits_{(s_1, s_2), \ldots, (s_{k-1}, s_k)} \langle u_{s_1} u_{s_2} \rangle \langle u_{s_3} u_{s_4} \rangle \ldots \langle u_{s_{k-1}} u_{s_k} \rangle, & \text{for} \quad k \text{ even}, \\ 0, & \text{for} \quad k \text{ odd} \end{cases} \tag{2.1}$$

where we have written u_i for $u(x_i)$. The sum above is taken over all nonequivalent decompositions of $\{1, 2, \ldots, k\}$ in pairs. Two decompositions are said to be equivalent if they are only distinguished by a permutation of the pairs and a permutation of the elements in the pairs. The number of terms in the sum is $1 \cdot 3 \cdot 5 \cdot \ldots \cdot (k-3) \cdot (k-1) = (k-1)!!$. The set $\{x_1, \ldots, x_k\} = \{x_i, i \in I\}$ is decomposed into the maximum ε-adjoining subsets $\{x_i, i \in I_1\}, \ldots, \{x_i, i \in I_p\}$. Hence, we must prove

$$\langle u_1 \ldots u_k \rangle = \prod_{j=1}^{p} \langle \prod_{i \in I_j} u_i \rangle , \tag{2.2}$$

where $I_j \doteq \{i_{j1}, \ldots, i_{jk_j}\}$ and hence $k = \sum_{i=1}^{p} k_i$.

Let k be odd. Then the left-hand side of (2.2) vanishes because of (2.1). From $k = \sum_{i=1}^{p} k_i$ we obtain that at least one of the k_i is odd and again because of (2.1) at least one factor on the right-hand side of (2.2) is zero. We have proved (2.2) with k odd. Let k be even and at least one of the k_i odd. Then the right-hand side of (2.2) vanishes. We consider (2.1) for the left-hand side of (2.2). In every decomposition of $\{1, 2, \ldots, k\}$ in pairs there exists a pair $\{i_l, i_{l+1}\}$ possessing the property that x_{i_l} and $x_{i_{l+1}}$ are not ε-adjoining, and therefore $\langle u_{i_l} u_{i_{l+1}} \rangle = 0$. Hence, we have $\langle u_1 \ldots u_k \rangle = 0$ and (2.2) is proved in this case. Let k and $k_1, \ldots, k_p$ be even. From (2.1)

$$\prod_{j=1}^{p} \langle \prod_{i \in I_j} u_i \rangle$$

$$= \prod_{j=1}^{p} \sum_{(s_1, s_2), \ldots, (s_{k_j-1}, s_{k_j})} \langle u_{i_{js_1}} u_{i_{js_2}} \rangle \ldots \langle u_{i_{js_{k_j-1}}} u_{i_{js_{k_j}}} \rangle \tag{2.3}$$

$$= \sum_{(q_1, q_2), \ldots, (q_{k-1}, q_k)} \langle u_{q_1} u_{q_2} \rangle \ldots \langle u_{q_{k-1}} u_{q_k} \rangle \tag{2.4}$$

is obtained. The sum in (2.4) is taken over all nonequivalent decompositions of $\{1, 2, \ldots, k\}$ in pairs for which $\{x_{q_i}, x_{q_{i+1}}\}$ is ε-adjoining for $i = 1, 3, \ldots, k-1$. We will prove this, and further suppose that a decomposition exists which is not contained in the sum of (2.4). The pertinent product is decomposed into subproducts so that a subproduct contains those factors whose pertinent point pairs form a maximum ε-adjoining subset. The contradiction to the decomposition in (2.3) follows from the uniqueness of the decomposition of a set in maximum ε-adjoining subsets. ◀

For applications to differential equations we need weakly correlated fields where the sample functions are bounded and differentiable.

Theorem 2.3. *Let ε be any real number, $\varepsilon > 0$. Then weakly correlated fields exist possessing sample continuous partial derivatives of any order. The random field $v(x, \omega) \doteq h(u(x, \omega)) - \langle h(u(x, \omega)) \rangle$ is a weakly correlated field if $\bar{u}(x, \omega) = u(x, \omega) - \langle u(x) \rangle$ is weakly correlated, and the moments $\langle h(u(x))^k \rangle$, $k = 1, 2, \ldots$, $x \in \mathcal{D} \subset \mathbb{R}^m$, exist for the real function $h(t)$ on $\mathbb{R}^1$.*

Proof. The correlation function constructed in Lemma 2.2 belongs to $C_0^\infty(\mathbb{R})$. Thus, the first part of this theorem follows from the connection between the derivatives of a random field and the derivatives of the correlation function. We obtain the second part of this theorem by means of Theorem 2.1. ◀

We now give the definition of a weakly correlated connected random field.

Definition 2.3. A *random vector field* $(f_{1\varepsilon}(x, \omega), f_{2\varepsilon}(x, \omega))$, $x \in \mathcal{D} \subset \mathbb{R}^m$, with $\langle f_{1\varepsilon}(x)\rangle \equiv \langle f_{2\varepsilon}(x)\rangle \equiv 0$ is said to be *weakly correlated connected* with correlation length ε if

$$\langle \prod_{i\in I} f_{r_i\varepsilon}(x_i)\rangle = \prod_{j=1}^{p} \langle \prod_{i\in I_j} f_{r_i\varepsilon}(x_i)\rangle \tag{2.5}$$

for all $k \geqq 1$, where $I_j = \{i_{j1}, i_{j2}, \dots, i_{jk_j}\}, j = 1, 2, \dots, p, \sum_{j=1}^{p} k_j = k, r_i = 1, 2$ and $\{x_i, i \in I_1\}, \dots, \{x_i, i \in I_p\}$ denotes the decomposition of $\{x_1, \dots, x_k\} = \{x_i, i \in I\}$ in maximum ε-adjoining subsets.

The definition of a weakly correlated connected vector field implies the weak correlation of the components of the vector field (e.g. $r_i = 1$ for $i = 1, 2, \dots, k$).

Theorem 2.4. *A random vector field $(f_{1\varepsilon}(x, \omega), f_{2\varepsilon}(x, \omega))$ is weakly correlated connected if the random fields $f_{1\varepsilon}(x, \omega)$ and $f_{2\varepsilon}(x, \omega)$ are independent and weakly correlated.*

Proof. We prove the relation (2.5) for $\langle \prod_{i\in I} f_{r_i\varepsilon}(x_i)\rangle$. Denote by $\{x_i, i \in I_1\}, \dots, \{x_i, i \in I_p\}$ the decomposition of $\{x_i, i \in I\}$ in maximum ε-adjoining subsets, $\{x_i, i \in J'\}$, $J' \subset I$ those points x_i for which $r_i = 1$, and $J'' = I \setminus J' \subset I$. Furthermore,

$$\{x_i, i \in J_1'\}, \dots, \{x_i, i \in J_p'\}, \bigcup_{j=1}^{p} J_j' = J'$$

and

$$\{x_i, i \in J_1''\}, \dots, \{x_i, i \in J_p''\}, \bigcup_{j=1}^{p} J_j'' = J''$$

are the decompositions of $\{x_i, i \in J'\}$ and $\{x_i, i \in J''\}$, respectively, where $\{x_i, i \in J_r'\} \cup \{x_i, i \in J_r''\} = \{x_i, i \in I_r\}$. It is possible that $J_r' = \emptyset$ or $J_r'' = \emptyset$, respectively. From the independence and the weak correlation of $f_{1\varepsilon}, f_{2\varepsilon}$ we obtain

$$\begin{aligned}
\langle \prod_{i\in I} f_{r_i\varepsilon}(x_i)\rangle &= \langle \prod_{i\in J'} f_{1\varepsilon}(x_i)\rangle \langle \prod_{i\in J''} f_{2\varepsilon}(x_i)\rangle \\
&= \prod_{s=1}^{p} \langle \prod_{i\in J_s'} f_{1\varepsilon}(x_i)\rangle \langle \prod_{i\in J_s''} f_{2\varepsilon}(x_i)\rangle \\
&= \prod_{s=1}^{p} \langle \prod_{i\in J_s'} f_{1\varepsilon}(x_i) \prod_{i\in J_s''} f_{2\varepsilon}(x_i)\rangle \\
&= \prod_{s=1}^{p} \langle \prod_{i\in I_s} f_{r_i\varepsilon}(x_i)\rangle
\end{aligned}$$

which completes the proof. ◀

A simple example of a weakly correlated connected vector field is given by $(\bar{h}(u(x, \omega)), \bar{g}(u(x, \omega)))$, where $\bar{h} \doteq h - \langle h \rangle$, $\bar{g} \doteq g - \langle g \rangle$ and $u(x, \omega)$ is a weakly correlated random field. The real functions h and g satisfy the properties of Theorem 2.3.

A generalization of the above definition to random vector fields $(f_{1\varepsilon}(x, \omega), \dots, f_{l\varepsilon}(x, \omega))$, $l \geqq 2$, can be performed in a straightforward fashion.

2.1.2. A limit theorem for linear functionals

Linear functionals of the form

$$r_{i\varepsilon}(x, \omega) = \frac{1}{\sqrt{\varepsilon^m}} \int_{\mathcal{D}} F_i(x, y) f_\varepsilon(y, \omega) \, dy , \qquad i = 1, 2, \dots , l , \tag{2.6}$$

arise in the solutions of random differential equation problems in which weakly correlated fields $f_\varepsilon(x, \omega)$ appear as coefficients or as the nonhomogeneous term. Let $\mathcal{D}$ be a bounded domain $\mathcal{D} \subset \mathbb{R}^m$ with smooth boundary and let $F_i(x, y)$, for $i = 1, 2, \dots , l$, be functions on $\mathcal{G} \times \mathcal{D} \subset \mathbb{R}^p \times \mathbb{R}^m$ where $F_i(x, y) \in \mathbf{L}_2(\mathcal{D})_y$ for all $x \in \mathcal{G}$. $\mathbf{L}_2(\mathcal{D})$ denotes the space of all square integrable functions on $\mathcal{D}$. We assume that $f_\varepsilon(x, \omega)$ is a weakly correlated field, whose sample functions are almost surely continuous. In general, continuity of the sample functions is necessary for the existence of an almost sure solution of a random differential equation

We will prove the convergence in distribution of the functionals in (2.6) to a Gaussian field $\xi_i(x, \omega)$. The proof uses the method of moments; and the next theorem establishes the convergence for the second moments.

Theorem 2.5. *Assume that* (i) *the deterministic functions* $F_i(x, y) \in \mathbf{L}_2(\mathcal{D})_y$ *for* $i = 1, 2$, (ii) *the weakly correlated field* $f_\varepsilon(x, \omega)$ *on* $\mathcal{D} \subset \mathbb{R}^m$ *possesses continuous sample functions, and* (iii) $\langle f_\varepsilon(x)^2 \rangle \leqq c_2 < \infty$. *Then for the second moments*

$$\lim_{\varepsilon \downarrow 0} \langle r_{1\varepsilon}(x_1) \, r_{2\varepsilon}(x_2) \rangle = \int_{\mathcal{D}} F_1(x_1, y) \, F_2(x_2, y) \, a(y) \, dy \tag{2.7}$$

holds where the $r_{i\varepsilon}$ *are defined by* (2.6), *and the function* $a(y)$ *is given by*

$$a(y) \doteq \lim_{\varepsilon \downarrow 0} \frac{1}{\varepsilon^m} \int_{\mathcal{K}_\varepsilon(0)} \langle f_\varepsilon(y) \, f_\varepsilon(y + z) \rangle \, dz , \tag{2.8}$$

where $\mathcal{K}_\varepsilon(x) = \{z \in \mathbb{R}^m : |z - x| \leqq \varepsilon\}$. $a(y)$ *is called the intensity of the weakly correlated field* $f_\varepsilon(x, \omega)$.

Remark 2.1. *Let* $f_\varepsilon(x, \omega)$ *be a wide-sense homogeneous field. Then we have for the intensity*

$$a(y) \equiv a = \lim_{\varepsilon \downarrow 0} \frac{1}{\varepsilon^m} \int_{\mathcal{K}_\varepsilon(0)} \langle f_\varepsilon(0) \, f_\varepsilon(z) \rangle \, dz .$$

If $R_\varepsilon(x, y) = \langle f_\varepsilon(x) \, f_\varepsilon(y) \rangle$ *is continuous at* (x, x) *then the intensity is a continuous function.*

The proof of the first statement follows immediately from the definition of the intensity. The continuity is the result of the inequalities

$$\left|\frac{1}{\varepsilon^m}\int\limits_{\mathcal{K}_\varepsilon(0)}[\langle f_\varepsilon(x_1)\,f_\varepsilon(x_1+y)\rangle - \langle f_\varepsilon(x_2)\,f_\varepsilon(x_2+y)\rangle]\,\mathrm{d}y\right|$$

$$\leqq \frac{1}{\varepsilon^m}\int\limits_{\mathcal{K}_\varepsilon(0)}[\sqrt{\langle f_\varepsilon^2(x_1+y)\rangle}\,\sqrt{\langle(f_\varepsilon(x_1)-f_\varepsilon(x_2))^2\rangle}$$

$$+\sqrt{\langle f_\varepsilon^2(x_2)\rangle}\,\sqrt{\langle(f_\varepsilon(x_1+y)-f_\varepsilon(x_2+y))^2\rangle}\,]\,\mathrm{d}y$$

$$\leqq 2\sqrt{c_2}\,V_1\sqrt{\eta}\,,$$

where $x_2 \in \mathcal{K}_\delta(x_1)$ und $\varepsilon \leqq \varepsilon_0$, because continuity in mean square of $f_\varepsilon(x, \omega)$ follows from the continuity of $R_\varepsilon(x, y)$ at (x, x). V_1 denotes the volume of the sphere with unit radius.

Proof of Theorem 2.5. This proof is given in three sections. First we give the proof for $F_i(x, y) \equiv \mathbf{1}_{\mathcal{B}_i}(y)$, for $i = 1, 2$, where the Lebesgue measure λ of the boundary $\partial\mathcal{B}_i$ of $\mathcal{B}_i$ is assumed to be zero and $\mathbf{1}_{\mathcal{B}}(y)$ is defined by $\mathbf{1}_{\mathcal{B}}(y) = \begin{cases} 1 & \text{for } y \in \mathcal{B} \\ 0 & \text{for } y \notin \mathcal{B} \end{cases}$. Then we establish the assertion of this theorem for continuous functions; and finally for $F_i(x, y) \in \mathbf{L}_2(\mathcal{D})_y$.

(1) We prove

$$\lim_{\varepsilon\downarrow 0}\frac{1}{\varepsilon^m}\int\limits_{\mathcal{B}_1}\int\limits_{\mathcal{B}_2}\langle f_\varepsilon(y_1)\,f_\varepsilon(y_2)\rangle\,\mathrm{d}y_1\,\mathrm{d}y_2 = \int\limits_{\mathcal{B}_1\cap\mathcal{B}_2} a(y)\,\mathrm{d}y \tag{2.9}$$

where $\mathcal{B}_1$ and $\mathcal{B}_2$ are subsets of $\mathcal{D}$ with $\lambda(\partial\mathcal{B}_1) = \lambda(\partial\mathcal{B}_2) = 0$. It is easy to see that

$$\frac{1}{\varepsilon^m}\int\limits_{\mathcal{B}_1}\int\limits_{\mathcal{B}_2} R_\varepsilon(y_1, y_2)\,\mathrm{d}y_1\,\mathrm{d}y_2 = \frac{1}{\varepsilon^m}\int\limits_{\mathcal{B}_1\setminus\partial\mathcal{B}_2}\left(\int\limits_{\mathcal{K}_\varepsilon(y_1)\cap\mathcal{B}_2} R_\varepsilon(y_1, y_2)\,\mathrm{d}y_2\right)\mathrm{d}y_1\,,$$

since $\lambda(\partial\mathcal{B}_2) = 0$ where $R_\varepsilon(y_1, y_2) \doteq \langle f_\varepsilon(y_1)\,f_\varepsilon(y_2)\rangle$. With the aid of

$$\lim_{\varepsilon\downarrow 0}\frac{1}{\varepsilon^m}\int\limits_{\mathcal{K}_\varepsilon(y_1)\cap\mathcal{B}_2} R_\varepsilon(y_1, y_2)\,\mathrm{d}y_2 = \begin{cases} a(y_1) & \text{for } y_1 \in \mathcal{B}_1 \cap \mathcal{B}_2 \setminus (\partial\mathcal{B}_1 \cup \partial\mathcal{B}_2)\,, \\ 0 & \text{for } y_1 \in \mathcal{B}_1 \setminus (\mathcal{B}_2 \cup \partial\mathcal{B}_1 \cup \partial\mathcal{B}_2) \end{cases}$$

we obtain

$$\lim_{\varepsilon\downarrow 0}\frac{1}{\varepsilon^m}\int\limits_{\mathcal{K}_\varepsilon(y_1)\cap\mathcal{B}_2} R_\varepsilon(y_1, y_2)\,\mathrm{d}y_2 = a(y_1)\,\mathbf{1}_{\mathcal{B}_1\cap\mathcal{B}_2}(y_1)$$

almost surely for all $y_1 \in \mathcal{B}_1$. Furthermore, we have

$$\left|\frac{1}{\varepsilon^m}\int\limits_{\mathcal{K}_\varepsilon(y_1)\cap\mathcal{B}_2} R_\varepsilon(y_1, y_2)\,\mathrm{d}y_2\right| \leqq V_1 c_2\,.$$

Applying Lebesgue's theorem, we can show that

$$\lim_{\varepsilon \downarrow 0} \int\limits_{\mathscr{B}_1 \setminus \partial \mathscr{B}_2} \left(\frac{1}{\varepsilon^m} \int\limits_{\mathscr{K}_\varepsilon(y_1) \cap \mathscr{B}_2} R_\varepsilon(y_1, y_2)\, \mathrm{d}y_2 \right) \mathrm{d}y_1 = \int\limits_{\mathscr{B}_1 \cap \mathscr{B}_2} a(y_1)\, \mathrm{d}y_1 .$$

The first part of this theorem is proved.

(2) We now prove (2.7), with respect to y for continuous functions $F_1(x, y)$ and $F_2(x, y)$ on the compact domain $\mathscr{D}$. Let $t_i(y)$, for $i = 1, 2$, be step functions on $\mathscr{D}$, i.e.

$$t_i(y) = \sum_{k=1}^{m_i} a_{ik} \mathbf{1}_{\mathscr{C}_{ik}}(y)$$

where $\mathscr{C}_{ik} \subset \mathscr{D}$ and $\lambda(\partial \mathscr{C}_{ik}) = 0$. It can be easily seen, using (2.9), that

$$\lim_{\varepsilon \downarrow 0} \frac{1}{\varepsilon^m} \int\limits_{\mathscr{D}} \int\limits_{\mathscr{D}} t_1(y_1)\, t_2(y_2)\, \langle f_\varepsilon(y_1) f_\varepsilon(y_2) \rangle\, \mathrm{d}y_2\, \mathrm{d}y_1$$

$$= \lim_{\varepsilon \downarrow 0} \sum_{k, l=1}^{m_1, m_2} a_{1k} a_{2l} \frac{1}{\varepsilon^m} \int\limits_{\mathscr{C}_{1k}} \int\limits_{\mathscr{C}_{2l}} R_\varepsilon(y_1, y_2)\, \mathrm{d}y_2\, \mathrm{d}y_1$$

$$= \sum_{k, l=1}^{m_1, m_2} a_{1k} a_{2l} \int\limits_{\mathscr{C}_{1k} \cap \mathscr{C}_{2l}} a(y)\, \mathrm{d}y = \int\limits_{\mathscr{D}} t_1(y)\, t_2(y)\, a(y)\, \mathrm{d}y .$$

This equation establishes (2.7) for step functions.

Now we consider step functions $t_1(y)$ and $t_2(y)$ associated with the continuous functions $F_i(y) \doteq F_i(x_i, y)$, with $|F_i(y)| \leq A_i$, for $i = 1, 2$, where

$$|t_i(y) - F_i(y)| < \eta , \quad \text{for all} \quad y \in \mathscr{D} .$$

Let

$$\bar{r}_{i\varepsilon}(\omega) \doteq \frac{1}{\sqrt{\varepsilon^m}} \int\limits_{\mathscr{D}} t_i(y) f_\varepsilon(y)\, \mathrm{d}y , \qquad r_{i\varepsilon} \doteq r_{i\varepsilon}(x_i)$$

then we have

$$|\langle r_{1\varepsilon} r_{2\varepsilon} \rangle - \int\limits_{\mathscr{D}} F_1(y)\, F_2(y)\, a(y)\, \mathrm{d}y|$$

$$\leq |\langle r_{1\varepsilon} r_{2\varepsilon} \rangle - \langle \bar{r}_{1\varepsilon} \bar{r}_{2\varepsilon} \rangle| + |\langle \bar{r}_{1\varepsilon} \bar{r}_{2\varepsilon} \rangle - \int\limits_{\mathscr{D}} t_1(y)\, t_2(y)\, a(y)\, dy|$$

$$+ | \int\limits_{\mathscr{D}} (t_1(y)\, t_2(y) - F_1(y)\, F_2(y))\, a(y)\, \mathrm{d}y| .$$

The first summand on the right-hand side of the inequality above can be estimated by

$$|\langle r_{1\varepsilon} r_{2\varepsilon} \rangle - \langle \bar{r}_{1\varepsilon} \bar{r}_{2\varepsilon} \rangle| \leq \eta(A_1 + A_2 + \eta) \frac{1}{\varepsilon^m} \int\limits_{\mathscr{D}} \int\limits_{\mathscr{K}_\varepsilon(y_1) \cap \mathscr{D}} |R_\varepsilon(y_1, y_2)|\, \mathrm{d}y_2\, \mathrm{d}y_1$$

$$\leq V_1 V(\mathscr{D})\, c_2 \eta (A_1 + A_2 + \eta)$$

and the third summand by

$$|\int_{\mathcal{D}} (t_1(y)\, t_2(y) - F_1(y)\, F_2(y))\, a(y)\, dy| \leqq V_1 V(\mathcal{D})\, c_2 \eta (A_1 + A_2 + \eta)\,,$$

where we have used $|a(y)| \leqq V_1 c_2$. $V(\mathcal{D})$ denotes the volume of $\mathcal{D}$. The second summand converges to zero as $\varepsilon \downarrow 0$ and we obtain

$$\lim_{\varepsilon \downarrow 0} |\langle r_{1\varepsilon} r_{2\varepsilon}\rangle - \int_{\mathcal{D}} F_1(y)\, F_2(y)\, a(y)\, dy| \leqq V_1 c_2 V(\mathcal{D})\, 2\eta (A_1 + A_2 + \eta)\,.$$

(2.7) is proved for continuous functions $F_1(x_1, y)$ and $F_2(x_2, y)$ because η is an arbitrary positive real number.

(3) In this part we assume that $F_i(y) \in \mathbf{L}_2(\mathcal{D})$ for $i = 1, 2$. Then a sequence $\{F_i^k(y)\}_{k=1,2,\ldots}$ of continuous functions $F_i^k(y)$ exists so that

$$\lim_{k \to \infty} ||F_i - F_i^k|| = 0 \quad \text{for} \quad i = 1, 2$$

where $||.||$ denotes the norm in $\mathbf{L}_2(\mathcal{D})$. Defining

$$r_{i\varepsilon}^k \doteq \frac{1}{\sqrt{\varepsilon^m}} \int_{\mathcal{D}} F_i^k(y)\, f_\varepsilon(y)\, dy$$

it follows that

$$\begin{aligned} &|\langle r_{1\varepsilon} r_{2\varepsilon}\rangle - \int_{\mathcal{D}} F_1(y)\, F_2(y)\, a(y)\, dy| \\ &\leqq |\langle r_{1\varepsilon} r_{2\varepsilon}\rangle - \langle r_{1\varepsilon}^k r_{2\varepsilon}^k\rangle| + |\langle r_{1\varepsilon}^k r_{2\varepsilon}^k\rangle - \int_{\mathcal{D}} F_1^k(y)\, F_2^k(y)\, a(y)\, dy| \\ &\qquad + |\int_{\mathcal{D}} (F_1^k(y)\, F_2^k(y) - F_1(y)\, F_2(y))\, a(y)\, dy|\,. \end{aligned} \tag{2.10}$$

The second summand of the right-hand side of (2.10) converges to zero as $\varepsilon \downarrow 0$ if k is a fixed number. For the third summand we can estimate

$$\begin{aligned} &|\int_{\mathcal{D}} (F_1^k(y)\, F_2^k(y) - F_1(y)\, F_2(y))\, a(y)\, dy| \\ &\leqq V_1 c_2 \,(||F_1^k||\, ||F_2^k - F_2|| + ||F_2||\, ||F_1^k - F_1||)\,. \end{aligned}$$

Finally, we obtain for the first summand

$$\begin{aligned} &|\langle r_{1\varepsilon} r_{2\varepsilon}\rangle - \langle r_{1\varepsilon}^k r_{2\varepsilon}^k\rangle| \\ &\leqq \frac{1}{\varepsilon^m} \left| \int_{\mathcal{D}}\int_{\mathcal{D}} (F_1(y_1)\, F_2(y_2) - F_1^k(y_1)\, F_2^k(y_2))\, R_\varepsilon(y_1, y_2)\, dy_2\, dy_1 \right| \\ &\leqq \frac{1}{\varepsilon^m} \int_{\mathcal{D}}\int_{\mathcal{D}} |F_1(y_1)|\, |F_2(y_2) - F_2^k(y_2)|\, |R_\varepsilon(y_1, y_2)|\, dy_2\, dy_1 \\ &\qquad + \frac{1}{\varepsilon^m} \int_{\mathcal{D}}\int_{\mathcal{D}} |F_2^k(y_2)|\, |F_1(y_1) - F_1^k(y_1)|\, |R_\varepsilon(y_1, y_2)|\, dy_2\, dy_1 \end{aligned}$$

$$\leqq \frac{1}{\varepsilon^m} \int\limits_{\mathcal{D}} |F_2(y_2) - F_2^k(y_2)| \int\limits_{\mathcal{K}_\varepsilon(y_2)\cap\mathcal{D}} |F_1(y_1)|\, |R_\varepsilon(y_1, y_2)|\, dy_1\, dy_2$$

$$+ \frac{1}{\varepsilon^m} \int\limits_{\mathcal{D}} |F_1(y_1) - F_1^k(y_1)| \int\limits_{\mathcal{K}_\varepsilon(y_1)\cap\mathcal{D}} |F_2^k(y_2)|\, |R_\varepsilon(y_1, y_2)|\, dy_2\, dy_1$$

$$\leqq \frac{1}{\varepsilon^m} ||F_2 - F_2^k|| \left\| \int\limits_{\mathcal{K}_\varepsilon(y_2)\cap\mathcal{D}} |F_1(y_1)|\, |R_\varepsilon(y_1, y_2)|\, dy_1 \right\|$$

$$+ \frac{1}{\varepsilon^m} ||F_1 - F_1^k|| \left\| \int\limits_{\mathcal{K}_\varepsilon(y_1)\cap\mathcal{D}} |F_2^k(y_2)|\, |R_\varepsilon(y_1, y_2)|\, dy_2 \right\| .$$

An estimation shows that

$$\frac{1}{\varepsilon^m} \left\| \int\limits_{\mathcal{K}_\varepsilon(y_1)\cap\mathcal{D}} |F_2^k(y_2)\, R_\varepsilon(y_1, y_2)|\, dy_2 \right\|$$

$$\leqq \frac{1}{\varepsilon^m} \left\| \left[\int\limits_{\mathcal{K}_\varepsilon(y_1)\cap\mathcal{D}} F_2^k(y_2)^2\, dy_2 \int\limits_{\mathcal{K}_\varepsilon(y_1)\cap\mathcal{D}} R_\varepsilon(y_1, y_2)^2\, dy_2 \right]^{1/2} \right\|$$

$$\leqq \frac{c_2 \sqrt{V_1}}{\varepsilon^{m/2}} \left\| \left[\int\limits_{\mathcal{K}_\varepsilon(y_1)\cap\mathcal{D}} F_2^k(y_2)^2\, dy_2 \right]^{1/2} \right\|$$

$$\leqq \frac{c_2 \sqrt{V_1}}{\varepsilon^{m/2}} \left[\int\limits_{\mathcal{D}} \int\limits_{\mathcal{K}_\varepsilon(y_1)\cap\mathcal{D}} F_2^k(y_2)^2\, dy_2\, dy_1 \right]^{1/2}$$

$$\leqq \frac{c_2 \sqrt{V_1}}{\varepsilon^{m/2}} \left[\int\limits_{\mathcal{D}} F_2^k(y_2)^2\, dy_2 \int\limits_{\mathcal{K}_\varepsilon(y_2)\cap\mathcal{D}} dy_1 \right]^{1/2} \leqq c_2 V_1\, ||F_2^k||$$

if the Fubini's theorem is applied. Then the inequality

$$|\langle r_{1\varepsilon} r_{2\varepsilon}\rangle - \langle r_{1\varepsilon}^k r_{2\varepsilon}^k\rangle| \leqq V_1 c_2\, [||F_2 - F_2^k||\, ||F_1|| + ||F_1 - F_1^k||\, ||F_2^k||]$$

follows and therefore

$$\lim_{k\to\infty} |\langle r_{1\varepsilon} r_{2\varepsilon}\rangle - \langle r_{1\varepsilon}^k r_{2\varepsilon}^k\rangle| = 0 .$$

According to (2.10), the proof is complete. ◀

We now give a lemma which is necessary for Theorem 2.6 which concerns convergence of the k-th moments.

Lemma 2.3. *Let* $F_i(y) \doteq F_i(x_i, y) \in \mathbf{L}_2(\mathcal{D})$ *and* $\langle |f_\varepsilon(x)|^k\rangle \leqq c_k < \infty$. *Then the convergence relation*

$$\lim_{\varepsilon\downarrow 0} \frac{1}{\sqrt{\varepsilon^{mk}}} \int\limits_{\mathcal{E}_k} \prod_{i=1}^{k} F_i(y_i) \left\langle \prod_{i=1}^{k} f_\varepsilon(y_i)\right\rangle dy_1 \dots dy_k = 0$$

is true for $k \geqq 3$, where the set $\mathscr{E}_k$ is defined by

$$\mathscr{E}_k \doteq \{\{y_1, \dots, y_k\} \in \mathscr{D}^k \colon \{y_1, \dots, y_k\}\ \varepsilon\text{-adjoining}\} .$$

Proof. An estimation leads to

$$\begin{aligned}
&\frac{1}{\varepsilon^m} \iint\limits_{\mathscr{E}_2} |F_1(y_1)\, F_2(y_2)|\, \mathrm{d}y_1\, \mathrm{d}y_2 \\
&\leqq \frac{1}{\varepsilon^m} \int\limits_{\mathscr{D}} \int\limits_{\mathscr{K}_\varepsilon(y_1) \cap \mathscr{D}} |F_1(y_1)\, F_2(y_2)|\, \mathrm{d}y_1\, \mathrm{d}y_2 \\
&\leqq \frac{1}{\varepsilon^m} \left[\int\limits_{\mathscr{D}} \int\limits_{\mathscr{K}_\varepsilon(y_1) \cap \mathscr{D}} F_1(y_1)^2\, \mathrm{d}y_2\, \mathrm{d}y_1 \int\limits_{\mathscr{D}} \int\limits_{\mathscr{K}_\varepsilon(y_2) \cap \mathscr{D}} F_2(y_2)^2\, \mathrm{d}y_1\, \mathrm{d}y_2 \right]^{1/2} \\
&\leqq V_1\, \|F_1\|\, \|F_2\| .
\end{aligned} \tag{2.11}$$

Furthermore, we have

$$\begin{aligned}
\int\limits_{\mathscr{K}_\varepsilon(x) \cap \mathscr{D}} |F_i(y)|\, \mathrm{d}y &\leqq [V_1 \varepsilon^m \int\limits_{\mathscr{K}_\varepsilon(x) \cap \mathscr{D}} F_i(y)^2\, \mathrm{d}y]^{1/2} \\
&\leqq [V_1 \varepsilon^m A_{i\varepsilon}]^{1/2}
\end{aligned} \tag{2.12}$$

where

$$A_{i\varepsilon} \doteq \sup_{x \in \mathscr{D}} \int\limits_{\mathscr{K}_\varepsilon(x) \cap \mathscr{D}} F_i^2(y)\, \mathrm{d}y \leqq A_\varepsilon .$$

The result $\lim\limits_{\varepsilon \downarrow 0} A_\varepsilon = 0$ follows from the absolute continuity of the Lebesgue integral. With the aid of (2.11) and (2.12), we can show that

$$\begin{aligned}
&\left| \frac{1}{\sqrt{\varepsilon^{mk}}} \int\limits_{\mathscr{D}} \int\limits_{\mathscr{K}_\varepsilon(y_1) \cap \mathscr{D}} \cdots \int\limits_{\mathscr{K}_\varepsilon(y_{k-1}) \cap \mathscr{D}} \prod_{i=1}^{k} F_i(y_i) \left\langle \prod_{i=1}^{k} f_\varepsilon(y_i) \right\rangle \mathrm{d}y_1 \dots \mathrm{d}y_k \right| \\
&\leqq \frac{c_k}{\sqrt{\varepsilon^{m(k-2)}}} (\varepsilon^m V_1 A_\varepsilon)^{(k-2)/2} \frac{1}{\varepsilon^m} \int\limits_{\mathscr{D}} \int\limits_{\mathscr{K}_\varepsilon(y_1) \cap \mathscr{D}} |F_1(y_1)\, F_2(y_2)|\, \mathrm{d}y_2\, \mathrm{d}y_1 \\
&\leqq c_k V_1^{k/2} A_\varepsilon^{k/2-1}\, \|F_1\|\, \|F_2\|
\end{aligned}$$

which completes the proof. ◀

The following theorem includes the convergence of the k-th moments of random fields of (2.6), where $k > 2$. In Theorem 2.5 we proved this convergence for $k = 2$.

Theorem 2.6. *For the k-th moments of the random fields defined by* (2.6) *the limit value of $\langle r_{i_1\varepsilon}(x_1) \dots r_{i_k\varepsilon}(x_k)\rangle$, as $\varepsilon \downarrow 0$, is given by*

$$\lim_{\varepsilon \downarrow 0} \langle r_{i_1\varepsilon}(x_1) \dots r_{i_k\varepsilon}(x_k)\rangle = \begin{cases} 0\,, & \text{for } k \text{ odd}, \\ \sum\limits_{(p_1, p_2), \dots, (p_{k-1}, p_k)} A_{p_1 p_2} \dots A_{p_{k-1} p_k}, & \text{for } k \text{ even} \end{cases}$$

where

$$A_{pq} \doteq \lim_{\varepsilon \downarrow 0} \langle r_{i_p\varepsilon}(x_p)\, r_{i_q\varepsilon}(x_q)\rangle$$

and $i_p \in N$, $x_p \in \mathcal{S}$ *for* $p = 1, 2, \dots, k$. *The sum above is taken over all nonequivalent decompositions of* $\{1, 2, \dots, k\}$ *in pairs. The same conditions as in Theorem 2.5 are assumed and, in addition, we assume* $\langle |f_\varepsilon(x)|^k \rangle \leq c_k < \infty$.

For example,

$$\begin{aligned}
&\lim_{\varepsilon \downarrow 0} \langle r_{1\varepsilon}(x_1)\, r_{1\varepsilon}^2(x_2)\, r_{2\varepsilon}(x_3) \rangle \\
&= 2 \lim_{\varepsilon \downarrow 0} \langle r_{1\varepsilon}(x_1)\, r_{1\varepsilon}(x_2) \rangle \lim_{\varepsilon \downarrow 0} \langle r_{1\varepsilon}(x_2)\, r_{2\varepsilon}(x_3) \rangle \\
&\quad + \lim_{\varepsilon \downarrow 0} \langle r_{1\varepsilon}(x_1)\, r_{2\varepsilon}(x_3) \rangle \lim_{\varepsilon \downarrow 0} \langle r_{1\varepsilon}^2(x_2) \rangle
\end{aligned}$$

follows from this theorem. The limit values A_{pq} are calculated by Theorem 2.6.

Proof. We first define the sets

$$\mathscr{B}(I_1, \dots, I_s) \doteq \Big\{ \{y_1, \dots, y_k\} \in \mathscr{D}^k : \{y_i, i \in I_r\} \text{ maximum } \varepsilon\text{-adjoining}, \quad r = 1, 2, \dots, s;\ \bigcup_{r=1}^{s} I_r = I \Big\}$$

for $s = 1, 2, \dots, k$. The relation

$$\mathscr{B}(I_1, \dots, I_s) = \mathscr{E}(I_1) \times \dots \times \mathscr{E}(I_s) \setminus \mathscr{F}(I_1, \dots, I_s) \tag{2.13}$$

holds where

$$\mathscr{F}(I_1, \dots, I_s) = \bigcup_{(p_1, p_2), (p_3, \dots, p_s)} \mathscr{E}(I_{p_1} \cup I_{p_2}) \times \mathscr{E}(I_{p_3}) \times \dots \times \mathscr{E}(I_{p_s})$$

and $\mathscr{E}(I_r)$ denotes the set previously defined in Lemma 2.3; that is

$$\mathscr{E}(I_r) = \big\{ \{x_i, i \in I_r\} \in \mathscr{D}^{k_r} : \{x_i, i \in I_r\} \ \varepsilon\text{-adjoining} \big\} .$$

The union above is taken over all nonequivalent decompositions (p_1, p_2), $(p_3, \dots, p_s)$ of $(1, 2, \dots, s)$ and I_r has k_r elements $\left(\sum_{r=1}^{s} k_r = k\right)$.

From (2.13) we also obtain the measurability of the sets $\mathscr{B}$ since the sets

$$\mathscr{E}(\{1, \dots, r\}) = \bigcup_{(p_1, \dots, p_r)} \bigcap_{j=1}^{r-1} \{\{y_1, \dots, y_r\} \in \mathscr{D}^r : |y_{p_j} - y_{p_{j+1}}| < \varepsilon\}$$

are measurable. In this formula the union is taken over all the permutations $(p_1, \dots, p_r)$ of $(1, 2, \dots, r)$. Hence, we have

$$\mathscr{D}^k = \bigcup_{s=1}^{k} \bigcup_{\{I_1, \dots, I_s\}} \mathscr{B}(I_1, \dots, I_s) ,$$

and in this equation the union is taken over all non-equivalent decompositions $\{I_1, \dots, I_s\}$ of $\{1, \dots, k\}$. The sets $\mathscr{B}(I_1, \dots, I_s)$ are disjoint sets.

The definition of a weakly correlated field $f_\varepsilon(x, \omega)$ implies that

$$\langle \prod_{i=1}^{k} r_{i\varepsilon} \rangle = \sum_{s=1}^{k} \sum_{I_1, \dots, I_s} \frac{1}{\sqrt{\varepsilon^{km}}} \int_{\mathscr{B}(I_1, \dots, I_s)} T(y_1, \dots, y_k)\, \mathrm{d}y_1 \dots \mathrm{d}y_k \tag{2.14}$$

where

$$T(y_1, \dots, y_k) = \prod_{j=1}^{s} \{\prod_{i\in I_j} F_i(y_i) \langle \prod_{i\in I_j} f_\varepsilon(y_i)\rangle\}$$

was written as an abbreviation. In (2.14) we have put $r_{i_t\varepsilon}(x_t) = r_{t\varepsilon}$. It is possible to restrict in (2.14) to $k_j \geqq 2$, for $j = 1, 2, \dots, s$, because all the other summands vanish since $\langle f_\varepsilon(y)\rangle = 0$. By means of (2.13), we get for a summand $A_s = A(I_1, \dots, I_s)$ from (2.14)

$$\begin{aligned} A_s &= \frac{1}{\sqrt{\varepsilon^{km}}} \int\limits_{\mathscr{E}(I_1)\times\dots\times\mathscr{E}(I_s)} \prod_{j=1}^{s} \{\prod_{j\in I_j} F_i(y_i) \langle \prod_{i\in I_j} f_\varepsilon(y_i)\rangle\} \, dy_1 \dots dy_k \\ &\quad - \frac{1}{\sqrt{\varepsilon^{km}}} \int\limits_{\mathscr{F}(I_1,\dots,I_s)} \prod_{j=1}^{s} \{\prod_{i\in I_j} F_i(y_i) \langle \prod_{i\in I_j} f_\varepsilon(y_i)\rangle\} \, dy_1 \dots dy_k \\ &= A_{s1} - A_{s2}\,. \end{aligned} \tag{2.15}$$

Furthermore, it follows that

$$A_{s1} = \prod_{j=1}^{s} \frac{1}{\sqrt{\varepsilon^{k_j m}}} \int\limits_{\mathscr{E}(I_j)} \langle \prod_{i\in I_j} f_\varepsilon(y_i)\rangle \prod_{i\in I_j} F_i(y_i) \, dy_i\,.$$

The relation

$$\lim_{\varepsilon\downarrow 0} A_{s1} = 0 \text{ for } k \text{ odd or for } k \text{ even and at least one } k_j \geqq 3$$

can be obtained from Lemma 2.3. If k is even and $k_j = 2$ for all j ($s = k/2$) we put $I_j = \{p_j, q_j\}$. Then we obtain from Theorem 2.5

$$\begin{aligned} \lim_{\varepsilon\downarrow 0} A_{s1} &= \prod_{j=1}^{k/2} \lim_{\varepsilon\downarrow 0} \frac{1}{\varepsilon^m} \int\limits_{\mathscr{E}(p_j, q_j)} F_{p_j}(y_{p_j}) F_{q_j}(y_{q_j}) \langle f_\varepsilon(y_{p_j}) f_\varepsilon(y_{q_j})\rangle \, dy_{p_j} \, dy_{q_j} \\ &= \prod_{j=1}^{k/2} \lim_{\varepsilon\downarrow 0} \langle r_{p_j\varepsilon} r_{q_j\varepsilon}\rangle\,. \end{aligned}$$

We can estimate the term A_{s2} from (2.15) by

$$\begin{aligned} |A_{s2}| \leqq \sum_{(r_1, r_2), (r_3, \dots, r_s)} \frac{1}{\sqrt{\varepsilon^{m(k_{r_1}+k_{r_2})}}} \int\limits_{\mathscr{E}(I_{r_1}\cup I_{r_2})} &|\langle \prod_{i\in I_{r_1}} f_\varepsilon(y_i)\rangle \\ \times \langle \prod_{i\in I_{r_2}} f_\varepsilon(y_i)\rangle \prod_{\substack{i\in I_{r_1}\\ i\in I_{r_2}}} F_i(y_i) \,|\, dy_i \\ \times \prod_{j=3}^{s} \frac{1}{\sqrt{\varepsilon^{mk_{r_j}}}} \int\limits_{\mathscr{E}(I_{r_j})} |\langle \prod_{i\in I_{r_j}} f_\varepsilon(y_i)\rangle \prod_{i\in I_{r_j}} F_i(y_i)| \, dy_i\,. \end{aligned}$$

If follows that

$$\lim_{\varepsilon\downarrow 0} A_{s2}(I_1, \dots, I_s) = 0$$

since the first factor of every summand converges to zero as $\varepsilon \downarrow 0$ because $k_{r_1} + k_{r_2} \geqq 4$ and the remaining factors also converge to zero as $\varepsilon \downarrow 0$ for $k_{r_j} \geqq 3$, or they are bounded for $k_{r_j} = 2$. Hence we obtain from (2.15)

$$\lim_{\varepsilon \downarrow 0} A(I_1, \dots, I_s) = \begin{cases} \prod_{j=1}^{k/2} \lim_{\varepsilon \downarrow 0} \langle r_{p_j\varepsilon} r_{q_j\varepsilon} \rangle, & \text{for } k \text{ even and } s = k/2, \quad k_j = 2 \quad \text{for all } j, \\ 0, & \text{otherwise}, \end{cases}$$

and finally from (2.14)

$$\lim_{\varepsilon \downarrow 0} \langle r_{1\varepsilon} \dots r_{k\varepsilon} \rangle = \begin{cases} \sum_{(p_1, q_1), \dots, (p_{k/2}, q_{k/2})} \prod_{j=1}^{k/2} \lim_{\varepsilon \downarrow 0} \langle r_{p_j\varepsilon} r_{q_j\varepsilon} \rangle & \text{for } k \text{ even}, \\ 0 & \text{for } k \text{ odd}. \end{cases}$$ ◀

The next theorem establishes convergence in distribution of the random field of (2.6) to a Gaussian field.

Theorem 2.7. *Let $(f_\varepsilon(x, \omega))_{\varepsilon \downarrow 0}$ be a sequence of weakly correlated fields on $\mathcal{D} \subset R^m$ where $f_\varepsilon(x, \omega)$ possesses continuous sample functions a.s. and $\langle |f_\varepsilon^k(x, \omega)| \rangle \leqq c_k < \infty$ for all $k \geqq 1$. The intensity of the weakly correlated field is defined by*

$$a(x) \doteq \lim_{\varepsilon \downarrow 0} \frac{1}{\varepsilon^m} \int_{\mathcal{K}_\varepsilon(0)} \langle f_\varepsilon(x) f_\varepsilon(x + y) \rangle \, \mathrm{d}y .$$

Then the convergence in distribution, that is

$$\lim_{\varepsilon \downarrow 0} \big(r_{1\varepsilon}(x, \omega), r_{2\varepsilon}(x, \omega), \dots, r_{l\varepsilon}(x, \omega)\big) = \big(\xi_1(x, \omega), \xi_2(x, \omega), \dots, \xi_l(x, \omega)\big)$$

is obtained for the random fields

$$r_{i\varepsilon}(x, \omega) = \frac{1}{\sqrt{\varepsilon^m}} \int_{\mathcal{D}} F_i(x, y) f_\varepsilon(y) \, \mathrm{d}y , \qquad i = 1, 2, \dots, l, \qquad x \in \mathcal{G} ,$$

where $F_i(x, y) \in \mathbf{L}_2(\mathcal{D})_y$. The random vector field $\big(\xi_1(x, \omega), \dots, \xi_l(x, \omega)\big)$ is a Gaussian vector field with mean

$$\langle \xi_i(x) \rangle = 0$$

and correlation relations

$$\langle \xi_i(x) \, \xi_j(y) \rangle = \int_{\mathcal{D}} F_i(x, z) \, F_j(y, z) \, a(z) \, \mathrm{d}z$$

for $i, j = 1, 2, \dots, l$.

The convergence in distribution of $r_\varepsilon(x, \omega) \doteq \big(r_{1\varepsilon}(x, \omega), \dots, r_{l\varepsilon}(x, \omega)\big)$ to $\xi(x, \omega) \doteq \big(\xi_1(x, \omega), \dots, \xi_l(x, \omega)\big)$ means the convergence of all finite-dimensional joint distribution functions of $r_\varepsilon(x, \omega)$ to the corresponding distribution function of $\xi(x, \omega)$, that is, for all positive integer $p \geqq 1$, $x_q \in \mathcal{G}$ and $y_q \in R^l$ for $q = 1, 2,$

..., p we have

$$\lim_{\varepsilon\downarrow 0} F_{\varepsilon;x_1,\dots,x_p}(y_1,\dots,y_p) = F_{x_1,\dots,x_p}(y_1,\dots,y_p). \tag{2.16}$$

$F_{\varepsilon;x_1,\dots,x_p}(y_1,\dots,y_p)$ denotes the joint distribution function of $\left(r_\varepsilon(x_1),\dots,r_\varepsilon(x_p)\right)$,

$$F_{\varepsilon;x_1,\dots,x_p}(y_1,\dots,y_p) \doteq \mathsf{P}\left(r_\varepsilon(x_1) < y_1,\dots,r_\varepsilon(x_p) < y_p\right),$$

and $F_{x_1,\dots,x_p}(y_1,\dots,y_p)$ the joint distribution function of $\left(\xi(x_1),\dots,\xi(x_p)\right)$, that is

$$F_{x_1,\dots,x_p}(y_1,\dots,y_p) \doteq \mathsf{P}\left(\xi(x_1) < y_1,\dots,\xi(x_p) < y_p\right).$$

Let $\mu_\varepsilon^p(x_1,\dots,x_p)$ be the measure defined on $\mathbb{R}^{lp}$ by the finite-dimensional distribution function $F_{\varepsilon;x_1,\dots,x_p}(y_1,\dots,y_p)$, and $\mu^p(x_1,\dots,x_p)$ the measure defined by $F_{x_1,\dots,x_p}(y_1,\dots,y_p)$. Then we obtain from (2.16) the weak convergence of the measures μ_ε^p to the measure μ^p, i.e. we obtain the limit relation

$$\lim_{\varepsilon\downarrow 0} \int_{\mathbb{R}^{lp}} g\,\mathrm{d}\mu_\varepsilon^p = \int_{\mathbb{R}^{lp}} g\,\mathrm{d}\mu^p$$

for all bounded continuous functions g on $\mathbb{R}^{lp}$. In general, the weak convergence of the measures $\mu_\varepsilon^p(x_1,\dots,x_p)$ to $\mu^p(x_1,\dots,x_p)$ as $\varepsilon\downarrow 0$ is equivalent to the convergence of $F_{\varepsilon;x_1,\dots,x_p}(y_1,\dots,y_p)$ to $F_{x_1,\dots,x_p}(y_1,\dots,y_p)$ at every continuity point of $F_{x_1,\dots,x_p}$. For Gaussian random vectors the set of continuity points coincides with the whole space.

Proof of Theorem 2.7. Using Theorem 2.5 and Theorem 2.6 we have for $i_q \in \{1,2,\dots,l\}$, $x_{j_q} \in \mathcal{G}$

$$\lim_{\varepsilon\downarrow 0} \langle r_{i_1\varepsilon}(x_{j_1}) \dots r_{i_p\varepsilon}(x_{j_p})\rangle = \langle \xi_{i_1}(x_{j_1}) \dots \xi_{i_p}(x_{j_p})\rangle$$

where the vector field $\left(\xi_1(x,\omega),\dots,\xi_l(x,\omega)\right)$ was defined by Theorem 2.7. Hence, all moments of $\left(r_{1\varepsilon}(x),\dots,r_{l\varepsilon}(x)\right)$ converge to the corresponding moments of $\left(\xi_1(x),\dots,\xi_l(x)\right)$.

We will prove (2.16) for an arbitrary positive integer $p \geqq 1$ and for every point $x_q \in \mathcal{G}$ and $y_q \in \mathbb{R}^l$, $q = 1,\dots,p$. The convergence of the joint distribution function follows from the convergence of the characteristic functions

$$\lim_{\varepsilon\downarrow 0} \left\langle \exp\left(i \sum_{s=1}^{p} t_s^{\mathsf{T}} r_\varepsilon(x_s)\right)\right\rangle = \left\langle \exp\left(i \sum_{s=1}^{p} t_s^{\mathsf{T}} \xi(x_s)\right)\right\rangle \tag{2.17}$$

for every $t_s \in \mathbb{R}^l$. Since $\xi(x_s)$, for $s = 1,\dots,p$, are Gaussian random vectors the right-hand side of (2.17) is a continuous function of the components of t_s. Furthermore, we note that

$$\left| \mathrm{e}^{ih} - \sum_{s=0}^{2n-1} \frac{(ih)^s}{s!} \right| \leqq \frac{h^{2n}}{(2n)!}$$

and obtain the inequality

$$\overline{\lim_{\varepsilon\downarrow 0}} \left| \langle \mathrm{e}^{iA_\varepsilon}\rangle - \sum_{s=0}^{2n-1} \frac{\langle (iA_0)^s\rangle}{s!} \right| \leqq \frac{\langle A_0^{2n}\rangle}{(2n)!} \tag{2.18}$$

where $A_\varepsilon \doteq \sum_{s=1}^{p} t_s^\mathsf{T} r_\varepsilon(x_s)$ and $A_0 \doteq \sum_{s=1}^{p} t_s^\mathsf{T} \xi(x_s)$ taking into consideration the convergence of the moments of $r_\varepsilon(x, \omega)$ to the corresponding moments of $\xi(x, \omega)$. It is now easy to see that

$$\overline{\lim_{\varepsilon \downarrow 0}} \left| \langle e^{iA_\varepsilon} \rangle - \sum_{s=0}^{\infty} \frac{\langle (iA_0)^s \rangle}{s!} \right| \leqq \frac{\langle A_0^{2n} \rangle}{(2n)!} + \left| \sum_{s=2n}^{\infty} \frac{\langle (iA_0)^s \rangle}{s!} \right| .$$

The series $\sum_{s=0}^{\infty} 1/s! \, \langle (iA_0)^s \rangle$ converges since A_0 is a Gaussian random variable with mean $\langle A_0 \rangle = 0$ and variance

$$\langle A_0^2 \rangle = \sum_{s,r=1}^{p} t_s^\mathsf{T} \langle \xi(x_s) \, \xi^\mathsf{T}(x_r) \rangle \, t_r .$$

The relation

$$\lim_{n \to \infty} \left[\frac{\langle A_0^{2n} \rangle}{(2n)!} + \left| \sum_{s=2n}^{\infty} \frac{\langle (iA_0)^s \rangle}{s!} \right| \right] = 0$$

completes the proof. ◀

Finally, we will formulate this limit theorem for the case of weakly correlated connected fields.

Theorem 2.8. *Let $(f_{1\varepsilon}(x, \omega), \ldots, f_{l\varepsilon}(x, \omega))_{\varepsilon \downarrow 0}$ be weakly correlated connected fields on $\mathcal{D} \subset \mathbb{R}^m$, where the sample functions of these random fields are continuous a.s. The random field $r_\varepsilon(x, \omega)$ is defined by*

$$r_\varepsilon(x, \omega) \doteq \sum_{i=1}^{l} r_{i\varepsilon}(x, \omega) ,$$

where

$$r_{i\varepsilon}(x, \omega) \doteq \frac{1}{\sqrt{\varepsilon^m}} \int_{\mathcal{D}} F_i(x, y) \, f_{i\varepsilon}(y, \omega) \, dy ,$$

and $F_i(x, y)$, for $i = 1, 2, \ldots, l$, denote functions on $\mathcal{G} \times \mathcal{D}$ with $F_i(x, y) \in \mathbf{L}_2(\mathcal{D})_y$ for all $x \in \mathcal{G}$. The intensity $a_{ij}(x)$ between $f_{i\varepsilon}(x, \omega)$ and $f_{j\varepsilon}(x, \omega)$ is given by

$$a_{ij}(x) \doteq \lim_{\varepsilon \downarrow 0} \frac{1}{\varepsilon^m} \int_{\mathcal{K}_\varepsilon(0)} \langle f_{i\varepsilon}(x) \, f_{j\varepsilon}(x + y) \rangle \, dy . \tag{2.19}$$

Then the following hold

(1) *for $\langle f_{i\varepsilon}^2(x) \rangle \leqq c_2 < \infty$, $i = 1, 2, \ldots, l$, $x \in \mathcal{D}$, the relation*

$$\lim_{\varepsilon \downarrow 0} \langle r_{i\varepsilon}(x_1) \, r_{j\varepsilon}(x_2) \rangle = \int_{\mathcal{D}} F_i(x_1, y) \, F_j(x_2, y) \, a_{ij}(y) \, dy$$

for $i, j = 1, 2, \ldots, l$;

(2) *for* $\langle |f_{i\varepsilon}^k(x)| \rangle \leqq c_k < \infty$, $i = 1, 2, \ldots, l$, $x \in \mathcal{D}$, *the convergence of the k-th moments*

$$\lim_{\varepsilon \downarrow 0} \langle r_{i_1\varepsilon}(x_1) \ldots r_{i_k\varepsilon}(x_k) \rangle$$

$$= \begin{cases} \sum\limits_{(j_1, j_2)\ldots(j_{k-1}, j_k)} \lim\limits_{\varepsilon \downarrow 0} \langle r_{i_{j_1}\varepsilon}(x_{j_1}) \, r_{i_{j_2}\varepsilon}(x_{j_2}) \rangle \ldots \lim\limits_{\varepsilon \downarrow 0} \langle r_{i_{j_{k-1}}\varepsilon}(x_{j_{k-1}}) \, r_{i_{j_k}\varepsilon}(x_{j_k}) \rangle & \text{for } k \text{ even,} \\ 0 & \text{for } k \text{ odd;} \end{cases}$$

(3) *for* $\langle |f_{i\varepsilon}^k(x)| \rangle \leqq c_k < \infty$, $i = 1, 2, \ldots, l$, $x \in \mathcal{D}$, $k = 1, 2, \ldots$, *the convergence in distribution*

$$\lim_{\varepsilon \downarrow 0} r_\varepsilon(x, \omega) = \xi(x, \omega) .$$

The random field $\xi(x, \omega)$ *is a Gaussian field with mean* $\langle \xi(x) \rangle \equiv 0$ *and correlation function*

$$\langle \xi(x) \, \xi(y) \rangle = \sum_{i,j=1}^{l} \int_{\mathcal{D}} F_i(x, z) \, F_j(y, z) \, a_{ij}(z) \, dz .$$

Remark 2.2. *Assuming that the random fields* $f_{i\varepsilon}(x, \omega)$, $i = 1, 2, \ldots, l$, *in the weakly correlated connected vector field* $(f_{1\varepsilon}(x, \omega), \ldots, f_{l\varepsilon}(x, \omega))$ *are independent, then* $a_{ij}(x) \equiv 0$ *for* $i \neq j$ *and the limit field* $\xi(x, \omega)$ *can be written as a sum of* l *independent Gaussian fields* $\xi_i(x, \omega)$,

$$\xi(x, \omega) = \sum_{i=1}^{l} \xi_i(x, \omega) .$$

The mean of $\xi_i(x, \omega)$ *is* $\langle \xi_i(x) \rangle \equiv 0$ *and the correlation function*

$$\langle \xi_i(x) \, \xi_i(y) \rangle = \int_{\mathcal{D}} F_i(x, z) \, F_i(y, z) \, a_{ii}(z) \, dz .$$

Remark 2.3. *Let* $R_{ii\varepsilon}(x, y) \doteq \langle f_{i\varepsilon}(x, \omega) \, f_{i\varepsilon}(y, \omega) \rangle$ *for* $i = 1, 2, \ldots, l$ *be continuous functions at* (x, x), $x \in \mathcal{D}$, *then the intensities satisfy the relation*

$$a_{ij}(x) = a_{ji}(x) \quad \text{for} \quad i, j = 1, 2, \ldots, l .$$

In order to show that this property is satisfied we use the fact that continuity in mean square follows from the continuity of $R_{ii\varepsilon}(x, y)$ at (x, x). Furthermore, we consider

$$\left| \frac{1}{\varepsilon^m} \int_{\mathcal{K}_\varepsilon(0)} \{ \langle f_{i\varepsilon}(x) \, f_{j\varepsilon}(x + y) \rangle - \langle f_{j\varepsilon}(x) \, f_{i\varepsilon}(x + y) \rangle \} \, dy \right|$$

$$\leqq \frac{1}{\varepsilon^m} \left[\sqrt{\langle f_{i\varepsilon}^2(x) \rangle} \int_{\mathcal{K}_\varepsilon(0)} \sqrt{\langle (f_{j\varepsilon}(x + y) - f_{j\varepsilon}(x))^2 \rangle} \, dy \right.$$

$$\left. + \sqrt{\langle f_{j\varepsilon}^2(x) \rangle} \int_{\mathcal{K}_\varepsilon(0)} \sqrt{\langle (f_{i\varepsilon}(x + y) - f_{i\varepsilon}(x))^2 \rangle} \, dy \right]$$

$$\leqq 2 V_1 \sqrt{c_2 \eta}$$

for $y \in \mathcal{K}_\varepsilon(0)$.

Proof of Theorem 2.8. We put as an abbreviation

$$r_{i\varepsilon}(x, \omega) = \frac{1}{\sqrt{\varepsilon^m}} \int_{\mathcal{D}} F_i(x, y) f_{i\varepsilon}(y, \omega) \, dy$$

for $i = 1, 2, \dots, l$, and obtain

$$\lim_{\varepsilon \downarrow 0} \langle r_{i\varepsilon}(x_1) \, r_{j\varepsilon}(x_2) \rangle = \int_{\mathcal{D}} F_i(x_1, y) \, F_j(x_2, y) \, a_{ij}(y) \, dy ,$$

$i, j = 1, 2, \dots, l$, using the proof of Theorem 2.5. The limit value $\lim_{\varepsilon \downarrow 0} \langle r_{i\varepsilon}(x_1) \, r_{i\varepsilon}(x_2) \rangle$, $i = 1, 2, \dots, l$, was calculated previously and the limit value $\lim_{\varepsilon \downarrow 0} \langle r_{i\varepsilon}(x_1) \, r_{j\varepsilon}(x_2) \rangle$, $i, j = 1, 2, \dots, l$, $i \neq j$, can be determined similarly if we take into account the definition of weakly correlated connected vector fields. The first statement of this theorem has been proved. The proof of the second statement follows similarly as the proof of Theorem 2.6.

To establish the third statement we prove the convergence of the moments and obtain convergence in distribution from Theorem 2.7. Thus,

$$\begin{aligned}\lim_{\varepsilon \downarrow 0} \langle r_\varepsilon(x_1) \, r_\varepsilon(x_2) \rangle &= \lim_{\varepsilon \downarrow 0} \sum_{i,j=1}^{l} \langle r_{i\varepsilon}(x_1) \, r_{j\varepsilon}(x_2) \rangle \\ &= \sum_{i,j=1}^{l} \int_{\mathcal{D}} F_i(x_1, y) \, F_j(x_2, y) \, a_{ij}(y) \, dy .\end{aligned}$$

For an odd k we have

$$\lim_{\varepsilon \downarrow 0} \langle r_\varepsilon(x_1) \dots r_\varepsilon(x_k) \rangle = 0 ,$$

and for k even

$$\begin{aligned}\lim_{\varepsilon \downarrow 0} \langle r_\varepsilon(x_1) \dots r_\varepsilon(x_k) \rangle &= \sum_{i_1, \dots, i_k = 1}^{l} \lim_{\varepsilon \downarrow 0} \langle r_{i_1\varepsilon}(x_1) \dots r_{i_k\varepsilon}(x_k) \rangle \\ &= \sum_{i_1, \dots, i_k = 1}^{l} \sum_{(j_1, j_2) \dots (j_{k-1}, j_k)} \lim_{\varepsilon \downarrow 0} \langle r_{i_{j_1}\varepsilon}(x_{j_1}) \, r_{i_{j_2}\varepsilon}(x_{j_2}) \rangle \cdot \dots \\ &\qquad \cdot \lim_{\varepsilon \downarrow 0} \langle r_{i_{j_{k-1}}\varepsilon}(x_{j_{k-1}}) \, r_{i_{j_k}\varepsilon}(x_{j_k}) \rangle \\ &= \sum_{(j_1, j_2) \dots (j_{k-1}, j_k)} \lim_{\varepsilon \downarrow 0} \sum_{i_{j_1}, i_{j_2} = 1}^{l} \langle r_{i_{j_1}\varepsilon}(x_{j_1}) \, r_{i_{j_2}\varepsilon}(x_{j_2}) \rangle \cdot \dots \\ &\qquad \cdot \lim_{\varepsilon \downarrow 0} \sum_{i_{j_{k-1}}, i_{j_k} = 1}^{l} \langle r_{i_{j_{k-1}}\varepsilon}(x_{j_{k-1}}) \, r_{i_{j_k}\varepsilon}(x_{j_k}) \rangle \\ &= \sum_{(j_1, j_2) \dots (j_{k-1}, j_k)} \lim_{\varepsilon \downarrow 0} \langle r_\varepsilon(x_{j_1}) \, r_\varepsilon(x_{j_2}) \rangle \dots \\ &\qquad \dots \lim_{\varepsilon \downarrow 0} \langle r_\varepsilon(x_{j_{k-1}}) \, r_\varepsilon(x_{j_k}) \rangle .\end{aligned}$$

The sum $\sum_{(j_1, j_2) \dots (j_{k-1}, j_k)}$ is taken over all the nonequivalent decompositions of $(1, 2, \dots, k)$ in pairs. ◀

2.1.3. A limit theorem for functions of linear functionals

In this section we generalize the limit theorems of the last section in order to apply then to random eigenvalue problems. Before the general theorem is proved investigations must be carried out for the simplest case. Let $f_\varepsilon(x, \omega)$ be a weakly correlated process on $\mathcal{D} \subset R^1$ where the sample functions of $f_\varepsilon(x, \omega)$ are continuous a.s., and let $\langle |f_\varepsilon(x)|^p \rangle \leqq c_p < \infty$ be for $\varepsilon \leqq \varepsilon_0$, $p = 1, 2, \ldots$ The random variable $\bar{r}_\varepsilon(\omega)$ is defined by

$$\bar{r}_\varepsilon(\omega) \doteq \int_{\mathcal{D}} F(y) f_\varepsilon(y, \omega) \, dy \,,$$

where $F \in \mathbf{L}_2(\mathcal{D})$. Assume that $\bar{r}_\varepsilon(\omega) \in \mathcal{J}$ a.s.

Let $d(y)$ be a real function on $\mathcal{J}$ satisfying the following conditions:

(1′) d can be written as

$$d(y) = d_0 + d_1 y + y^2 g(y) \,,$$

where $g(y)$ is a bounded function on $\mathcal{K}_\delta(0) \doteq \{y : |y| < \delta\}$ for a suitable $\delta > 0$, $|g(y)| \leqq c_0$.

(2′) All moments of $\hat{d}_\varepsilon(\omega) \doteq d(\bar{r}_\varepsilon(\omega))$ exist and $\langle |\hat{d}_\varepsilon|^p \rangle \leqq \bar{c}_p < \infty$ for $\varepsilon \leqq \varepsilon_0$.

Using this assumptions we will show that

$$\frac{1}{\sqrt{\varepsilon}} \left(d(\bar{r}_\varepsilon(\omega)) - d_0 \right)$$

converges in distribution to a Gaussian random variable as $\varepsilon \downarrow 0$. This statement can be established by the convergence of the moments. It is easy to see that

$$\frac{1}{\sqrt{\varepsilon^k}} \langle \left(d(\bar{r}_\varepsilon) - d_0 \right)^k \rangle = \frac{1}{\sqrt{\varepsilon^k}} \langle \left(d_1 \bar{r}_\varepsilon + \bar{r}_\varepsilon^2 g(\bar{r}_\varepsilon) \right)^k \rangle$$

$$= d_1^k \frac{1}{\sqrt{\varepsilon^k}} \langle \bar{r}_\varepsilon^k \rangle + \frac{1}{\sqrt{\varepsilon^k}} \sum_{i=1}^{k} \binom{k}{i} \langle (d_1 \bar{r}_\varepsilon)^{k-i} \left(\bar{r}_\varepsilon^2 g(\bar{r}_\varepsilon) \right)^i \rangle \,.$$

Let $a(y)$ be the intensity of the weakly correlated process $f_\varepsilon(x, \omega)$ then we deduce for the first summand of the last equation

$$\lim_{\varepsilon \downarrow 0} \frac{d_1^k}{\sqrt{\varepsilon^k}} \langle \bar{r}_\varepsilon^k \rangle = \begin{cases} (k-1)!! \, \sigma^k & \text{for} \quad k \text{ even} \,, \\ 0 & \text{for} \quad k \text{ odd} \end{cases}$$

where

$$\sigma^2 = d_1^2 \lim_{\varepsilon \downarrow 0} \frac{1}{\varepsilon} \langle \bar{r}_\varepsilon^2 \rangle = d_1^2 \int_{\mathcal{D}} F^2(y) \, a(y) \, dy \,.$$

Furthermore, we will show that the second summand converges to zero as $\varepsilon \downarrow 0$. For this, the estimate

$$\langle \bar{r}_\varepsilon^{k+i} g(\bar{r}_\varepsilon)^i \rangle \leqq \langle \bar{r}_\varepsilon^{2(k+i)} \rangle^{1/2} \langle g(\bar{r}_\varepsilon)^{2i} \rangle^{1/2}$$

is obtained using Cauchy's inequality. The convergence of the summand to zero follows from the boundedness of $\langle g(\bar{r}_\varepsilon)^{2i}\rangle$ and $\langle \bar{r}_\varepsilon^{2(k+i)}\rangle = O(\sqrt{\varepsilon^{2(k+i)}})$. Let P_ε be the distribution law of $\bar{r}_\varepsilon(\omega)$ on $\mathbb{R}^1$. Then the boundedness of $\langle g(\bar{r}_\varepsilon)^{2i}\rangle$ follows from the estimate

$$\begin{aligned}
\langle g(\bar{r}_\varepsilon)^{2i}\rangle &= \int\limits_{\{y:|y|<\delta\}} g(y)^{2i}\,\mathrm{d}\mathsf{P}_\varepsilon(y) + \int\limits_{\{y:|y|\geqq\delta\}} g(y)^{2i}\,\mathrm{d}\mathsf{P}_\varepsilon(y) \\
&\leqq c_0^{2i} + \frac{1}{\delta^{4i}} \int\limits_{-\infty}^{\infty} (d(y) - d_0 - d_1 y)^{2i}\,\mathrm{d}\mathsf{P}_\varepsilon(y) \\
&\leqq c_0^{2i} + \frac{1}{\delta^{4i}} \langle (d(\bar{r}_\varepsilon) - d_0 - d_1\bar{r}_\varepsilon)^{2i}\rangle ,
\end{aligned}$$

using the condition (2′).

Hence, we showed the convergence

$$\lim_{\varepsilon\downarrow 0} \frac{1}{\sqrt{\varepsilon^k}} \langle \big(d(\bar{r}_\varepsilon) - d_0\big)^k\rangle = \langle \xi^k\rangle ,$$

where ξ is a Gaussian random variable with mean $\langle\xi\rangle = 0$ and variance

$$\langle \xi^2\rangle = \sigma^2 = d_1^2 \int_{\mathscr{D}} F^2(y)\, a(y)\,\mathrm{d}y .$$

The convergence in distribution

$$\lim_{\varepsilon\downarrow 0} \frac{1}{\sqrt{\varepsilon}} \big(d(\bar{r}_\varepsilon) - d_0\big) = \xi$$

can be obtained from the convergence of the moments.

The assumption (1′) is equivalent to $d(y) \in \mathbf{C}^2\big(\mathscr{K}_\delta(0)\big)$ for a suitable $\delta > 0$. The proof of this statement follows from the Taylor expansion

$$d(y) = d(0) + \frac{\mathrm{d}d}{\mathrm{d}y}(0)\, y + \frac{1}{2}\frac{\mathrm{d}^2 d}{\mathrm{d}y^2}(\bar{y})\, y^2$$

for $y \in \mathscr{K}_\delta(0)$, where we put $g(y) \doteq 1/2\ \mathrm{d}^2 d/\mathrm{d}y^2\,(\bar{y})$. The assumption $|g(y)| \leqq c_0$ is derived from continuity. The function $g(y)$ for $y \notin \mathscr{K}_\delta(0)$ is defined by

$$g(y) \doteq \frac{1}{y^2}\left(d(y) - d(0) - \frac{\mathrm{d}d}{\mathrm{d}y}(0)\, y\right).$$

We now give a generalization of this simple case so that it can be applied to random eigenvalue problems. Let $\big(f_{1\varepsilon}(x,\omega), \ldots, f_{l\varepsilon}(x,\omega)\big)$ be a weakly correlated connected vector field on $\mathscr{D} \subset \mathbb{R}^m$ whose sample functions are continuous and $\langle |f_{i\varepsilon}^p(x)|\rangle \leqq c_p < \infty$ for $i = 1, 2, \ldots, l$, $x \in \mathscr{D}$ and $p = 1, 2, \ldots$ Let $F_j(x, y)$ for $j = 1, 2, \ldots, t$, be real functions and $F_j(x, y) \in \mathbf{L}_2(\mathscr{D})_y$ for all $x \in \mathscr{G}$. The random fields $\bar{r}_{ij\varepsilon}(x,\omega)$ are defined by

$$\bar{r}_{ij\varepsilon}(x,\omega) \doteq \int_{\mathscr{D}} F_j(x, y)\, f_{i\varepsilon}(y,\omega)\,\mathrm{d}y , \qquad j = 1, 2, \ldots, t , \quad i = 1, 2, \ldots, l .$$

We assume that

$$\left(\bar{r}_{11\varepsilon}(x, \omega), \dots, \bar{r}_{l1\varepsilon}(x, \omega), \dots, \bar{r}_{1t\varepsilon}(x, \omega), \dots, \bar{r}_{lt\varepsilon}(x, \omega)\right) \in \mathcal{J} \subset \mathbb{R}^{lt} \quad \text{a.s.}$$

where $\mathcal{J} = \mathbb{R}^{lt}$ is also possible.

Let

$$d_k(y_{11}, \dots, y_{l1}, y_{12}, \dots, y_{l2}, \dots, y_{1t}, \dots, y_{lt}), \qquad k = 1, 2, \dots, s,$$

be functions on $\mathcal{J}$ with the following properties:

(1) d_k possesses the representation

$$d_k(y_{11}, \dots, y_{lt}) = d_{k0} + \sum_{i=1}^{l} \sum_{j=1}^{t} d_{kij} y_{ij} + \Big(\sum_{i=1}^{l} \sum_{j=1}^{t} y_{ij}^2\Big)^\alpha g_k(y_{11}, \dots, y_{lt}) \tag{2.20}$$

where $\frac{1}{2} < \alpha$ and the real function g_k is bounded on

$$\mathcal{K}_\delta^{lt}(0) \doteq \{(y_{11}, \dots, y_{lt}) \colon \sum_{i=1}^{l} \sum_{j=1}^{t} y_{ij}^2 \leqq \delta^2\}$$

for a $\delta > 0$, $|g_k(y_{11}, \dots, y_{lt})| \leqq c_0$.

(2) All moments of

$$\hat{d}_{k\varepsilon}(x_{11}, \dots, x_{l1}, \dots, x_{1t}, \dots, x_{lt}, \omega) \doteq d_k\left(\bar{r}_{11\varepsilon}(x_{11}, \omega), \dots, \bar{r}_{lt\varepsilon}(x_{lt}, \omega)\right)$$

exist, and consequently

$$\langle |\hat{d}_{k\varepsilon}(x_{11}, \dots, x_{lt})|^p \rangle \leqq \bar{c}_p(x_{11}, \dots, x_{lt}) < \infty$$

for all $\varepsilon > 0$ and $p = 1, 2, \dots$

Especially, condition (1) is satisfied if the function $d_k \in \mathbf{C}^2\left(\mathcal{K}_\delta^{lt}(0)\right)$ for $k = 1, 2, \dots, s$. In this case we have by means of the Taylor expansion

$$d_k(y) = d_k(0) + \sum_{i=1}^{l} \sum_{j=1}^{t} \frac{\partial d_k}{\partial y_{ij}}(0)\, y_{ij} + \sum_{i,p=1}^{l} \sum_{j,q=1}^{t} \frac{\partial^2 d_k}{\partial y_{ij}\, \partial y_{pq}}(\bar{y})\, y_{ij} y_{pq}$$

where $y \doteq (y_{11}, \dots, y_{lt}) \in \mathcal{K}_\delta^{lt}(0)$. From this, (2.20) is obtained with $\alpha = 1$ if

$$g_k(y) \doteq \begin{cases} \dfrac{1}{2} \displaystyle\sum_{i,p=1}^{l} \sum_{j,q=1}^{t} \frac{\partial^2 d_k}{\partial y_{ij}\, \partial y_{pq}}(\bar{y}) \frac{y_{ij} y_{pq}}{|y|^2}, & \text{for} \quad y \in \mathcal{K}_\delta^{lt}(0) \setminus \{0\}, \\ \dfrac{1}{|y|^2}\left(d_k(y) - d_k(0) - \displaystyle\sum_{i=1}^{l} \sum_{j=1}^{t} \frac{\partial d_k}{\partial y_{ij}}(0)\, y_{ij}\right), & \text{for} \quad y \notin \mathcal{K}_\delta^{lt}(0), \end{cases}$$

and $g_k(0) \doteq 0$. The condition $|g_k| \leqq c_0$ for $y \in \mathcal{K}_\delta^{lt}(0)$ follows from the boundedness of the second derivatives of d_k on $\mathcal{K}_\delta^{lt}(0)$ and $|y_{ij} y_{pq}|/\left|y\right|^2| \leqq 1$.

With these conditions we can formulate Theorem 2.9.

Theorem 2.9. *Let d_k, $k = 1, 2, \dots, s$, be real functions on $\mathcal{J}$ satisfying conditions (1) and (2) above. Then, the sequence of random vectors*

$$R_\varepsilon(\underline{x}, \omega) \doteq \frac{1}{\sqrt{\varepsilon^m}} \left(d_1(\bar{r}_{11\varepsilon}, \dots, \bar{r}_{lt\varepsilon}) - d_{10}, \dots, d_s(\bar{r}_{11\varepsilon}, \dots, \bar{r}_{lt\varepsilon}) - d_{s0}\right)$$

converges in distribution to a Gaussian vector field $\xi(\underline{x}, \omega) = (\xi_1(\underline{x}, \omega), \dots, \xi_s(\underline{x}, \omega))$ *as* $\varepsilon \downarrow 0$,

$$\lim_{\varepsilon \downarrow 0} R_\varepsilon(\underline{x}, \omega) = \xi(\underline{x}, \omega) ,$$

where $\bar{r}_{ij\varepsilon} \doteq \bar{r}_{ij\varepsilon}(x_{ji}, \omega)$, $\underline{x} = (x_{11}, \dots, x_{lt})$. *The mean vector of* ξ *is zero and the elements of the correlation matrix are given by*

$$\langle \xi_p(\underline{x})\, \xi_q(\underline{y}) \rangle = \sum_{i,u=1}^{l} \sum_{j,v=1}^{t} d_{pij} d_{quv} \int_{\mathcal{D}} F_j(x_{ij}, z)\, F_v(y_{uv}, z)\, a_{iu}(z)\, dz$$

for $p, q = 1, 2, \dots, s$. *The function* a_{ik} *denotes the intensity between the random fields* $f_{i\varepsilon}$ *and* $f_{k\varepsilon}$ (cf. (2.19)).

Proof. This proof is again based on the convergence of the moments. First we show

$$\lim_{\varepsilon \downarrow 0} \langle \tilde{d}_{i_1\varepsilon}(\underline{x}^1)\, \tilde{d}_{i_2\varepsilon}(\underline{x}^2) \dots \tilde{d}_{i_k\varepsilon}(\underline{x}^k) \rangle = \langle \xi_{i_1}(\underline{x}^1)\, \xi_{i_2}(\underline{x}^2) \dots \xi_{i_k}(\underline{x}^k) \rangle , \tag{2.21}$$

where

$$\tilde{d}_{i\varepsilon}(\underline{x}) \doteq \frac{1}{\sqrt{\varepsilon^m}} [d_i(\bar{r}_{11\varepsilon}(x_{11}, \omega), \dots, \bar{r}_{lt\varepsilon}(x_{lt}, \omega)) - d_{i0}]$$

and $\underline{x} = (x_{11}, \dots, x_{lt})$, $\underline{x}^i = (x_{11i}, \dots, x_{lti})$, $i_p \in \{1, 2, \dots, s\}$, $p = 1, 2, \dots, k$. From (2.20) we have

$$\langle \prod_{p=1}^{k} d_{i_p\varepsilon}(\underline{x}^p) \rangle = \frac{1}{\sqrt{\varepsilon^{mk}}} \langle \prod_{p=1}^{k} [\sum_{u=1}^{l} \sum_{v=1}^{t} d_{i_puv} \bar{r}^p_{uv\varepsilon} + |\bar{r}^p_\varepsilon|^{2\alpha}\, g_{i_p}(\bar{r}^p_{11\varepsilon}, \dots, \bar{r}^p_{lt\varepsilon})] \rangle$$

if we define $\bar{r}^p_{uv\varepsilon} \doteq \bar{r}_{uv\varepsilon}(x_{uvp})$, $\bar{r}^p_\varepsilon \doteq (\bar{r}^p_{11\varepsilon}, \dots, \bar{r}^p_{lt\varepsilon})$.

Furthermore, we have

$$\langle \prod_{p=1}^{k} \tilde{d}_{i_p\varepsilon}(\underline{x}^p) \rangle = \frac{1}{\sqrt{\varepsilon^{mk}}} \langle \prod_{p=1}^{k} [\sum_{u=1}^{l} \sum_{v=1}^{t} d_{i_puv} \bar{r}^p_{uv\varepsilon}] \rangle + A_\varepsilon . \tag{2.22}$$

A_ε denotes a sum in which each summand is a product of k factors and among these k factors there is at least one factor of the form $|\bar{r}^p_\varepsilon|^{2\alpha}\, g_{i_p}(\bar{r}^p_\varepsilon)$. For the first summand on the right-hand side of (2.22) we calculate

$$\frac{1}{\sqrt{\varepsilon^{mk}}} \langle \prod_{p=1}^{k} [\sum_{u=1}^{l} \sum_{v=1}^{t} d_{i_puv} \bar{r}^p_{uv\varepsilon}] \rangle = \left\langle \prod_{p=1}^{k} \left[\sum_{u=1}^{l} \frac{1}{\sqrt{\varepsilon^m}} \int_{\mathcal{D}} \bar{F}_{up}(y)\, f_{u\varepsilon}(y, \omega)\, dy \right] \right\rangle$$

where

$$\bar{F}_{up}(y) \doteq \sum_{v=1}^{t} d_{i_puv} F_v(x_{uvp}, y) .$$

Applying Theorem 2.8 it follows that

$$\lim_{\varepsilon \downarrow 0} \frac{1}{\sqrt{\varepsilon^{mk}}} \langle \prod_{p=1}^{k} [\sum_{u=1}^{l} \sum_{v=1}^{t} d_{i_puv} \bar{r}^p_{uv\varepsilon}] \rangle$$

$$= \sum_{(j_1, j_2) \dots (j_{k-1}, j_k)} \langle \xi_{i_{j_1}}(\underline{x}^{j_1})\, \xi_{i_{j_2}}(\underline{x}^{j_2}) \rangle \dots \langle \xi_{i_{j_{k-1}}}(\underline{x}^{j_{k-1}})\, \xi_{i_{j_k}}(\underline{x}^{j_k}) \rangle , \tag{2.23}$$

with

$$\langle \xi_{i_p}(\underline{x}^p)\, \xi_{i_q}(\underline{x}^q)\rangle = \sum_{i,u=1}^{l} \int_{\mathcal{D}} \bar{F}_{ip}(y)\, \bar{F}_{uq}(y)\, a_{iu}(y)\, dy$$

$$= \sum_{i,u=1}^{l} \sum_{j,v=1}^{t} d_{i_p ij} d_{i_q uv} \int_{\mathcal{D}} F_j(x_{ijp}, y)\, F_v(x_{uvq}, y)\, a_{iu}(y)\, dy .$$

The sum on the right-hand side of (2.23) is taken over all the non-equivalent decompositions $\{(j_1, j_2), (j_3, j_4), \dots, (j_{k-1}, j_k)\}$ of $(1, 2, \dots, k)$ in pairs. It follows that

$$\lim_{\varepsilon\downarrow 0} \frac{1}{\sqrt{\varepsilon^{mk}}} \langle \prod_{p=1}^{k} [\sum_{u=1}^{l} \sum_{v=1}^{t} d_{i_p uv} \bar{r}^p_{uv\varepsilon}]\rangle = \langle \xi_{i_1}(\underline{x}^1)\, \xi_{i_2}(\underline{x}^2) \dots \xi_{i_k}(\underline{x}^k)\rangle$$

if the Gaussian vector field $\xi(\underline{x}, \omega) = \big(\xi_1(\underline{x}, \omega), \dots, \xi_s(\underline{x}, \omega)\big)$ is defined by the moments of Theorem 2.9. To prove (2.21) we still have to show the convergence

$$\lim_{\varepsilon\downarrow 0} A_\varepsilon = 0 .$$

The summands of A_ε are of the form

$$\bar{A}_\varepsilon \doteq \frac{1}{\sqrt{\varepsilon^{mk}}} \langle \prod_{q=1}^{k_1} [\sum_{u=1}^{l} \sum_{v=1}^{t} \tilde{d}_{quv} \tilde{\bar{r}}^q_{uv\varepsilon}] \prod_{q=1}^{k_2} [|\tilde{\tilde{\bar{r}}}^q_\varepsilon|^{2\alpha}\, \tilde{g}_q(\tilde{\tilde{\bar{r}}}^q_\varepsilon)]\rangle \tag{2.24}$$

where $\tilde{d}_{quv} \doteq d_{i_{j_q}uv}$, $\tilde{\bar{r}}^q_{uv\varepsilon} \doteq \bar{r}^{j_q}_{uv\varepsilon}$, $\tilde{\tilde{\bar{r}}}^q_\varepsilon \doteq \bar{r}^{j'_q}_\varepsilon$, $\tilde{g}_q \doteq g_{j'_q}$. We put $\{j_1, \dots, j_{k_1}\} \cup \{j'_1, \dots, j'_{k_2}\} = \{1, 2, \dots, k\}$ and hence $k_1 + k_2 = k$. We have $1 \leqq k_2 \leqq k$ and for $k_2 = k$ the first factor of (2.24) is equal to one ($k_1 = 0$). It is now easy to see that

$$\bar{A}_\varepsilon = \sum_{u_1,\dots,u_{k_1}=1}^{l} \sum_{v_1,\dots,v_{k_1}=1}^{t} \tilde{d}_{1u_1v_1} \dots \tilde{d}_{k_1u_{k_1}v_{k_1}}$$

$$\times \frac{1}{\sqrt{\varepsilon^{mk}}} \langle \tilde{\bar{r}}^1_{u_1v_1\varepsilon} \dots \tilde{\bar{r}}^{k_1}_{u_{k_1}v_{k_1}\varepsilon} \prod_{q=1}^{k_2} |\tilde{\tilde{\bar{r}}}^q_\varepsilon|^{2\alpha}\, \tilde{g}_q\rangle .$$

By means of Cauchy's inequality, we obtain

$$\frac{1}{\sqrt{\varepsilon^{mk}}} \langle \prod_{p=1}^{k_1} \tilde{\bar{r}}^p_{u_pv_p\varepsilon} \prod_{q=1}^{k_2} |\tilde{\tilde{\bar{r}}}^q_\varepsilon|^{2\alpha}\, \tilde{g}_q\rangle$$

$$\leqq \left[\frac{1}{\varepsilon^{mk}} \langle \{ \prod_{p=1}^{k_1} \tilde{\bar{r}}^p_{u_pv_p\varepsilon} \prod_{q=1}^{k_2} |\tilde{\tilde{\bar{r}}}^q_\varepsilon|^{2\alpha} \}^2\rangle \langle \prod_{q=1}^{k_2} \tilde{g}^2_q\rangle\right]^{1/2}, \tag{2.25}$$

and

$$\langle \prod_{q=1}^{k_2} \tilde{g}^2_q\rangle \leqq [\prod_{q=1}^{k_2} \langle \tilde{g}^{2k_2}_q\rangle]^{1/k_2} .$$

We now prove the boundedness of the moments

$$\langle g^{2p}_q(\bar{r}^q_\varepsilon)\rangle = \langle g^{2p}_q\big(\bar{r}_{11\varepsilon}(x_{11q}), \dots, \bar{r}_{lt\varepsilon}(x_{ltq})\big)\rangle .$$

To do this, let $\mathsf{P}_{\varepsilon q}$ be the distribution law of the random vector $(\bar{r}_{11\varepsilon}(x_{11q}), \ldots, \bar{r}_{lt\varepsilon}(x_{ltq}))$. Then we can estimate

$$\begin{aligned}
\langle g_q^{2p}(\bar{r}_\varepsilon^q)\rangle &= \int\limits_{\{y:|y|<\delta\}} g_q^{2p}(y)\, \mathrm{d}\mathsf{P}_{\varepsilon q}(y) + \int\limits_{\{y:|y|\geqq\delta\}} g_q^{2p}(y)\, \mathrm{d}\mathsf{P}_{\varepsilon q}(y) \\
&\leqq c_0^{2p} + \frac{1}{\delta^{4p\alpha}} \int\limits_{\mathbb{R}^{tl}} \Big(d_q(y) - d_{q0} - \sum_{i=1}^{l}\sum_{j=1}^{t} d_{qij} y_{ij}\Big)^{2p} \mathrm{d}\mathsf{P}_{\varepsilon q}(y) \\
&\leqq c_0^{2p} + \frac{1}{\delta^{4p\alpha}} \langle \Big(d_q(\bar{r}_\varepsilon^q) - d_{q0} - \sum_{i=1}^{l}\sum_{j=1}^{t} d_{qij} \bar{r}_{ij\varepsilon}^q\Big)^{2p}\rangle \\
&\leqq c_0^{2p} + \frac{1}{\delta^{4p\alpha}} \sum_{u=0}^{2p} \binom{2p}{u} (-1)^u \langle d_q(\bar{r}_\varepsilon^q)^{2p-u} [d_{q0} + \sum_{i=1}^{l}\sum_{j=1}^{t} d_{qij} \bar{r}_{ij\varepsilon}^q]^u\rangle \\
&\leqq c_0^{2p} + \frac{1}{\delta^{4p\alpha}} \sum_{u=0}^{2p} \binom{2p}{u} [\langle d_q(\bar{r}_\varepsilon^q)^{4p-2u}\rangle \langle \{d_{q0} + \sum_{i=1}^{l}\sum_{j=1}^{t} d_{qij} \bar{r}_{ij\varepsilon}^q\}^{2u}\rangle]^{1/2}
\end{aligned}$$

by means of condition (1) for the functions g_k, $k = 1, 2, \ldots, s$. Using condition (2), the boundedness of $\langle g_q^{2p}(\bar{r}_\varepsilon^q)\rangle$ for $\varepsilon > 0$ is proved. Now the convergence of $\bar{A}_\varepsilon$ follows from (2.25) if we can show

$$\lim_{\varepsilon\downarrow 0} \frac{1}{\varepsilon^{mk}} \langle [\prod_{p=1}^{k_1} \tilde{\bar{r}}^p_{u_p v_p \varepsilon} \prod_{q=1}^{k_2} |\tilde{\tilde{\bar{r}}}_\varepsilon^q|^{2\alpha}]^2\rangle = 0\,. \tag{2.26}$$

With the aid of Hölder's inequality the estimate

$$\begin{aligned}
&\frac{1}{\varepsilon^{mk}} \langle [\prod_{p=1}^{k_1} \tilde{\bar{r}}^p_{u_p v_p \varepsilon} \prod_{q=1}^{k_2} |\tilde{\tilde{\bar{r}}}_\varepsilon^q|^{2\alpha}]^2\rangle \\
&\leqq \frac{1}{\varepsilon^{mk}} [\prod_{p=1}^{k_1} \langle (\tilde{\bar{r}}^p_{u_p v_p \varepsilon})^{2(k_1+k_2)}\rangle \prod_{q=1}^{k_2} \langle |\tilde{\tilde{\bar{r}}}_\varepsilon^q|^{4\alpha(k_1+k_2)}\rangle]^{1/(k_1+k_2)}
\end{aligned}$$

follows and we have for $\frac{1}{2} < \alpha \leqq 1$

$$\langle |\tilde{\tilde{\bar{r}}}_\varepsilon^q|^{4\alpha(k_1+k_2)}\rangle \leqq \langle |\tilde{\tilde{\bar{r}}}_\varepsilon^q|^{4(k_1+k_2)}\rangle^\alpha\,,$$

and for $\alpha > 1$

$$\langle |\tilde{\tilde{\bar{r}}}_\varepsilon^q|^{4\alpha(k_1+k_2)}\rangle \leqq \langle |\tilde{\tilde{\bar{r}}}_\varepsilon^q|^{4[\alpha](k_1+k_2)} + |\tilde{\tilde{\bar{r}}}^q|^{4([\alpha]+1)(k_1+k_2)}\rangle\,.$$

By means of Theorem 2.8, we obtain

$$\langle \prod_{i=1}^{p} r_{u_i v_i \varepsilon}(x_i)\rangle = O(\sqrt{\varepsilon^{mp}})$$

for $1 \leqq v_i \leqq t$, $1 \leqq u_i \leqq l$, $p = 2, 3, \ldots$, and then

$$\begin{aligned}
&\frac{1}{\varepsilon^{mk}} \langle [\prod_{p=1}^{k_1} \tilde{r}_{u_p v_p \varepsilon} \prod_{q=1}^{k_2} |\tilde{\tilde{\bar{r}}}_\varepsilon^q|^{2\alpha}]^2\rangle \\
&\leqq \begin{cases} \dfrac{1}{\varepsilon^{mk}} O(\varepsilon^{m(k_1+2k_2\alpha)}) = O(\varepsilon^{mk_2(2\alpha-1)})\,, & \text{for } \tfrac{1}{2} < \alpha \leqq 1\,, \\[2ex] \dfrac{1}{\varepsilon^{mk}} O(\varepsilon^{m(k_1+2k_2[\alpha])}) = O(\varepsilon^{mk_2(2[\alpha]-1)})\,, & \text{for } \alpha > 1\,. \end{cases}
\end{aligned}$$

Hence, (2.26) is proved, and also the convergence of the moments (2.21). Convergence in distribution follows analogously as in the proof of Theorem 2.7; and this completes the proof. ◀

2.2. Limit distributions for eigenvalues and eigenvectors of random matrices

In this section we deal with limit theorems for the distribution laws of eigenvalues and eigenvectors of random matrices, which are obtained by the method of Ritz for eigenvalue problems for random differential operators. First we give some facts on the method of Ritz for eigenvalue problems. These facts inform us of the type of eigenvalue problems for random matrices which can be considered. A second part contains some considerations on random eigenvalue problems for matrices. Then we prove the most important theorem of this section. This theorem deals with the limit distribution law for random eigenvalues and eigenvectors. In a third section we consider some applications of this main theorem.

2.2.1. The method of Ritz

In our subsequent investigations we will refer to some results for eigenvalue problems; especially in the application of the method of Ritz to eigenvalue problems. These results can be found in Michlin [1] and Zeidler [2].

Let $\mathscr{X}$ be a real separable Hilbert space with dim $\mathscr{X} = \infty$. The scalar product is denoted by $(\!(.\,,.)\!)$ and the norm on $\mathscr{X}$ by $||.||$. A bilinear form $a(u, v)$ is a mapping of $\mathscr{X} \times \mathscr{X}$ into $\mathbb{R}$:

$$(u, v) \in \mathscr{X} \times \mathscr{X} \to a(u, v) \in \mathbb{R},$$

which is linear in both arguments. We say that

a is bounded if, and only if, $|a(u, v)| \leqq c||u|| \, ||v||$ for some $c > 0$ and all $u, v \in \mathscr{X}$;

a is positive if, and only if, $a(u, u) > 0$ for all $u \neq 0$ from $\mathscr{X}$;

a is positive definite if, and only if, $a(u, u) \geqq s||u||^2$ for some $s > 0$ and all $u \in \mathscr{X}$;

a is symmetric if, and only if, $a(u, v) = a(v, u)$ for all $u, v \in \mathscr{X}$;

a is compact if, and only if, the convergence $a(u_n, v_n) \to a(u, v)$ follows from the weak convergence of u_n to u and v_n to v.

A sequence $\{\varphi_i\}$ of elements from $\mathscr{X}$ is said to be a basis of $\mathscr{X}$ if a finite set of the φ_i is always independent and

$$\mathscr{X} = \overline{\bigcup_n \mathscr{X}_n} \quad \text{where} \quad \mathscr{X}_n = \text{lin}\,\{\varphi_1, \ldots, \varphi_n\} \quad \text{for } n = 1, 2, \ldots$$

To investigate eigenvalue problems for ordinary differential operators and partial differential operators we consider the eigenvalue problem

$$a(u, v) = \bar{\lambda} b(u, v) \tag{2.27}$$

for bilinear forms. $\bar{\lambda} \in \mathbb{R}$, $u \in \mathscr{X}$, $u \neq 0$ is a solution of (2.27) if this equation is satisfied for all $v \in \mathscr{X}$. The number $\bar{\lambda}$ denotes an eigenvalue and $u \in \mathscr{X}$ an eigenelement of the eigenvalue problem (2.27). The Ritz's eigenvalue problem associated with the eigenvalue problem (2.27) above is defined by

$$a({}^n u, \varphi_j) = {}^n\bar{\lambda} b({}^n u, \varphi_j) , \qquad j = 1, 2, \ldots , n , \tag{2.28}$$

where ${}^n u = \sum_{k=1}^{n} {}^n u_k \varphi_k$. These equations are algebraic equations for the determination of ${}^n\bar{\lambda}$ and ${}^n u_k$, $k = 1, 2, \ldots , n$:

$$\sum_{k=1}^{n} {}^n u_k [a(\varphi_k, \varphi_j) - {}^n\bar{\lambda} b(\varphi_k, \varphi_j)] = 0 , \qquad j = 1, 2, \ldots , n .$$

The eigenvalues ${}^n\bar{\lambda}$ are computed from

$$\det [a(\varphi_k, \varphi_j) - {}^n\bar{\lambda} b(\varphi_k, \varphi_j)] = 0 ,$$

and then the vector $({}^n u_1, {}^n u_2, \ldots , {}^n u_n)^\mathsf{T}$ is determined from the homogeneous equation system which belongs to the eigenvalue ${}^n\bar{\lambda}$.

The following theorem is concerned with the existence of eigenvalues and eigenelements of (2.27) and the connection between the solutions ${}^n\bar{\lambda}$, ${}^n u$ of (2.28) and $\bar{\lambda}$, u of (2.27).

Theorem 2.10. *Assume that the bilinear forms a and b of (2.27) are bounded and symmetric. Furthermore, let a be positive and compact and let b be positive definite. The following statements holds*

(1) *The variational problems*

$$\min_{u \in \mathscr{X}} b(u, u) = \bar{\lambda}^{-1}$$

with the conditions $a(u, u) = 1$, $a(u_i, u) = 0$ *for* $i = 1, 2, \ldots , m - 1$ *lead to the eigensolutions* $(\bar{\lambda}_m, u_m)$ *of (2.27) for* $m = 1, 2, \ldots$ *The condition* $a(u_i, u) = 0$ *is not applicable for* $m = 1$.

(2) *Every eigenvalue* $\bar{\lambda}_m$ *is of finite multiplicity, i.e. the corresponding space of eigenelements is of finite dimension. We have*

$$0 < \ldots \leqq \bar{\lambda}_2 \leqq \bar{\lambda}_1 \quad \text{and} \quad \lim_{m \to \infty} \bar{\lambda}_m = 0 .$$

(3) *For every* $u \in \mathscr{X}$,

$$u = \sum_{k=1}^{\infty} a(u, u_k)\, u_k \quad \text{in } \mathscr{X} ,$$

and $a(u_i, u_j) = \delta_{ij}$ *for* $i, j = 1, 2, \ldots$

(4) *The eigenvalues* ${}^n\bar{\lambda}$ *of* (2.28) *converge to the eigenvalues of* (2.27) *as* $n \to \infty$, *i.e.*

$$\lim_{n\to\infty} {}^n\bar{\lambda}_i = \bar{\lambda}_i ,$$

if $0 < {}^n\bar{\lambda}_n \leqq {}^n\bar{\lambda}_{n-1} \leqq \dots \leqq {}^n\bar{\lambda}_1$ *are the eigenvalues of* (2.28). *Assuming* $\lim_{n\to\infty} {}^n\bar{\lambda} = \bar{\lambda} \neq 0$, *then the sequence* $\{{}^n u\}$ *of eigenelements of* (2.28) *corresponding to* $\{{}^n\bar{\lambda}\}$, *with* $b({}^n u, {}^n u) = 1$, *contains a subsequence* $\{{}^{n'} u\}$ *which converges in* $\mathscr{X}$, $\lim_{n'\to\infty} {}^{n'}u = u$. $(\bar{\lambda}, u)$ *is an eigensolution of* (2.27).

By means of statement (4) we can show that the convergence

$$\lim_{n\to\infty} {}^n u_i = u_i \quad \text{in } \mathscr{X} \tag{2.29}$$

follows from the property that the eigenvalue $\bar{\lambda}_i$ is simple. We establish (2.29) using the fact that a subsequence $\{{}^{n''}u_i\}$ which converges in $\mathscr{X}$ can be obtained from every subsequence $\{{}^{n'}u_i\}$.

Theorem 2.10 can be stated for random bilinear forms if the assumptions are satisfied for almost all $\omega \in \Omega$. Consequently, the statements are true for almost all $\omega \in \Omega$, and the subsequence which is found in statement (4) is dependent on ω. In the case of a simple eigenvalue we also obtain (2.29) almost surely.

We now consider eigenvalue problems for random differential operators

$$M(\omega)\, u = \lambda N(\omega)\, u , \tag{2.30}$$

for $a \leqq x \leqq b$, with the boundary conditions

$$U_i[u] = 0 , \qquad i = 1, 2, \dots , 2m . \tag{2.31}$$

The differential operators $M(\omega)$ and $N(\omega)$ are defined by

$$M(\omega)\, u \doteq \sum_{i=0}^{m} (-1)^i \,[f_i(x, \omega)\, u^{(i)}]^{(i)} ,$$

$$N(\omega)\, u \doteq \sum_{i=0}^{m'} (-1)^i \,[g_i(x, \omega)\, u^{(i)}]^{(i)} ,$$

where

$$f_i(x, \omega), \quad i = 0, 1, \dots , m , \qquad g_i(x, \omega) , \quad i = 0, 1, \dots , m' ,$$

are sufficiently smooth stochastic processes. Assume that the order $2m'$ of the differential operator N is less than the order $2m$ of M, i.e. $m' < m$. Let the boundary conditions be of the form

$$U_i[u] \doteq \sum_{j=0}^{2m-1} [\alpha_{ij} u^{(j)}(a) + \beta_{ij} u^{(j)}(b)] = 0$$

for $i = 1, 2, \dots , 2m$ where α_{ij} and β_{ij} denote real non-random numbers. A function u belongs to the domain of definition of $M(\omega)$ if $u \in \mathbf{C}^{2m}[a, b]$ and the boundary conditions (2.31) are satisfied. Then we write $u \in \mathrm{D}(M(\omega))$.

The boundary conditions can be classified as required and natural boundary conditions. The required boundary conditions are obtained by elimination of the m-th and higher derivatives from as many boundary conditions (3.31) as possible. The remaining boundary conditions which are linearly independent of the required boundary conditions are called natural boundary conditions.

Let $M(\omega)$ and $N(\omega)$ be symmetric, i.e.

$$(\!(M(\omega)\,u, v)\!) = (\!(u, M(\omega)\,v)\!)\,, \qquad (\!(N(\omega)\,u, v)\!) = (\!(u, N(\omega)\,v)\!)\,,$$

for every $u, v \in \mathrm{D}(M(\omega))$. Furthermore, let $M(\omega)$ be positive definite on $\mathbf{L}_2(a, b)$, and $N(\omega)$ positive, i.e.

$$(\!(M(\omega)\,u, u)\!) \geqq s(\omega)\,(\!(u, u)\!)\,, \qquad (\!(N(\omega)\,u, u)\!) > 0 \quad \text{a.s.} \tag{2.32}$$

for every $u \in \mathrm{D}(M(\omega))$, $u \neq 0$, where $s(\omega) > 0$ a.s. The scalar product in $\mathbf{L}_2(a, b)$ is denoted by $(\!(.\,, .)\!)$.

The generalized eigenvalue problem belonging to (2.30) and (2.31) leads to

$$a(u, v) = \grave{\lambda} b(u, v)\,, \tag{2.33}$$

where $\lambda = \bar{\lambda}^{-1}$ and

$$a(u, v) \doteq (\!(N(\omega)\,u, v)\!)\,, \qquad b(u, v) \doteq (\!(M(\omega)\,u, v)\!)\,.$$

A solution $(\bar{\lambda}, u)$ of (2.33) is an element u from $\mathcal{H}_M$ and a $\bar{\lambda} \in R$ which satisfy (2.33) for every $v \in \mathcal{H}_M$. $\mathcal{H}_M$ denotes the energetic space of the operator M, i.e. $\mathcal{H}_M$ is the completion of $\mathrm{D}_{(M(\omega))}$ in the metric $[\![u, v]\!] \doteq (\!(M(\omega)\,u, v)\!)$. The energetic space is a Hilbert space with scalar product $[\![u, v]\!] = (\!(M(\omega)\,u, v)\!)$ for u, $v \in \mathrm{D}(M(\omega))$. In general, the functions from $\mathcal{H}_M$ satisfy only the required boundary conditions.

Using the conditions for $M(\omega)$ and $N(\omega)$, the bilinear forms a and b are a.s. bounded and a.s. symmetric. Furthermore, by means of inequality (2.32) a is positive and b is positive definite because $\mathcal{X} = \mathcal{H}_M$. To get the results of Theorem 2.10 we must prove that $a(u, v)$ is compact a.s.

We will deal with the proof of the compactness for the eigenvalue problem (2.30) with the boundary conditions

$$u(a) = u'(a) = \dots = u^{(m-1)}(a) = u(b) = u'(b) = \dots = u^{(m-1)}(b) = 0$$

where $0 \leqq g_i(x, \omega) \leqq G$ on $[a, b]$ a.s. In this case we have $\mathcal{H}_M = \mathring{\mathbf{W}}_2^m(a, b)$. First, the strong convergence in $\mathring{\mathbf{W}}_2^j(a, b)$, for $j < m$, follows from the weak convergence $u_n \rightharpoonup u$ in $\mathring{\mathbf{W}}_2^m(a, b)$ since the imbedding operator of $\mathring{\mathbf{W}}_2^m(a, b)$ in $\mathring{\mathbf{W}}_2^j(a, b)$ is compact, and consequently this property holds. Using

$$|a(u, v)| \leqq \sum_{i=0}^{m'} |(\!(g_i u^{(i)}, v^{(i)})\!)| \leqq \sum_{i=0}^{m'} \sqrt{(\!(g_i u^{(i)}, u^{(i)})\!)(\!(g_i v^{(i)}, v^{(i)})\!)}$$
$$\leqq G(m' + 1)\sqrt{\|u\|_{\mathring{\mathbf{W}}_2^{m'}}, \|v\|_{\mathring{\mathbf{W}}_2^{m'}}}\,,$$

we have

$$|a(u_n, v_n) - a(u, v)| \leqq |a(u_n - u, v_n)| + |a(u, v_n - v)| \to 0$$

as $n \to \infty$ from the weak convergence $u_n \to u$ and $v_n \to v$ in the sense of $\mathring{\mathbf{W}}_2^m(a, b)$. Hence, the compactness of $a(u, v)$ (ω) a.s. is proved in this case.

We consider the random eigenvalue problem

$$-(f_1(x, \omega)\, u')' + f_0(x, \omega)\, u = \lambda g_0(x, \omega)\, u \tag{2.34}$$

for $a \leqq x \leqq b$, with the boundary conditions

$$\alpha u'(a) - \beta u(a) = 0\,, \qquad \gamma u'(b) + \delta u(b) = 0$$

where $\alpha > 0$, $\gamma > 0$, $\beta \geqq 0$, $\delta \geqq 0$, $\beta + \delta > 0$. Let

$$f_1(x, \omega) \geqq 0\,, \qquad f_0(x, \omega) \geqq 0\,, \qquad \int_a^b \frac{\mathrm{d}x}{f_1(x, \omega)} < A(\omega) < \infty \quad \text{a.s.}$$

and

$$0 < G_0(\omega) \leqq g_0(x, \omega) \leqq G_1(\omega) \quad \text{a.s.}$$

By a calculation we obtain

$$\begin{aligned} &(\!(M(\omega)\, u, v)\!) \\ &= (\!(f_1 u', v')\!) + (\!(f_0 u, v)\!) + \frac{\delta}{\gamma} f_1(b)\, u(b)\, v(b) + \frac{\beta}{\alpha} f_1(a)\, u(a)\, v(a) \end{aligned}$$

and from this the symmetry of $M(\omega)$ a.s., and for $\beta f_1(a) > 0$ a.s. the inequality

$$(\!(M(\omega)\, u, u)\!) \geqq c_1[(\!(f_1 u', u')\!) + u^2(a)]$$

where $c_1 > 0$, Furthermore, we have

$$\begin{aligned} u^2(x) &\leqq 2u^2(a) + 2\left(\int_a^x u'(t)\, \mathrm{d}t\right)^2 \\ &\leqq 2u^2(a) + 2\left(\int_a^x \frac{\mathrm{d}t}{f_1(t)} \cdot \int_a^x f_1(t)\, u'^2(t)\, \mathrm{d}t\right) \\ &\leqq c_2\,[(\!(f_1 u', u')\!) + u^2(a)]\,; \end{aligned}$$

and therefore $M(\omega)$ is positive definite,

$$(\!(M(\omega)\, u, u)\!) \geqq c(\omega)\, (\!(u, u)\!)\,,$$

with $c(\omega) > 0$ a.s. For $\alpha = 0$ or $\gamma = 0$ the bilinear form $b(u, v)$ is also positive definite. It is easy to see that $a(u, v) = (\!(g_0\, u, v)\!)$ is positive.

We prove the compactness of $a(u, v)$ on $\mathscr{H}_M$. We have

$$b(u, u) = [\![u, u]\!] \geqq (\!(f_1 u', u')\!) + \frac{\beta}{\alpha} f_1(a)\, u^2(a)$$

for $0 < f_1(a) < \infty$ a.s. Then from $u \in \mathscr{S}$ where $\mathscr{S} = \{u \in \mathscr{H}_M : b(u, u) \leqq c^2\}$ the inequalities

$$(\!(f_1 u', u')\!) \leqq C^2\,, \qquad |u(a)| \leqq C\sqrt{\frac{\alpha}{\beta f_1(a)}} \tag{2.35}$$

are obtained. Defining

$$K(x, t) = \begin{cases} \dfrac{1}{\sqrt{f_1(t)}}, & \text{for} \quad a \leqq t < x, \\ 0, & \text{otherwise,} \end{cases}$$

we have

$$u(x) = u(a) + \int_a^b K(x, t) \sqrt{f_1(t)}\, u'(t)\, \mathrm{d}t,$$

where $\int_a^b \int_a^b K(x, t)^2 \, \mathrm{d}x \, \mathrm{d}t < \infty$ a.s. Because of the inequalities (2.35) a sequence $(u_n(x))$, $u_n \in \mathcal{S}$, can be found such that $\int_a^b K(x, t) \sqrt{f_1(t)}\, u_n'(t)\, \mathrm{d}t$ converges in $\mathbf{L}_2(a, b)$, and a subsequence $(u_{n'}(x))$ of $(u_n(x))$ such that $(u_{n'}(a))$ converges. The convergence of $(u_{n'})$ in $\mathbf{L}_2(a, b)$ follows from

$$⟪u, u⟫ \leqq 2(b - a)\, u^2(a) + 2 \int_a^b \Big(\int_a^b K(x, t) \sqrt{f_1(t)}\, u'(t)\, \mathrm{d}t\Big)^2 \mathrm{d}x\,.$$

Hence, the bilinear form $a(u, v)$ is compact, and the statements of Theorem 2.10 hold for the eigenvalue problem (2.34). Now we investigate the conditions of Theorem 2.10 for the eigenvalue problem

$$\big(f_2(x, \omega)\, u''\big)'' = \lambda\big(g(x, \omega)\, u - g_1 u''\big) \tag{2.36}$$

with the boundary conditions

$$u(a) = u(b) = u''(a) = u''(b) = 0\,,$$

where

$$f_2(x, \omega) \geqq c(\omega) > 0\,,$$

$$g(x, \omega) \geqq \bar{c}(\omega) > 0 \quad \text{and} \quad g_1 = \text{const} > 0\,.$$

In Section 2.2.4 we will deal in detail with this eigenvalue problem as an application of the general theory.

We have

$$b(u, v) = ⟪M(\omega)\, u, v⟫ = ⟪f_2 u'', v''⟫\,,$$

$$a(u, v) = ⟪N(\omega)\, u, v⟫ = ⟪g_1 u', v'⟫ + ⟪gu, v⟫\,,$$

and $⟪N(\omega)\, u, u⟫ > 0$ for $u \neq 0$. To prove the positive definiteness of $b(u, u)$ on $\mathbf{L}_2(a, b)$ we use Poincaré's inequality

$$\int_a^b u^2(x)\, \mathrm{d}x \leqq \frac{1}{2}\,(b - a)^2 \int_a^b u'^2(x)\, \mathrm{d}x + \frac{1}{b - a}\left(\int_a^b u(x)\, \mathrm{d}x\right)^2. \tag{2.37}$$

For the proof of this inequality

$$\big(u(x) - u(y)\big)^2 = u^2(x) + u^2(y) - 2u(x)\, u(y) \leqq (b - a) \int_a^b u'^2(t)\, \mathrm{d}t$$

is obtained from

$$u(x) - u(y) = \int_y^x u'(t)\,\mathrm{d}t\,. \qquad (2.38)$$

Then Poincaré's inequality follows by integration over x and y from a to b in each case. Using

$$\int_a^b u'(x)\,\mathrm{d}x = 0 \quad \text{and} \quad \int_a^b u^2(x)\,\mathrm{d}x \leqq (b-a)^2 \int_a^b u'^2(x)\,\mathrm{d}x\,,$$

we have

$$\langle\!\langle M(\omega)u, u\rangle\!\rangle \geqq c\langle\!\langle u'', u''\rangle\!\rangle \geqq \frac{2c}{(b-a)^4}\langle\!\langle u, u\rangle\!\rangle$$

if we apply the inequality (2.37) to $u'(x)$. Hence, the bilinear form $b(u, u)$ is positive definite on $\mathbf{L}_2(a, b)$.

We now show the compactness of $a(u, v)$. Integrating (2.38) for the function u' over y from a to b, and using $\int_a^b u'(y)\,\mathrm{d}y = 0$, we have

$$(b-a)\,u'(x) = \int_a^b \int_y^x u''(t)\,\mathrm{d}t\,\mathrm{d}y = \int_a^b K(x, t)\sqrt{f_2(t, \omega)}\,u''(t)\,\mathrm{d}t\,,$$

where

$$K(x, t) = \begin{cases} \dfrac{t-a}{\sqrt{f_2(t, \omega)}}\,, & \text{for} \quad t < x\,, \\ \dfrac{t-b}{\sqrt{f_2(t, \omega)}}\,, & \text{for} \quad t > x\,. \end{cases}$$

Because of this relation, from a set $\mathcal{S}$ with $\langle\!\langle f_2 u'', u''\rangle\!\rangle \leqq C$ for all $u \in \mathcal{S}$ a sequence $(u_n(x))$ can be found so that $(u_n'(x))$ converges in $\mathbf{L}_2(a, b)$. With the aid of

$$u(x) = \int_a^x u'(t)\,\mathrm{d}t$$

a subsequence $(u_{n'}(x))$ can be found from the sequence $(u_n(x))$, and $(u_{n'}(x))$ converges in $\mathbf{L}_2(a, b)$. Hence, the bilinear form $a(u, v)$ is compact.

Theorem 2.10 has been stated in that manner so that it can be also applied to eigenvalue problems of partial elliptic differential operators.

2.2.2. Eigenvalue problems for random matrices

This section is concerned with eigenvalue problems of the form

$$\big(A_0 + B(\omega)\big)\,U = \Lambda\big(C_0 + D(\omega)\big)\,U \qquad (2.39)$$

where

$$A_0 = (a_{ij})_{1 \leqq i, j \leqq n}\,, \qquad C_0 = (c_{ij})_{1 \leqq i, j \leqq n}$$

denote non-random symmetric matrices and

$$B(\omega) = \big(b_{ij}(\omega)\big)_{1 \leqq i, j \leqq n}\,, \qquad D(\omega) = \big(d_{ij}(\omega)\big)_{1 \leqq i, j \leqq n}$$

are random symmetric matrices. $U = (u_i)^{\mathrm{T}}_{1 \leq i \leq n}$ is a vector and Λ a real number (because the matrices A_0, B, C_0, D are symmetric matrices). Let $\langle B(\omega) \rangle = \langle D(\omega) \rangle = 0$. This condition does not restrict the generality of the eigenvalue problem since we can write the matrix $(A_0 + \langle B \rangle) + (B(\omega) - \langle B \rangle)$ for $A_0 + B(\omega)$ and $(C_0 + \langle D \rangle) + (D(\omega) - \langle D \rangle)$ for $C_0 + D(\omega)$. Furthermore, the matrix $C_0 + D(\omega)$ is assumed to be positive definite; that is

$$\langle\!\langle (C_0 + D(\omega))\, V, V \rangle\!\rangle \geqq s_0(\omega)\, |V|^2$$

for all $V \in \mathbb{R}^n$, where $s_0(\omega) > 0$ a.s. Let the k-th moment of $s_0^{-1}(\omega)$ exist for $k = 1, 2, \dots$,

$$\langle s_0^{-k} \rangle < \infty\,, \qquad k = 1, 2, \dots$$

The scalar product of vectors is denoted by $\langle\!\langle \cdot\,, \cdot \rangle\!\rangle$ and the norm by $|\cdot|$,

$$\langle\!\langle U, V \rangle\!\rangle \doteq \sum_{i=1}^{n} u_i v_i\,, \qquad |U|^2 \doteq \langle\!\langle U, U \rangle\!\rangle\,,$$

where

$$U = (u_i)_{1 \leq i \leq n}\,, \qquad V = (v_i)_{1 \leq i \leq n}\,.$$

Under the above assumptions the eigenvalue problem (2.39) possesses exactly n real eigenvalues Λ_i, $i = 1, 2, \dots, n$. It is customary to order them according to increasing magnitude so that

$$\Lambda_1(\omega) \leqq \Lambda_2(\omega) \leqq \dots \leqq \Lambda_n(\omega)$$

for almost all $\omega \in \Omega$. Using Courant's principle as in Theorem 1.6 the measurability of the eigenvalues $\Lambda_i(\omega)$ is obtained from the property that these eigenvalues are also eigenvalues of the eigenvalue problem

$$\big(A_0 + B(\omega)\big)\, \big(C_0 + D(\omega)\big)^{-1}\, V = \Lambda V\,.$$

The eigenvector $U(\omega)$ corresponding to the eigenvalue $\Lambda(\omega)$ and $\bar{U}(\omega)$ to $\bar{\Lambda}$ with $\Lambda(\omega) \neq \bar{\Lambda}(\omega)$ are orthogonal relative to $C_0 + D(\omega)$, i.e.

$$\langle\!\langle \big(C_0 + D(\omega)\big)\, U(\omega), \bar{U}(\omega) \rangle\!\rangle = 0\,.$$

We now formulate a result which states the existence of the moments of random eigenvalues and eigenvectors.

Lemma 2.4. *Let b_{ij}, $i, j = 1, 2, \dots, n$, and $s_0^{-1}(\omega)$ be random variables for which all the moments exist ($s_0(\omega)$ is the random variable introduced in the definition of a positive definite matrix). Then all the moments exist of the eigenvalues $\Lambda_i(\omega)$ and the eigenvectors $U_i(\omega)$, $i = 1, 2, \dots, n$, of the eigenvalue problem* (2.39). *The eigenvectors are orthonormalized by $\langle\!\langle \big(C_0 + D(\omega)\big)\, U_i, U_j \rangle\!\rangle = \delta_{ij}$.*

Proof. The proof for the eigenvalues is a simple conclusion from the maximum principle. We have

$$\Lambda_{n-i} = \max_U \langle\!\langle (A_0 + B)\, U, U \rangle\!\rangle\,, \quad \text{for} \quad i = 0, 1, \dots, n-1\,,$$

with the conditions for U

$$\mathbb{C}(C_0 + D) U, U\mathbb{D} = 1 , \qquad \mathbb{C}(C_0 + D) U, U_s\mathbb{D} = 0$$
$$\text{for} \quad s = n, n-1 , \dots , n - i + 1 .$$

The vectors $U_1, \dots , U_n$ denote the eigenvectors of the eigenvalue problem (2.39) associated with the eigenvalues $\Lambda_1, \dots , \Lambda_n$. Hence, the estimate

$$\begin{aligned} |\mathbb{C}(A_0 + B) U, U\mathbb{D}| &= | \sum_{i,j=1}^{n} (a_{ij} + b_{ij}) u_i u_j| \\ &\leqq [\sum_{i,j=1}^{n} (a_{ij} + b_{ij})^2]^{1/2} \sum_{i=1}^{n} u_i^2 \\ &\leqq |A_0 + B| \, |U|^2 \leqq \frac{1}{s_0} |A_0 + B| \mathbb{C}(C_0 + D) U, U\mathbb{D} \end{aligned}$$

is obtained for all $U \in \mathbb{R}^n$. From this it follows that

$$|\mathbb{C}(A_0 + B) U, U\mathbb{D}| \leqq \frac{1}{s_0} |A_0 + B|$$

for those $U \in \mathbb{R}^n$ over which the maximum is taken, and therefore

$$|\Lambda_{n-i}| \leqq \frac{1}{s_0} |A_0 + B| \quad \text{for} \quad i = 0, 1, \dots , n - 1 .$$

Lemma 2.4 is proved for the eigenvalues.

The eigenvectors U associated with the eigenvalue Λ are determined by

$$\big(A_0 + B(\omega) - \Lambda(C_0 + D(\omega))\big) U = 0 ,$$

or by

$$[(A_0 + B(\omega)) \big(C_0 + D(\omega)\big)^{-1} - \Lambda I] U' = 0 .$$

where $U' = \big(C_0 + D(\omega)\big) U$. The inverse matrix $\big(C_0 + D(\omega)\big)^{-1}$ exists because the assumption of positive definiteness. The elements of $\big(C_0 + D(\omega)\big)^{-1}$ are random variables. Using Theorem 1.7, the components of U' are measurable, and hence also the components of U if the parameters appearing in U are assumed to be measurable.

Let

$$A_k = \{\omega \in \Omega : \text{rank} \big(A_0 + B(\omega) - \Lambda\big(C_0 + D(\omega)\big)\big) = n - k\} ,$$

$k = 1, 2, \dots , n - 1$. From (2.39) we obtain k linear independent vectors $V_1(\omega), \dots , V_k(\omega)$ on A_k which can be transformed to orthonormal vectors $\overline{U}_1(\omega), \dots , \overline{U}_k(\omega)$ with respect to $[\![U, V]\!] \doteq \mathbb{C}\big(C_0 + D(\omega)\big) U, V\mathbb{D}$. Hence, we have

$$\mathbb{C}\big(C_0 + D(\omega)\big) \overline{U}_i, \overline{U}_j\mathbb{D} = \delta_{ij} , \quad \text{for} \quad i, j = 1, 2, \dots , k ,$$

and the components of $\overline{U}_i(\omega)$ are random variables on A_k. Let A_{ik} be the set A_k associated with the i-th eigenvalue $\Lambda_i(\omega)$ and we denote the above calculated vectors $\overline{U}_j(\omega), j = 1, 2, \dots , k$, by $U_{ik1}(\omega), \dots , U_{ikk}(\omega)$. For the first eigenvalue the sets $A_{11}, \dots, A_{1l_1}$ are obtained with $\mathsf{P}(A_{1i}) \neq 0$, $i = 1, 2, \dots, l_1$. Then we define for A_{1s}

$$U_1(\omega) \doteq U_{1s1}(\omega)\,, \qquad U_2(\omega) \doteq U_{1s2}(\omega), \dots, \qquad U_s(\omega) \doteq U_{1ss}(\omega)\,,$$

and consider the eigenvalue $\Lambda_{s+1}(\omega)$ from which the following eigenvectors are determined. By this method we can construct n eigenvectors for all $\omega \in \Omega$ whose components are random variables with

$$\langle\!\langle (C_0 + D(\omega))\, U_i, U_j \rangle\!\rangle = \delta_{ij}\,, \quad \text{for} \quad i, j = 1, 2, \dots, n\,.$$

It is clear that the eigenvectors associated with the different eigenvalues are orthogonal. Hence, the existence of the moments of the eigenvectors follows from the inequality

$$\langle |U_i|^{2k} \rangle \leqq \left\langle \frac{1}{s_0^k} \langle\!\langle (C_0 + D(\omega))\, U_i, U_i \rangle\!\rangle^k \right\rangle = \left\langle \frac{1}{s_0^k} \right\rangle;$$

and this completes the proof of Lemma 2.4. ◀

The dependence of the eigenvalues and eigenvectors on the values b_{ij} and d_{ij} $i, j = 1, 2, \dots, n$, yields the following lemma.

Lemma 2.5. *Let Λ_0 be a simple eigenvalue of the unperturbed eigenvalue problem*

$$A_0 U_0 = \Lambda_0 C_0 U_0$$

associated with the eigenvalue problem (2.39). *The vector U_0 denotes the eigenvector associated with the eigenvalue Λ_0, and we have*

$$\langle\!\langle C_0 U_0, U_0 \rangle\!\rangle = 1\,,$$

and $U_0 = (u_{i0})^{\mathsf{T}}_{1 \leqq i \leqq n}$. Assuming

$$\sum_{i,j=1}^{n} b_{ij}^2 \leqq \gamma^2\,, \qquad \sum_{i,j=1}^{n} d_{ij}^2 \leqq \gamma^2$$

(γ sufficiently small) the expansions

$$\Lambda = \Lambda_0 + \sum_{i,j=1}^{n} (\alpha_{ij} b_{ij} + \beta_{ij} d_{ij}) + \sum_{i,j=1}^{n} (b_{ij}^2 + d_{ij}^2)\, R_\Lambda(b_{11}, \dots, b_{nn}, d_{11}, \dots, d_{nn})\,,$$

$$u_k = u_{k0} + \sum_{i,j=1}^{n} (\alpha_{kij} b_{ij} + \beta_{kij} d_{ij}) + \sum_{i,j=1}^{n} (b_{ij}^2 + d_{ii}^2)\, R_k(b_{11}, \dots, b_{nn}, d_{11}, \dots, d_{nn})$$

can be derived for the eigenvalue Λ and the eigenvector $U = (u_k)^{\mathsf{T}}_{1 \leqq k \leqq n}$ of the eigenvalue problem (2.39) *with $\Lambda|_{b_{ij}=d_{ij}=0} = \Lambda_0$ and $U|_{b_{ij}=d_{ij}=0} = U_0$, where*

$$|R_\Lambda| \leqq c_\Lambda \quad \text{and} \quad |R_k| \leqq c_k \quad \text{for} \quad k = 1, 2, \dots, n\,.$$

Proof. The eigenvalue Λ can be calculated from the determinantal equation

$$\psi(\Lambda, b_{ij}, d_{ij}) \doteq \det\left(A_0 + B - \Lambda(C_0 + D)\right) = \sum_{k=0}^{n} (-\Lambda)^{n-k} J_k = 0$$

corresponding to (2.39), and Λ_0 from

$$\psi(\Lambda_0, 0, 0) = \det(A_0 - \Lambda_0 C_0) = \sum_{k=0}^{n} (-\Lambda_0)^{n-k} J_{k0} = 0 ,$$

where $J_{k|B=D=0} = J_{k0}$. The eigenvalue Λ_0 is simple and therefore we get

$$\sum_{k=0}^{n-1} (n-k)(-\Lambda_0)^{n-k-1} J_{k0} \neq 0 .$$

Hence, using

$$\psi(\Lambda_0, 0, 0) = 0$$

and

$$\frac{\partial\psi}{\partial\Lambda}(\Lambda_0, 0, 0) = -\sum_{k=0}^{n-1} (n-k)(-\Lambda_0)^{n-k-1} J_{k0} \neq 0$$

we can solve the equation

$$\psi(\Lambda, b_{ij}, d_{ij}) = 0 ,$$

if the conditions $\sum_{i,j=1}^{n} b_{ij}^2 \leqq \gamma^2$, $\sum_{i,j=1}^{n} d_{ij}^2 \leqq \gamma^2$ are satisfied. The solution Λ is an analytic function in the variables b_{ij} and d_{ij}, $i, j = 1, 2, \dots, n$; and we have

$$\Lambda = \Lambda_0 + \sum_{i,j=0}^{n} (\alpha_{ij} b_{ij} + \beta_{ij} d_{ij}) + \overline{R}_\Lambda(b_{ij}, d_{ij}) .$$

The term $\overline{R}_\Lambda(b_{ij}, d_{ij})$ denotes a power series, and this series begins with summands of the second order in terms of b_{ij} and d_{ij}. It is easy to see that for $\sum_{i,j=1}^{n} b_{ij}^2 \leqq \gamma^2$, $\sum_{i,j=1}^{n} d_{ij}^2 \leqq \gamma^2$ a bounded function $R_\Lambda(b_{ij}, d_{ij})$, with $|R_\Lambda| \leqq c$, exists such that

$$\overline{R}_\Lambda(b_{ij}, d_{ij}) = \sum_{r,s=1}^{n} (b_{rs}^2 + d_{rs}^2) R_\Lambda(b_{ij}, d_{ij}) .$$

This lemma is proved for eigenvalues.

We consider the eigenvector U corresponding to the eigenvalue Λ of (2.39). The components of the eigenvector U can be calculated from the equations

$$\varphi_i(u_1, \dots, u_n; b_{uv}, d_{uv}) \doteq \sum_{j=1}^{n} \left(a_{ij} + b_{ij} - \Lambda(c_{ij} + d_{ij})\right) u_j = 0 , \quad i = 1, 2, \dots, n. \tag{2.40}$$

From the simplicity of the eigenvalue Λ_0 it follows that

$$\operatorname{rank}\left(\frac{\partial\varphi_i}{\partial u_j}(u_{10}, \dots, u_{n0}; 0, 0)\right) = \operatorname{rank}(a_{ij} - \Lambda_0 c_{ij}) = n - 1 .$$

$U_0 = (u_{i0})^{\mathrm{T}}_{1 \leq i \leq n}$ is the eigenvector associated with the eigenvalue Λ_0, where $\langle\!\langle CU_0, U_0 \rangle\!\rangle = 1$. Let

$$\det (a_{ij} - \Lambda_0 c_{ij})_{\substack{1 \leq i, j \leq n \\ i \neq p, j \neq q}} \neq 0 .$$

Hence, the system of equations (2.40), with $i \neq p$, can be solved for $u_1, \dots, u_{q-1}, u_{q+1}, \dots, u_n$ in a neighbourhood of $(u_{10}, \dots, u_{n0}; 0, 0)$,

$$u_i = \psi_i(u_q; b_{uv}, d_{uv}) , \qquad i = 1, \dots, q-1 , \quad q+1, \dots, n .$$

The parameter u_q is determined by the normalization condition

$$\begin{aligned} \chi(u_q; b_{uv}, d_{uv}) &\doteq \langle\!\langle (C_0 + D)\, U, U \rangle\!\rangle - 1 \\ &= \sum_{\substack{i, j=1 \\ i, j \neq q}}^{n} (c_{ij} + d_{ij})\, \psi_i \psi_j + 2u_q \sum_{\substack{i=1 \\ i \neq q}}^{n} (c_{iq} + d_{iq})\, \psi_i \\ &\quad + (c_{qq} + d_{qq})\, u_q^2 - 1 = 0 . \end{aligned} \tag{2.41}$$

We remark that

$$u_{j0} = u_{q0} \frac{\partial \psi_j}{\partial u_q} (u_{q0}; 0, 0) , \qquad j = 1, \dots, q-1, q+1, \dots, n ,$$

follows from the system of equations

$$\sum_{\substack{j=1 \\ j \neq q}}^{n} (a_{ij} - \Lambda_0 c_{ij})\, u_{j0} = - (a_{iq} - \Lambda_0 c_{iq})\, u_{q0} , \qquad i = 1, \dots, p-1, p+1, \dots, n ,$$

$$\sum_{\substack{j=1 \\ j \neq q}}^{n} (a_{ij} - \Lambda_0 c_{ij}) \frac{\partial \psi_j}{\partial u_q} (u_{q0}; 0, 0) = -(a_{iq} - \Lambda_0 c_{iq}) , \qquad i = 1, \dots, p-1, p+1, \dots, n .$$

Using

$$\chi(u_{q0}; 0, 0) = \langle\!\langle C_0 U_0, U_0 \rangle\!\rangle - 1 = 0 ,$$

and

$$\begin{aligned} \frac{\partial \chi}{\partial u_q} (u_{q0}; 0, 0) &= \frac{2}{u_{q0}} \left[\sum_{\substack{i, j=1 \\ i, j \neq q}}^{n} c_{ij} u_{i0} u_{j0} + \sum_{\substack{i=1 \\ i \neq q}}^{n} c_{iq} u_{i0} u_{q0} + c_{qq} u_{q0}^2 \right] \\ &= \frac{2}{u_{q0}} \sum_{i, j=1}^{n} c_{ij} u_{i0} u_{j0} = \frac{2}{u_{q0}} , \end{aligned}$$

(2.41) can be solved for u_q in a neighbourhood of $(u_{q0}; 0, 0)$; and thus

$$u_q = \psi_q(b_{uv}, d_{uv}) .$$

Hence, the components of the eigenvector $U = (u_i)^{\mathrm{T}}_{1 \leq i \leq n}$ calculated from (2.40) can be written as

$$u_i = \psi_i\big(\psi_q(b_{uv}, d_{uv}); b_{uv}, d_{uv}\big) , \qquad i = 1, \dots, q-1, q+1, \dots, n ,$$

$$u_q = \psi_q(b_{uv}, d_{uv}) ,$$

where

$$\langle\!\langle (C_0 + D)\, U, U \rangle\!\rangle = \sum_{i,j=1}^{n} (c_{ji} + d_{ij})\, u_i u_j = 1 .$$

Since the functions φ_i, $i = 1, 2, \ldots, n$, are analytic functions of the given variables, the solutions ψ_i, $i \neq q$, are also analytic functions of these variables. The same statement can be made for the function ψ_q. Then we obtain the expansions

$$u_k = u_{k0} + \sum_{i,j=1}^{n} (\alpha_{kij} b_{ij} + \beta_{kij} d_{ij}) + \bar{R}_k(b_{ij}, d_{ij})$$

in a domain determined by the conditions $\sum_{i,j=1}^{n} b_{ij}^2 \leqq \gamma^2$, $\sum_{i,j=1}^{n} d_{ij}^2 \leqq \gamma^2$. The function $\bar{R}_k(b_{ij}, d_{ij})$ is a power series which begins with summands of the second order. ◀

Before we state the main theorem of this section we calculate the linear terms in the elements of the matrices B and D of (2.39) of the expansions of the eigenvalues and eigenvectors. To this end, we prove the following lemma.

Lemma 2.6. *Let Λ_{i0}, $i = 1, 2, \ldots, n$, $\Lambda_{10} \leqq \Lambda_{20} \leqq \ldots \leqq \Lambda_{n0}$, be the eigenvalues of the eigenvalue problem $A_0 U_0 = \Lambda_0 C_0 U_0$, and U_{i0}, $i = 1, 2, \ldots, n$, be the eigenvectors associated with the eigenvalues Λ_{i0} where*

$$\langle\!\langle C_0 U_{i0}, U_{j0} \rangle\!\rangle = \delta_{ij} , \qquad i, j = 1, 2, \ldots, n .$$

The h-th eigenvalue Λ_{h0} is assumed to be simple. Let Λ_h be the h-th eigenvalue of the eigenvalue problem

$$(A_0 + B)\, U = \Lambda (C_0 + D)\, U ,$$

and let U_h be the eigenvector associated with Λ_h, where $\langle\!\langle (C_0 + D)\, U_h, U_h \rangle\!\rangle = 1$ and $\Lambda_{h|B=D=0} = \Lambda_{h0}$, $U_{h|B=D=0} = U_{h0}$. Then the expansions

$$\Lambda_h = \Lambda_{h0} + \Lambda_{h1} + \sum_{k=2}^{\infty} \Lambda_{hk} .$$

$$U_h = U_{h0} + U_{h1} + \sum_{k=2}^{\infty} U_{hk}$$

proved in Lemma 2.5 possesses the linear terms

$$\Lambda_{h1} = \langle\!\langle B U_{h0}, U_{h0} \rangle\!\rangle - \Lambda_{h0} \langle\!\langle D U_{h0}, U_{h0} \rangle\!\rangle ,$$

$$U_{h1} = \sum_{\substack{j=1 \\ j \neq h}}^{n} \frac{1}{\Lambda_{j0} - \Lambda_{h0}} [\Lambda_{h0} \langle\!\langle D U_{h0}, U_{j0} \rangle\!\rangle - \langle\!\langle B U_{h0}, U_{j0} \rangle\!\rangle]\, U_{j0}$$
$$- \frac{1}{2} \langle\!\langle D U_{h0}, U_{h0} \rangle\!\rangle\, U_{h0} .$$

Proof. First, the eigenvalue Λ_h is simple for $\sum_{i,j=1}^{n} b_{ij}^2 \leqq \gamma^2$, $\sum_{i,j=1}^{n} d_{ij}^2 \leqq \gamma^2$. We have the expansion

$$\Lambda_h = \Lambda_{h0} + \Lambda_{h1} + \sum_{k=2}^{\infty} \Lambda_{hk}\,, \tag{2.42}$$

where Λ_{hk}, $k = 1, 2, \ldots$, denotes the homogeneous part of the k-th order in the variables b_{ij} and d_{ij}, $i, j = 1, 2, \ldots, n$. In the same way, the expansion of the eigenvector U_h is

$$U_h = U_{h0} + U_{h1} + \sum_{k=2}^{\infty} U_{hk}\,, \tag{2.43}$$

where U_{hk} denotes the homogeneous part of the k-th order of U_h, U_{h0} is the eigenvector associated with the eigenvalue Λ_{h0}. where $(\!(C_0U_{h0}, U_{h0})\!) = 1$. Using the expansions (2.42) and (2.43) we obtain from (2.39) for the terms of 0-th order

$$A_0U_{h0} = \Lambda_{h0}C_0U_{h0}\,,$$

and for the terms of first order

$$(A_0 - \Lambda_{h0}C_0)\,U_{h1} = \Lambda_{h0}DU_{h0} + \Lambda_{h1}C_0U_{h0} - BU_{h0}\,. \tag{2.44}$$

From the normalization condition $(\!((C_0 + D)\,U_h, U_h)\!) = 1$ we can get for the terms of 0-th order

$$(\!(C_0U_{h0}, U_{h0})\!) = 1\,,$$

and for the terms of first order

$$(\!(C_0U_{h0}, U_{h1})\!) = -\tfrac{1}{2}(\!(DU_{h0}, U_{h0})\!)\,. \tag{2.45}$$

(2.44) can be solved if and only if the condition

$$(\!(\Lambda_{h0}DU_{h0} + \Lambda_{h1}C_0U_{h0} - BU_{h0}, U_{h0})\!) = 0$$

is satisfied, i.e. if we have

$$\Lambda_{h1} = (\!(BU_{h0}, U_{h0})\!) - \Lambda_{h0}(\!(DU_{h0}, U_{h0})\!)\,. \tag{2.46}$$

This gives the calculation of the term Λ_{h1}.

We determine U_{h1} from (2.44) and the normalization condition (2.45). Let $\Lambda_{10}, \Lambda_{20}, \ldots, \Lambda_{n0}$ be the eigenvalues, and let $U_{10}. U_{20}, \ldots, U_{n0}$ be the eigenvectors of the unperturbed problem $A_0U_{\cdot 0} = \Lambda_{\cdot 0}C_0U_{\cdot 0}$. Defining

$$V_h \doteq \Lambda_{h0}DU_{h0} + \Lambda_{h1}C_0U_{h0} - BU_{h0}\,,$$

Eq. (2.44) can be written in the form

$$\Lambda_{h0}U_{h1} - C_0^{-1}A_0U_{h1} = -C_0^{-1}V_h\,. \tag{2.47}$$

For the term $C_0^{-1}A_0U_{h1}$ the equation

$$C_0^{-1}A_0U_{h1} = \sum_{i=1}^{n} (\!(A_0U_{h1}, U_{i0})\!)\,U_{i0} = \sum_{i=1}^{n} \Lambda_{i0}(\!(C_0U_{h1}, U_{i0})\!)\,U_{i0} \tag{2.48}$$

is satisfied. Furthermore, the scalar product of (2.47) with C_0U_{j0} gives the relation

$$\langle\!\langle C_0U_{h1}, U_{j0}\rangle\!\rangle = \frac{\langle\!\langle V_h, U_{j0}\rangle\!\rangle}{\Lambda_{j0} - \Lambda_{h0}}, \quad \text{for} \quad j \neq h\,;$$

and the equation $\langle\!\langle V_h, U_{h0}\rangle\!\rangle = 0$ is obtained from (2.46). Then from (2.47) and (2.48) we have

$$U_{h1} = -\frac{1}{\Lambda_{h0}} C_0^{-1}V_h + \frac{1}{\Lambda_{h0}} \sum_{\substack{j=1\\ j\neq h}}^{n} \frac{\Lambda_{j0}}{\Lambda_{j0} - \Lambda_{h0}} \langle\!\langle V_h, U_{j0}\rangle\!\rangle\, U_{j0} + a' U_{h0}\,,$$

where a' is an arbitrary constant. Using

$$C_0^{-1}V_h = \sum_{j=1}^{n} \langle\!\langle V_h, U_{j0}\rangle\!\rangle\, U_{j0}\,,$$

the equation

$$U_{h1} = \sum_{\substack{j=1\\ j\neq h}}^{n} \frac{\langle\!\langle V_n, U_{j0}\rangle\!\rangle}{\Lambda_{j0} - \Lambda_{h0}} U_{j0} + aU_{h0}$$

can be obtained. The normalization condition (2.45) leads to

$$a = -\tfrac{1}{2}\langle\!\langle DU_{h0}, U_{h0}\rangle\!\rangle\,.$$

Finally, the linear term of the expansion (2.43) is

$$U_{h1} = \sum_{\substack{j=1\\ j\neq h}}^{n} \frac{1}{\Lambda_{j0} - \Lambda_{h0}} [\Lambda_{h0}\langle\!\langle DU_{h0}, U_{j0}\rangle\!\rangle - \langle\!\langle BU_{h0}, U_{j0}\rangle\!\rangle]\, U_{j0} - \frac{1}{2}\langle\!\langle DU_{h0}, U_{h0}\rangle\!\rangle\, U_{h0} \tag{2.49}$$

and this completes the proof of Lemma 2.6. ◀

Let $\big(f_{1\varepsilon}(x, \omega), f_{2\varepsilon}(x, \omega), \dots, f_{l\varepsilon}(x, \omega)\big)$ be a weakly correlated connected vector field on $\mathscr{D} \subset \mathbb{R}^m$, where the sample functions of this random vector field are continous a.s. Define the random variables

$$\bar{r}_{ij}^{\varepsilon}(\omega) = \int_{\mathscr{D}} F_{ij}(x)\, f_{i\varepsilon}(x, \omega)\, \mathrm{d}x = \langle\!\langle F_{ij}, f_{i\varepsilon}(., \omega)\rangle\!\rangle\,, \qquad i = 1, 2, \dots, l\,, \quad j = 1, 2, \dots, t\,.$$

$F_{ij}(x)$ denote functions from $\mathbf{L}_2(\mathscr{D})$. For further considerations the elements of the matrices $B(\omega)$ and $D(\omega)$ have the form

$$\begin{aligned} b_{ij}(\omega) &= \sum_{u=1}^{l} \sum_{v=1}^{t} \beta_{ij,uv}\bar{r}_{uv}^{\varepsilon}(\omega)\,,\\ d_{ij}(\omega) &= \sum_{u=1}^{l} \sum_{v=1}^{t} \delta_{ij,uv}\bar{r}_{uv}^{\varepsilon}(\omega)\,, \end{aligned} \tag{2.50}$$

where $\beta_{ij,uv}$ and $\delta_{ij,uv}$ are constants.

Now we state the main theorem of this section.

Theorem 2.11. *Consider the random eigenvalue problem*

$$(A_0 + B(\omega))\, U = \Lambda(C_0 + D(\omega))\, U\,. \tag{2.39}$$

Let the elements of the matrices $B(\omega)$ *and* $D(\omega)$ *have the form* (2.50). *The eigenvalues of* (2.39) *are denoted by* $\Lambda_{i\varepsilon}(\omega)$, $i = 1, 2, \ldots, n$, *with* $\Lambda_{1\varepsilon}(\omega) \leqq \Lambda_{2\varepsilon}(\omega) \leqq \ldots \leqq \Lambda_{n\varepsilon}(\omega)$ *and the associated eigenvectors by* $U_{1\varepsilon}(\omega), U_{2\varepsilon}(\omega), \ldots, U_{n\varepsilon}(\omega)$, where $\langle\!\langle (C_0 + D(\omega))\, U_{i\varepsilon}, U_{j\varepsilon} \rangle\!\rangle = \delta_{ij}$ *for* $i, j = 1, 2, \ldots, n$. *Let* $C_0 + D(\omega)$ *be a.s. positive definite,*

$$\langle\!\langle (C_0 + D(\omega))\, U, U \rangle\!\rangle \geqq s_0(\omega)\, |U|^2 \quad a.s.,$$

for all $U \in \mathbb{R}^n$, $s_0(\omega) > 0$ *a.s. Assume that all moments of* $s_0^{-1}(\omega)$ *exist,*

$$\langle s_0^{-k} \rangle < \infty\,, \quad \textit{for} \quad k = 1, 2, \ldots$$

The unperturbed problem

$$A_0 U_0 = \Lambda_0 C_0 U_0$$

corresponding to the eigenvalue problem (2.39) *possesses the eigenvalues* $\Lambda_{10} \leqq \Lambda_{20} \leqq \ldots \leqq \Lambda_{n0}$ *and the eigenvectors* $U_{10}, U_{20}, \ldots, U_{n0}$ *associated with the eigenvalues where* $U_{j0} = (u_{ji0})^{\mathsf{T}}_{1 \leqq i \leqq n}$ *and* $\langle\!\langle C_0 U_{i0}, U_{j0} \rangle\!\rangle = \delta_{ij}$ *for* $i, j = 1, 2, \ldots, n$. *Let the eigenvalues*

$$\Lambda_{i_1 0} < \Lambda_{i_2 0} < \ldots < \Lambda_{i_s 0}$$

be simple.

(1) *Then the convergence in distribution follows:*

$$\lim_{\varepsilon \downarrow 0} \frac{1}{\sqrt{\varepsilon^m}} (\Lambda_{i_1\varepsilon} - \Lambda_{i_1 0}, \Lambda_{i_2\varepsilon} - \Lambda_{i_2 0}, \ldots, \Lambda_{i_s\varepsilon} - \Lambda_{i_s 0}) = (\xi_{i_1}, \xi_{i_2}, \ldots, \xi_{i_s})\,.$$

The random vector $(\xi_{i_1}, \xi_{i_2}, \ldots, \xi_{i_s})$ *denotes a Gaussian random vector with means* $\langle \xi_g \rangle = 0$ *and second moments*

$$\langle \xi_g \xi_h \rangle = \sum_{u,p=1}^{l} \sum_{v,q=1}^{t} d_{gguv} d_{hhpq} \langle\!\langle F_{uv} F_{pq}, a_{up} \rangle\!\rangle$$

for $g, h \in \{i_1, \ldots, i_s\}$, *where*

$$d_{ghuv} = \sum_{i,j=1}^{n} (\beta_{ij,uv} - \Lambda_{h0} \delta_{ij,uv})\, u_{gio} u_{hjo}\,.$$

(2) *For the random eigenvectors we have*

$$\lim_{\varepsilon \downarrow 0} \frac{1}{\sqrt{\varepsilon^m}} (U_{i_1\varepsilon} - U_{i_1 0}, U_{i_2\varepsilon} - U_{i_2 0}, \ldots, U_{i_s\varepsilon} - U_{i_s 0}) = (L_{i_1}, L_{i_2}, \ldots, L_{i_s})$$

in distribution. The random vector $(L_{i_1}, L_{i_2}, \ldots, L_{i_s})$ *has a Gaussian distribution with means* $\langle L_g \rangle = 0$ *and second moments*

$$\langle L_g L_h^{\mathsf{T}} \rangle = \sum_{u,p=1}^{l} \sum_{v,q=1}^{t} D_{guv} D_{hpq}^{\mathsf{T}} \langle\!\langle F_{uv} F_{pq}, a_{up} \rangle\!\rangle$$

for $g, h \in \{i_1, \ldots, i_s\}$, *where*

$$D_{guv} = \sum_{\substack{i=1 \\ i \neq g}}^{n} \frac{1}{\Lambda_{i0} - \Lambda_{g0}} d_{giuv} U_{i0} + d_{guv} U_{g0}$$

and

$$d_{guv} = \frac{1}{2} \sum_{i,j=1}^{n} \delta_{ij,\,uv} u_{gi0} u_{gj0} \,.$$

(3) *Similarly, the eigenvalues and eigenvectors converge in distribution*

$$\lim_{\varepsilon \downarrow 0} \frac{1}{\sqrt{\varepsilon^m}} (\Lambda_{i_1\varepsilon} - \Lambda_{i_10}, \ldots, \Lambda_{i_s\varepsilon} - \Lambda_{i_s0}, U_{i_1\varepsilon} - U_{i_10}, \ldots, U_{i_s\varepsilon} - U_{i_s0})$$
$$= (\xi_{i_1}, \ldots, \xi_{i_s}, L_{i_1}, \ldots, L_{i_s})$$

to a zero mean Gaussian distribution. The second moments are given by

$$\langle \xi_g L_h \rangle = \sum_{u,p=1}^{l} \sum_{v,q=1}^{t} d_{gguv} D_{hpq} \langle\!\langle F_{uv} F_{pq}, a_{up} \rangle\!\rangle$$

for $g, h \in \{i_1, \ldots, i_s\}$ *and by the second moments above.*

The function $a_{up}(x)$ *denotes the intensity between the weakly correlated fields* $f_{u\varepsilon}(x, \omega)$ *and* $f_{p\varepsilon}(x, \omega)$; *that is*

$$a_{up}(x) = \lim_{\varepsilon \downarrow 0} \frac{1}{\varepsilon^m} \int_{\mathcal{K}_\varepsilon(0)} \langle f_{u\varepsilon}(x) f_{p\varepsilon}(x+y) \rangle \, dy \,.$$

Proof. The convergence statements are proved by means of Theorem 2.9. Consider an eigenvalue $\Lambda_{h\varepsilon}$ possessing the property that the eigenvalue $\Lambda_{h0} = \Lambda_{h|B=D=0}$ is simple. The vector $U_{h\varepsilon}$ denotes the eigenvector associated with $\Lambda_{h\varepsilon}$. Then we have

$$\Lambda_{h\varepsilon} = \Lambda_{h\varepsilon}(\bar{r}^\varepsilon_{uv}) = \Lambda_{h\varepsilon}(\bar{r}^\varepsilon_{11}, \ldots, \bar{r}^\varepsilon_{l1}, \bar{r}^\varepsilon_{12}, \ldots \bar{r}^\varepsilon_{l2}, , \ldots, \bar{r}^\varepsilon_{1t}, \ldots, \bar{r}^\varepsilon_{lt}) \,,$$
$$U_{h\varepsilon} = U_{h\varepsilon}(\bar{r}^\varepsilon_{uv}) \,.$$

The connection between the elements b_{ij}, d_{ij} and $\bar{r}^\varepsilon_{uv}$ is given by (2.50). By means of the Cauchy inequality it follows that

$$\sum_{i,j=1}^{n} b_{ij}^2 = \sum_{i,j=1}^{n} \Big(\sum_{u=1}^{l} \sum_{v=1}^{t} \beta_{ij,\,uv} \bar{r}^\varepsilon_{uv}\Big)^2 \leqq \sum_{i,j=1}^{n} \sum_{u=1}^{l} \sum_{v=1}^{t} \beta^2_{ij,\,uv} \sum_{u=1}^{l} \sum_{v=1}^{t} (\bar{r}^\varepsilon_{uv})^2 \,,$$
$$\sum_{i,j=1}^{n} d_{ij}^2 \leqq \sum_{i,j=1}^{n} \sum_{u=1}^{l} \sum_{v=1}^{t} \delta^2_{ij,\,uv} \sum_{u=1}^{l} \sum_{v=1}^{t} (\bar{r}^\varepsilon_{uv})^2 \,.$$

From this we obtain

$$\sum_{i,j=1}^{n} (b_{ij}^2 + d_{ij}^2) = \sum_{u=1}^{l} \sum_{v=1}^{t} (\bar{r}^\varepsilon_{uv})^2 \, R(\bar{r}^\varepsilon_{pq})$$

where $R(\bar{r}^\varepsilon_{pq})$ satisfies the inequality

$$|R(\bar{r}^\varepsilon_{pq})| \leqq \sum_{i,j=1}^{n} \sum_{u=1}^{l} \sum_{v=1}^{t} (\beta^2_{ij,\,uv} + \delta^2_{ij,\,uv}) \,.$$

Furthermore, from the expansions of Lemma 2.5 we have

$$\Lambda_{h\varepsilon} = \Lambda_{h0} + \sum_{u=1}^{l} \sum_{v=1}^{t} \bar{\alpha}_{uv} \bar{r}^{\varepsilon}_{uv} + \sum_{u=1}^{l} \sum_{v=1}^{t} (\bar{r}^{\varepsilon}_{uv})^2 \, \bar{R}_h(\bar{r}^{\varepsilon}_{pq}) ,$$

$$u_{hk\varepsilon} = u_{hk0} + \sum_{u=1}^{l} \sum_{v=1}^{t} \bar{\alpha}_{kh,\,uv} \bar{r}^{\varepsilon}_{uv} + \sum_{u=1}^{l} \sum_{v=1}^{t} (\bar{r}^{\varepsilon}_{uv})^2 \, \bar{R}_{hk}(\bar{r}^{\varepsilon}_{pq})$$

for $\sum_{u=1}^{l} \sum_{v=1}^{t} (\bar{r}^{\varepsilon}_{uv})^2 \leqq \bar{\delta}^2$ where $|\bar{R}_h(\bar{r}^{\varepsilon}_{pq})| \leqq \bar{c}_h$ and $|\bar{R}_{hk}(\bar{r}^{\varepsilon}_{pq})| \leqq \bar{c}_{hk}$. Thus the first condition of Theorem 2.9 is established.

With the aid of Lemma 2.4, the existence of the moments of $b_{ij}(\omega)$ and $d_{ij}(\omega)$ implies the existence of the moments of $\bar{r}^{\varepsilon}_{uv}(\omega)$ as well as the existence of the moments of $\Lambda_{h\varepsilon}$ and $U_{h\varepsilon}$. Now we can apply Theorem 2.9. We obtain the convergence in distribution of the eigenvalues and the eigenvectors. Applying the results of Theorem 2.9 we write the linear terms of the expansions of $\Lambda_{h\varepsilon}$ and $U_{h\varepsilon}$ in a different way than in Lemma 2.6. By means of the formula for $\Lambda_{h1\varepsilon}$ given by Lemma 2.6 and (2.50), it follows that

$$\begin{aligned}\Lambda_{h1\varepsilon} &= \sum_{i,j=1}^{n} (b_{ij} - \Lambda_{h0} d_{ij}) \, u_{hi0} u_{hj0} \\ &= \sum_{u=1}^{l} \sum_{v=1}^{t} \Big[\sum_{i,j=1}^{n} (\beta_{ij,\,uv} - \Lambda_{h0} \delta_{ij,\,uv}) \, u_{hi0} u_{hj0} \Big] \, \bar{r}^{\varepsilon}_{uv} \\ &= \sum_{u=1}^{l} \sum_{v=1}^{t} d_{hhuv} \bar{r}^{\varepsilon}_{uv} ,\end{aligned}$$

where $U_{h0} = (u_{hi0})^{\mathrm{T}}_{1 \leqq i \leqq n}$ and d_{hguv} is defined by

$$d_{hguv} \doteq \sum_{i,j=1}^{n} (\beta_{ij,\,uv} - \Lambda_{h0} \delta_{ij,\,uv}) \, u_{hi0} u_{gj0} \, .$$

Similar calculations show that

$$\begin{aligned}U_{h1\varepsilon} &= - \sum_{\substack{p=1 \\ p \neq h}}^{n} \frac{1}{\Lambda_{ph0}} \Big[\sum_{i,j=1}^{n} (b_{ij} - \Lambda_{h0} d_{ij}) \, u_{hi0} u_{pj0} \Big] \, U_{p0} \\ &\quad - \frac{1}{2} \sum_{i,j=1}^{n} d_{ij} u_{hi0} u_{hj0} U_{h0} \\ &= - \sum_{u=1}^{l} \sum_{v=1}^{t} \left[\sum_{\substack{p=1 \\ p \neq h}}^{n} \frac{1}{\Lambda_{ph0}} d_{hpuv} U_{p0} + d_{huv} U_{h0} \right] \bar{r}^{\varepsilon}_{uv} ,\end{aligned}$$

where $\Lambda_{ph0} \doteq \Lambda_{p0} - \Lambda_{h0}$ and

$$d_{huv} \doteq \tfrac{1}{2} \sum_{i,j=1}^{n} \delta_{ij,\,uv} u_{hi0} u_{hj0} \, .$$

Hence we can write

$$U_{h1\varepsilon} = \sum_{u=1}^{l} \sum_{v=1}^{t} D_{huv} \bar{r}^{\varepsilon}_{uv}$$

if the term D_{huv} is defined by

$$D_{huv} \doteq - \sum_{\substack{p=1 \\ p \neq h}}^{n} \frac{1}{\Lambda_{ph0}} d_{hpuv} U_{p0} - d_{huv} U_{h0} .$$

By the use of Theorem 2.9 we have proved the limit distribution of Theorem 2.11. ◀

Remark 2.4. *Assume that the weakly correlated connected vector field* $(f_{1\varepsilon}(x, \omega), \dots, f_{l\varepsilon}(x, \omega))$ *is wide-sense homogeneously connected, i.e.*

$$\langle f_{u\varepsilon}(x) f_{p\varepsilon}(y) \rangle = R_{up}(y - x) \quad \text{for} \quad u, p = 1, 2, \dots, l .$$

Then we obtain

$$a_{up} = \text{const} \quad \text{for } u, p = 1, 2, \dots, l ,$$

from the definition of the intensity. Furthermore, in this case the simpler relations

$$\langle \xi_g \xi_h \rangle = \sum_{u,p=1}^{l} \langle\!\langle e_{gu}, e_{hp} \rangle\!\rangle a_{up} ,$$

$$\langle L_g L_h^{\mathsf{T}} \rangle = \sum_{u,p=1}^{l} \langle\!\langle E_{gu} E_{hp}^{\mathsf{T}}, 1 \rangle\!\rangle a_{up} ,$$

$$\langle \xi_g L_h \rangle = \sum_{u,p=1}^{l} \langle\!\langle e_{gu} E_{hp}, 1 \rangle\!\rangle a_{up}$$

are found where

$$e_{gu} \doteq \sum_{v=1}^{t} d_{gguv} F_{uv} \quad \text{and} \quad E_{gu} \doteq \sum_{v=1}^{t} D_{guv} F_{uv} .$$

We give a special form of Theorem 2.11 for the further applications to concrete eigenvalue problems.

Theorem 2.12. *Assume that*

$$b_{ij} = \langle\!\langle f_{1\varepsilon} \eta_i, \eta_j \rangle\!\rangle \quad \text{and} \quad d_{ij} = \langle\!\langle f_{2\varepsilon} \psi_i, \psi_j \rangle\!\rangle$$

for $i, j = 1, 2, \dots, n$ *(as a result of the application of the method of Ritz to eigenvalue problems for differential operators) are contained in a problem of the form* (2.39). *Put*

$$u_{g\varepsilon}(x, \omega) = \sum_{i=1}^{n} u_{gi\varepsilon}(\omega) \varphi_i(x)$$

and

$$\bar{u}_{g0}(x) = \sum_{i=1}^{n} u_{gi0} \eta_i(x) , \qquad \bar{\bar{u}}_{g0}(x) = \sum_{i=1}^{n} u_{gi0} \psi_i(x) ,$$

where $U_{g\varepsilon}(\omega) = (u_{gi\varepsilon}(\omega))^{\mathsf{T}}_{1 \le i \le n}$ *denotes the eigenvector of* (2.39) *associated with the eigenvalue* $\Lambda_g(\omega)$, *and* $U_{g0} = (u_{gi0})^{\mathsf{T}}_{1 \le i \le n}$ *the eigenvector associated with the eigenvalue* Λ_{g0} *of the unperturbed problem of the eigenvalue problem* (2.39).

Then the second moments of the limit random variables of the eigenvalues are given by

$$\langle \xi_g \xi_h \rangle = \langle\!\langle a_{11} \bar{u}^2_{g0}, \bar{u}^2_{h0} \rangle\!\rangle - \Lambda_{g0} \langle\!\langle a_{21} \bar{\bar{u}}^2_{g0}, \bar{u}^2_{h0} \rangle\!\rangle$$
$$- \Lambda_{h0} \langle\!\langle a_{12} \bar{\bar{u}}^2_{h0}, \bar{u}^2_{g0} \rangle\!\rangle + \Lambda_{g0} \Lambda_{h0} \langle\!\langle a_{22} \bar{\bar{u}}^2_{g0}, \bar{\bar{u}}^2_{h0} \rangle\!\rangle \quad (2.51)$$

if $a_{ij}(x)$ *is the intensity between* $f_{i\varepsilon}$ *and* $f_{j\varepsilon}$ *for* $i, j = 1, 2$. *For the random fields* $u_{g\varepsilon}(x, \omega)$ *with* $g \in \{i_1, \dots, i_s\}$ *the convergence in distribution*

$$\lim_{\varepsilon \downarrow 0} \frac{1}{\sqrt{\varepsilon^m}} \left(u_{i_1 \varepsilon}(x, \omega) - u_{i_1 0}(x), \dots, u_{i_s \varepsilon}(x, \omega) - u_{i_s 0}(x) \right)$$
$$= \left(l_{i_1}(x, \omega), \dots, l_{i_s}(x, \omega) \right)$$

obtains, where

$$u_{g0}(x) = \sum_{i=1}^{n} u_{gi0} \varphi_i(x) .$$

The first and second moments of the Gaussian vector field $\left(l_{i_1}(x, \omega), \dots, l_{i_s}(x, \omega) \right)$ *are determined by*

$$\langle l_g(x) \rangle = 0 ,$$

$$\langle l_g(x)\, l_h(y) \rangle = \sum_{\substack{i=1 \\ i \neq g}}^{n} \sum_{\substack{j=1 \\ j \neq h}}^{n} \frac{1}{\Lambda_{ig0} \Lambda_{jh0}} [\Lambda_{g0} \Lambda_{h0} \langle\!\langle a_{22} \bar{\bar{u}}_{g0} \bar{\bar{u}}_{h0}, \bar{\bar{u}}_{i0} \bar{\bar{u}}_{j0} \rangle\!\rangle$$
$$- \Lambda_{g0} \langle\!\langle a_{21} \bar{\bar{u}}_{g0} \bar{u}_{h0}, \bar{\bar{u}}_{i0} \bar{u}_{j0} \rangle\!\rangle$$
$$- \Lambda_{h0} \langle\!\langle a_{12} \bar{\bar{u}}_{h0} \bar{u}_{g0}, \bar{\bar{u}}_{j0} \bar{u}_{i0} \rangle\!\rangle$$
$$+ \langle\!\langle a_{11} \bar{u}_{g0} \bar{u}_{h0}, \bar{u}_{i0} \bar{u}_{j0} \rangle\!\rangle]\, u_{i0}(x)\, u_{j0}(y)$$
$$- \frac{1}{2} \sum_{\substack{i=1 \\ i \neq g}}^{n} \frac{1}{\Lambda_{ig0}} [\Lambda_{g0} \langle\!\langle a_{22} \bar{\bar{u}}_{g0} \bar{\bar{u}}^2_{h0}, \bar{\bar{u}}_{i0} \rangle\!\rangle - \langle\!\langle a_{12} \bar{u}_{g0} \bar{\bar{u}}^2_{h0}, \bar{u}_{i0} \rangle\!\rangle]$$
$$\times\, u_{i0}(x)\, u_{h0}(y)$$
$$- \frac{1}{2} \sum_{\substack{j=1 \\ j \neq h}}^{n} \frac{1}{\Lambda_{jh0}} [\Lambda_{h0} \langle\!\langle a_{22} \bar{\bar{u}}_{h0} \bar{\bar{u}}^2_{g0}, \bar{\bar{u}}_{j0} \rangle\!\rangle - \langle\!\langle a_{21} \bar{u}_{h0} \bar{\bar{u}}^2_{g0}, \bar{u}_{j0} \rangle\!\rangle]$$
$$\times\, u_{j0}(y)\, u_{g0}(x)$$
$$+ \frac{1}{4} \langle\!\langle a_{22} \bar{\bar{u}}^2_{g0}, \bar{\bar{u}}^2_{h0} \rangle\!\rangle\, u_{g0}(x)\, u_{h0}(y) . \quad (2.52)$$

In connection with Theorem 2.11 *we have*

$$\langle l_g(x)\, l_h(y) \rangle = \Phi^{\mathsf{T}}(x)\, \langle L_g L_h^{\mathsf{T}} \rangle\, \Phi(y)$$

where,

$$\Phi(x) = \left(\varphi_i(x) \right)^{\mathsf{T}}_{1 \leq i \leq n} .$$

Proof. Equation (2.46) leads to

$$\Lambda_{g1} = \langle\!\langle f_{1\varepsilon} \bar{u}_{g0}, \bar{u}_{g0} \rangle\!\rangle - \Lambda_{g0} \langle\!\langle f_{2\varepsilon} \bar{\bar{u}}_{g0}, \bar{\bar{u}}_{g0} \rangle\!\rangle ,$$

and the statement of this theorem for the eigenvalues is proved. (2.49) implies

$$u_{g1\varepsilon}(x, \omega) \doteq \sum_{i=1}^{n} u_{gi1\varepsilon}(\omega)\, \varphi_i(x)$$

$$= \sum_{\substack{i=1\\ i\neq g}}^{n} \frac{1}{\Lambda_{ig0}} [\Lambda_{g0} \langle\!\langle f_{2\varepsilon} \bar{\bar{u}}_{g0}, \bar{\bar{u}}_{i0} \rangle\!\rangle - \langle\!\langle f_{1\varepsilon} \bar{u}_{g0}, \bar{u}_{i0} \rangle\!\rangle]\, u_{i0}(x)$$

$$- \frac{1}{2} \langle\!\langle f_{2\varepsilon} \bar{\bar{u}}_{g0}, \bar{\bar{u}}_{g0} \rangle\!\rangle\, u_{g0}(x) .$$

Hence, the statement for random fields $u_{g\varepsilon}(x, \omega)$, $g \in \{i_1, \dots, i_s\}$, is also clear, and the proof of Theorem 2.12 is complete. ◀

2.2.3. Simulation results

In this section we compare our theoretical results with results by simulation. This comparison is done for a concrete eigenvalue problem. First, we investigate the simulation of weakly correlated processes.

A random process $\bar{f}(x, \omega)$ is considered which is defined by

$$\bar{f}(x, \omega) = \zeta_i(\omega) + \frac{x - a_i}{a_{i+1} - a_i} \left(\zeta_{i+1}(\omega) - \zeta_i(\omega)\right)$$

for $x \in [a_i, a_{i+1}]$, $i = 0, 1, \dots, N-1$, where $a_i = \alpha + i(\beta - \alpha)/N$ and ζ_i denotes a random variable for which all moments $\langle \zeta_i^k \rangle$, $k = 1, 2, \dots$, exist, $i = 0, 1, \dots, N$. Let $\zeta_0, \zeta_1, \dots, \zeta_N$ be independent with $\langle \zeta_i \rangle = 0$, $\langle \zeta_i^2 \rangle = \sigma_i^2$. The definition of the process $\bar{f}(x, \omega)$ implies that the random variables $\bar{f}(x, \omega)$ and $\bar{f}(y, \omega)$ are independent if x and y satisfy the inequality $|x - y| \geqq 2(\beta - \alpha)/N$. The correlation function is given for $x \in [a_i, a_{i+1}]$, by

$$\langle \bar{f}(x)\, \bar{f}(y) \rangle = \begin{cases} \sigma_i^2\left(1 - s_i(y) - s_i(x)\right) + (\sigma_i^2 + \sigma_{i+1}^2)\, s_i(x)\, s_i(y) , & \text{for } y \in [a_i, a_{i+1}] , \\ \sigma_{ij+1} s_j(y) \left(1 - s_i(x)\right) + \sigma_{i+1j} s_i(x) \left(1 - s_j(y)\right) , & \text{for } y \in [a_j, a_{j+1}] \text{ and } i \neq j , \\ 0 , & \text{otherwise} , \end{cases} \tag{2.53}$$

where

$$\sigma_{ij} = \langle \zeta_i \zeta_j \rangle , \qquad s_i(x) = \frac{x - a_i}{a_{i+1} - a_i} .$$

Furthermore, for $x \in [a_i, a_{i+1}]$ we evaluate

$$\int_{\alpha}^{\beta} \langle \bar{f}(x) \bar{f}(y) \rangle \, dy = \int_{a_{i-1}}^{a_i} \langle \bar{f}(x) \bar{f}(y) \rangle \, dy + \int_{a_i}^{a_{i+1}} \langle \bar{f}(x) \bar{f}(y) \rangle \, dy + \int_{a_{i+1}}^{a_{i+2}} \langle \bar{f}(x) \bar{f}(y) \rangle \, dy = \sigma_i^2 (a_{i+1} - x) + \sigma_{i+1}^2 (x - a_i)$$

and with $\varepsilon = 2(\beta - \alpha)/N$,

$$\frac{1}{\varepsilon} \int_{\alpha}^{\beta} \langle \bar{f}(x) \bar{f}(y) \rangle \, dy = \frac{N}{2(\beta - \alpha)} [\sigma_i^2 (a_{i+1} - x) + \sigma_{i+1}^2 (x - a_i)] = \frac{1}{2} [\sigma_i^2 (i + 1 - \bar{x}) + \sigma_{i+1}^2 (\bar{x} - i)]$$

where we have put $x = \alpha + \bar{x}(\beta - \alpha)/N$ for $i \leqq x \leqq i + 1$. Especially, the equation

$$\frac{1}{\varepsilon} \int_{\alpha}^{\beta} \langle \bar{f}(x) \bar{f}(y) \rangle \, dy = \frac{\sigma^2}{2}$$

is obtained in the case, where $\sigma_i^2 = \sigma^2$ for $i = 0, 1, \dots, N$.

Hence, the above defined process $\bar{f}(x, \omega)$ is a weakly correlated process with correlation length $\varepsilon = 2(\beta - \alpha)/N$ and this process has intensity

$$a(x) = \tfrac{1}{2} [\sigma_i^2 (i + 1 - \bar{x}) + \sigma_{i+1}^2 (\bar{x} - i)]$$

for $x \in [a_i, a_{i+1}]$, $i = 0, 1, \dots, N - 1$ and $x = \alpha + \bar{x}(\beta - \alpha)/N$. This weakly correlated process is applied to the simulation of eigenvalue problems of random matrices. Similarly, we can define weakly correlated processes having continuous differentiable sample functions if we connect the points $\zeta_i(\omega)$ by suitable polynomials.

We now consider the simple eigenvalue problem

$$(A_0 + B(\omega)) \, U = \Lambda U \, , \tag{2.54}$$

where A_0 and $B(\omega)$ are defined by

$$A_0 = \begin{pmatrix} 2 & -2 \\ -2 & -1 \end{pmatrix}, \qquad B(\omega) = \begin{pmatrix} b_{11}(\omega) & 0 \\ 0 & b_{22}(\omega) \end{pmatrix},$$

respectively. Let b_{11} and b_{22} denote the random variables given by

$$b_{11}(\omega) = \int_{\alpha}^{\beta} \bar{f}_\varepsilon(x, \omega) \, dx \, , \qquad b_{22}(\omega) = \int_{\alpha}^{\beta} x \bar{f}_\varepsilon(x, \omega) \, dx$$

where $\bar{f}_\varepsilon(x, \omega)$ is the weakly correlated process defined above. For simplification we put $\sigma_i^2 = \sigma^2$ for $i = 0, 1, \dots, N$.

The equations

$$\begin{aligned}
\langle b_{11}\rangle &= \langle b_{22}\rangle = 0\,,\\
\langle b_{11}^2\rangle &= \frac{\sigma^2}{N}(\beta-\alpha)^2\left(1-\frac{1}{2N}\right),\\
\langle b_{11}b_{22}\rangle &= \frac{\sigma^2}{2N}(\beta-\alpha)(\beta^2-\alpha^2)\left(1-\frac{1}{2N}\right),\\
\langle b_{22}^2\rangle &= \frac{\sigma^2}{3N}(\beta-\alpha)(\beta^3-\alpha^3)+\frac{1}{4N^2}(\beta-\alpha)^2\\
&\quad\times\sigma^2\left(-\alpha^2-\beta^2+\frac{1}{9N^2}(\beta-\alpha)^2\right)
\end{aligned} \tag{2.55}$$

can be derived by means of (2.53). Theorem 2.5 leads to the approximate calculation of these moments. From

$$\lim_{N\to\infty}\frac{N}{2(\beta-\alpha)}\langle b_{11}^2\rangle = a\int_\alpha^\beta \mathrm{d}x = \frac{\sigma^2}{2}(\beta-\alpha)\,,$$

$$\lim_{N\to\infty}\frac{N}{2(\beta-\alpha)}\langle b_{11}b_{22}\rangle = a\int_\alpha^\beta x\,\mathrm{d}x = \frac{\sigma^2}{4}(\beta^2-\alpha^2)\,,$$

$$\lim_{N\to\infty}\frac{N}{2(\beta-\alpha)}\langle b_{22}^2\rangle = a\int_\alpha^\beta x^2\,\mathrm{d}x = \frac{\sigma^2}{6}(\beta^3-\alpha^3)$$

we obtain for a large natural N

$$\begin{aligned}
&\langle b_{11}^2\rangle \approx \frac{\sigma^2}{N}(\beta-\alpha)^2\,,\qquad \langle b_{22}^2\rangle \approx \frac{\sigma^2}{3N}(\beta-\alpha)(\beta^3-\alpha^3)\\
&\langle b_{11}b_{22}\rangle \approx \frac{\sigma^2}{2N}(\beta-\alpha)(\beta^2-\alpha^2)\,.
\end{aligned} \tag{2.56}$$

The approximately calculated second moments coincide with the exact second moments (2.55) up to terms of the order $O(N^{-2})$.

The integrals in the definition of b_{11} and b_{22} are calculated by Simpson's rule:

$$\begin{aligned}
&\int_\alpha^\beta g(x)\bar{f}_\varepsilon(x,\omega)\,\mathrm{d}x\\
&\approx \frac{\beta-\alpha}{3N}\,[g(\alpha)\,\zeta_0(\omega)+g(\beta)\,\zeta_N(\omega)\\
&\qquad + 4\sum_{i=0}^{N/2-1} g(a_{2i+1})\,\zeta_{2i+1}(\omega) + 2\sum_{i=1}^{N/2-1} g(a_{2i})\,\zeta_{2i}(\omega)]\,,
\end{aligned} \tag{2.57}$$

for N even. These random variables are denoted by $\bar{b}_{11}$ and $\bar{b}_{22}$, respectively. We have

$$\begin{aligned}
\langle \bar{b}_{11}^2 \rangle &= \frac{10\sigma^2}{9N} (\beta - \alpha)^2 \left(1 - \frac{1}{5N}\right), \\
\langle \bar{b}_{11} \bar{b}_{22} \rangle &= \frac{5\sigma^2}{9N} (\beta - \alpha)(\beta^2 - \alpha^2) \left(1 - \frac{1}{5N}\right), \\
\langle \bar{b}_{22}^2 \rangle &= \frac{20\sigma^2}{54N} (\beta - \alpha)(\beta^3 - \alpha^3) - \frac{\sigma^2}{9N^2} (\beta - \alpha)^2 \\
&\quad \times \left[\beta^2 + \alpha^2 + \frac{4}{3N} (\beta - \alpha)^2\right].
\end{aligned} \tag{2.58}$$

The approximate Gaussian distribution of $\bar{b}_{11}(\omega)$ and $\bar{b}_{22}(\omega)$ also follows from the central limit theorem because of the independence of $\zeta_0(\omega), \zeta_1(\omega), \ldots, \zeta_N(\omega)$.

The eigenvalues of problem (2.54) are of the form

$$\Lambda_{1/2} = \tfrac{1}{2} \left(1 + b_{11} + b_{22} \mp \sqrt{(3 + b_{11} - b_{22})^2 + 16}\right),$$

and $\Lambda_. - \Lambda_{.0}$ is of the form

$$\Lambda_{1/2} - \Lambda_{1/20} = \tfrac{1}{2} \left(b_{11} + b_{22} \mp \sqrt{(3 + b_{11} - b_{22})^2 + 16} \pm 5\right). \tag{2.59}$$

It is $\Lambda_{10} = -2$, $\Lambda_{20} = 3$, and these eigenvalues are simple. Theorem 2.11 leads to the limit relation

$$\lim_{N \to \infty} \sqrt{\frac{N}{2(\beta - \alpha)}} (\Lambda_1 - \Lambda_{10}, \Lambda_2 - \Lambda_{20}) = (\xi_1, \xi_2)$$

in distribution where (ξ_1, ξ_2) denotes a zero mean Gaussian random vector. The eigenvectors associated with the averaged problem $A_0 U_0 = \Lambda_0 U_0$ are

$$U_{10} = \frac{1}{\sqrt{5}} (1, 2)^{\mathrm{T}}, \qquad U_{20} = \frac{1}{\sqrt{5}} (2, -1)^{\mathrm{T}}.$$

In order to apply Theorem 2.11, we note that

$$\bar{r}_{11}^{\varepsilon}(\omega) = b_{11}(\omega), \qquad \bar{r}_{12}^{\varepsilon}(\omega) = b_{22}(\omega),$$

and, therefore,

$$\beta_{11,11} = \beta_{22,12} = 1, \qquad \beta_{ij,uv} = 0 \text{ otherwise.}$$

From this we have

$$\begin{aligned}
d_{1111} &= u_{110}^2 = \tfrac{1}{5}, \qquad & d_{2211} &= u_{210}^2 = \tfrac{4}{5}, \\
d_{1112} &= u_{120}^2 = \tfrac{4}{5}, \qquad & d_{2212} &= u_{220}^2 = \tfrac{1}{5};
\end{aligned}$$

and therefore

$$\langle \xi_1^2 \rangle = \frac{a}{25}\left(\int_\alpha^\beta \mathrm{d}x + 8\int_\alpha^\beta x\,\mathrm{d}x + 16\int_\alpha^\beta x^2\,\mathrm{d}x\right)$$

$$= \frac{a}{25}\left(\beta - \alpha + 4(\beta^2 - \alpha^2) + \frac{16}{3}(\beta^3 - \alpha^3)\right),$$

$$\langle \xi_1\xi_2 \rangle = \frac{a}{25}\left(4\int_\alpha^\beta \mathrm{d}x + 17\int_\alpha^\beta x\,\mathrm{d}x + 4\int_\alpha^\beta x^2\,\mathrm{d}x\right)$$

$$= \frac{a}{25}\left(4(\beta - \alpha) + \frac{17}{2}(\beta^2 - \alpha^2) + \frac{4}{3}(\beta^3 - \alpha^3)\right),$$

$$\langle \xi_2^2 \rangle = \frac{a}{25}\left(16\int_\alpha^\beta \mathrm{d}x + 8\int_\alpha^\beta x\,\mathrm{d}x + \int_\alpha^\beta x^2\,\mathrm{d}x\right)$$

$$= \frac{a}{25}\left(16(\beta - \alpha) + 4(\beta^2 - \alpha^2) + \frac{1}{3}(\beta^3 - \alpha^3)\right).$$

Hence, the eigenvalues $\Lambda_{1\varepsilon}$, $\Lambda_{2\varepsilon}$ have an approximate Gaussian distribution with the first and second moments given by

$$\langle \Lambda_{i\varepsilon} \rangle \approx \Lambda_{i0},$$

$$\langle (\Lambda_{i\varepsilon} - \Lambda_{i0})(\Lambda_{j\varepsilon} - \Lambda_{j0}) \rangle \approx \frac{2(\beta - \alpha)}{N} \langle \xi_i \xi_j \rangle$$

for $i, j = 1, 2$ where $a = \sigma^2/2$.

We will compare the approximately Gaussian distributed eigenvalues with the results obtained by simulation. Let $\alpha = 0$, $\beta = 1$. We have

$$\langle (\Lambda_{1\varepsilon} - \Lambda_{10})^2 \rangle \approx \frac{\sigma^2}{N} 0.413\bar{3}, \qquad \langle (\Lambda_{2\varepsilon} - \Lambda_{20})^2 \rangle \approx \frac{\sigma^2}{N} 0.813\bar{3},$$
$$\langle (\Lambda_{1\varepsilon} - \Lambda_{10})(\Lambda_{2\varepsilon} - \Lambda_{20}) \rangle \approx \frac{\sigma^2}{N} 0.553\bar{3}, \tag{2.60}$$

where σ^2 denotes the variance of the random variable $\zeta_i(\omega)$. The generation of independent uniformly distributed random variables is based on the algorithm

$$\bar{x}_{n+1} = (57\bar{x}_n + 645497) \bmod 999997, \qquad x_{n+1} = \frac{\bar{x}_{n+1}}{999997},$$

where x_{n+1} is the $(n + 1)$-th realization. We put $x_0 = 141593$. The number $y_{n+1} = 2x_{n+1}s - s$ gives a realization of a uniformly distributed random variable on $[-s, s]$. Furthermore, the number

$$z = \sigma_0 \sqrt{-2 \ln u_1} \cos(2\pi u_2)$$

is a realization of a mean zero Gaussian random variable with variance σ_0^2. In this formula u_1 and u_2 denote realizations of a uniformly distributed random variable on [0, 1]. From these realizations we can obtain realizations of the $b_{ii}(\omega)$ using (2.57), and by means of (2.59) realizations of $b_{ii}(\omega)$ lead to realizations of $\Lambda_{i\varepsilon} - \Lambda_{i0}$. We denote the number of the calculated realizations by n. In the following, the random variables $\zeta_i(\omega)$, $i = 0, 1, \ldots, N$, are assumed to have a uniform distribution on $[-s, s]$, or a mean zero Gaussian distribution with variance σ_0^2. These cases are characterized by the parameters s or σ_0, respectively.

We now deal with the comparison of the first and the second moments of b_{ii} and $\bar{\Lambda}_{i\varepsilon} \doteq \Lambda_{i\varepsilon} - \Lambda_{i0}$ obtained by the limit theorems and by simulation. Put $N = 10$. For this comparison, the calculations above imply $\langle b_{ii} \rangle = 0$, $\langle \bar{\Lambda}_{j\varepsilon} \rangle = 0$, and second moments

$$\sigma_{ij}^b = \langle b_{ii} b_{jj} \rangle , \qquad \sigma_{ij}^\Lambda = \langle \bar{\Lambda}_{i\varepsilon} \bar{\Lambda}_{i\varepsilon} \rangle$$

are given in Table 2.1. We have $\sigma^2(s) = s^2/3$ and $\sigma^2(\sigma_0) = \sigma_0^2$. The correlation coefficient is defined by $\varrho = \sigma_{12}/\sqrt{\sigma_{11}\sigma_{22}}$.

Table 2.1

σ_{ij}^b/σ^2	Eq. (2.55)	Eq. (2.56)	Eq. (2.58)	$\sigma_{ij}^\Lambda/\sigma^2$	Eq. (2.60)
$i = 1,\ j = 1$	0.095000	0.100000	0.108889	$i = 1,\ j = 1$	0.041333
$i = 1,\ j = 2$	0.047500	0.050000	0.054444	$i = 1,\ j = 2$	0.055333
$i = 2,\ j = 2$	0.030836	0.033333	0.035778	$i = 2,\ j = 2$	0.081333
ϱ^b	0.877611	0.866025	0.872278	ϱ^Λ	0.954338

Table 2.2

	$N = 10$				$N = 20$		
	$s = 1$			$\sigma_0 = 1$	$s = 1$		$\sigma_0 = 1$
n	300	1000	3000	1000	1000	3000	1000
m_1^b	−0.0234	−0.0077	−0.0010	0.0045	−0.0001	−0.0013	0.0005
m_2^b	−0.0107	−0.0045	−0.0019	0.0028	0.0003	0.0003	0.0001
s_{11}^b/σ^2	0.1115	0.1129	0.1108	0.1072	0 0564	0.0535	0.0517
s_{12}^b/σ^2	0.0570	0.0561	0.0548	0.0545	0.0267	0.0261	0.0258
s_{22}^b/σ^2	0.0370	0.0359	0.0354	0.0364	0.0172	0.0171	0.0176
r^b	0.8882	0.8810	0.8762	0.8721	0.8590	0.8617	0.8539
m_1^Λ	−0.0137	−0.0055	−0.0021	0.0020	0	−0.0002	0.0001
m_2^Λ	−0.0205	−0.0067	−0.0008	0.0053	0.0002	0.0003	0.0005
s_{11}^Λ/σ^2	0.0464	0.0455	0.0446	0.0450	0.0218	0.0214	0.0216
s_{12}^Λ/σ^2	0.0626	0.0620	0.0607	0.0600	0.0300	0.0290	0.0286
s_{22}^Λ/σ^2	0.0910	0.0916	0.0898	0.0875	0.0453	0.0433	0.0421
r^Λ	0.9623	0.9600	0.9583	0.9563	0.9526	0.9528	0.9498

Let estimates for $\langle b_{ii}\rangle$ and $\langle \bar{\Lambda}_{i\varepsilon}\rangle$ be denoted by m_i^b and m_i^Λ, and for σ_{ij}^b and σ_{ij}^Λ by the values s_{ij}^b, s_{ij}^Λ, respectively. The mean is estimated by $m = 1/n \sum_{p=1}^{n} x_p$ from a sample $\{x_1, \ldots, x_n\}$ and the second moments σ_{ij} by $s_{ij} = 1/(n-1) \sum_{p=1}^{n} (x_{ip} - m_i) \times (x_{jp} - m_j)$ from $\{(x_{i1}, x_{j1}), (x_{i2}, x_{j2}), \ldots, (x_{in}, x_{jn})\}$. The correlation coefficient is estimated by $r = s_{12}/\sqrt{s_{11}s_{22}}$. Table 2.2 shows the results of a simulation corresponding to the first and the second moments. We observe that in general the values of the theoretical results of Table 2.1 are better for $\sigma_0 = 1$ than in the case $s = 1$, which the inequality $\sigma^2(s = 1) = 1/3 < \sigma^2(\sigma_0 = 1) = 1$ yields. In the case of the uniform distribution the same results are obtained for the second moments s_{ij}/σ^2, if we put $s \neq 1$. This statement follows for s_{ij}^b/σ^2 from (2.56) and (2.58), and for s_{ij}^Λ/σ^2 from (2.59) taking into consideration only the linear terms. This remark is also true for the Gaussian distribution with $\sigma_0^2 \neq 1$. In general, the results obtained for $N = 20$ are better than the results for $N = 10$, by comparing the values of Table 2.1.

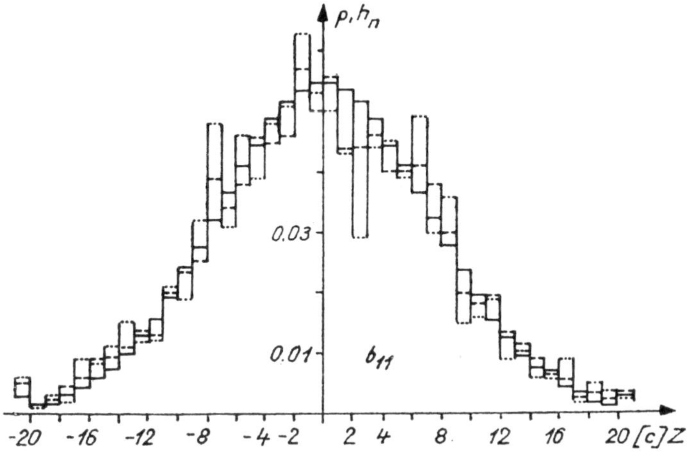

Fig. 2.1a. Comparison between theoretical probability and relative frequency of b_{11} ($q = 20$; $c = 0.025$; $s = 1$)

We now turn to remarks concerning the distribution. In Fig. 2.1a the relative frequency $h_n(z)$ of b_{11} and the theoretical probability $p(z)$ is represented for the case $N = 10$. We define

$$h_n(z) = \begin{cases} \dfrac{1}{n} H_{-q-1}\,, & \text{for } -\infty < z \leqq -qc\,, \\ \dfrac{1}{n} H_i\,, & \text{for } ic < z \leqq (i+1)\,c\,, \\ & \qquad i = -q, -q+1, \ldots, -1, 0, 1, \ldots, q-1\,, \\ \dfrac{1}{n} H_{q+1}\,, & \text{for } qc < z < \infty\,, \end{cases}$$

where H_i denotes the number of realizations of b_{11} in the given interval. The function $p(z)$ is determined by

$$p(z) = \begin{cases} \int_{-\infty}^{-qc} \varphi_{11}(t)\, dt\,, & \text{for } -\infty < z \leqq -qc\,, \\ \int_{ic}^{(i+1)c} \varphi_{11}(t)\, dt\,, & \text{for } ic < z \leqq (i+1)\, c\,, \\ & \qquad i = -q, \dots, -1, 0, 1, \dots, q-1\,, \\ \int_{qc}^{\infty} \varphi_{11}(t)\, dt\,, & \text{for } qc < z < \infty\,, \end{cases}$$

with

$$\varphi_{11}(t) = \sqrt{\frac{N}{2\pi}}\,\frac{1}{\sigma} \exp\left(-\frac{t^2 N}{2\sigma^2}\right).$$

In Fig. 2.1a the distribution of ζ_i is displayed for a uniform distribution on $[-1, 1]$, i.e. $s = 1$. We have used for $p(z)$ the line —, for $h_{1000}(z)$ the line ... and for $h_{3000}(z)$ the line --. It can be seen that $h_{3000}(z)$ is better than $h_{1000}(z)$, relative to $p(z)$ and we will consider this later in detail. Fig. 2.1b and Fig. 2.1c show the comparison just considered for the eigenvalues $\bar{\Lambda}_{1\varepsilon}$ and $\bar{\Lambda}_{2\varepsilon}$. We cannot find an essential distinction between the kind of correspondence of $h_n(z)$ and $p(z)$ for $b_{11}(\omega)$ and for the eigenvalues.

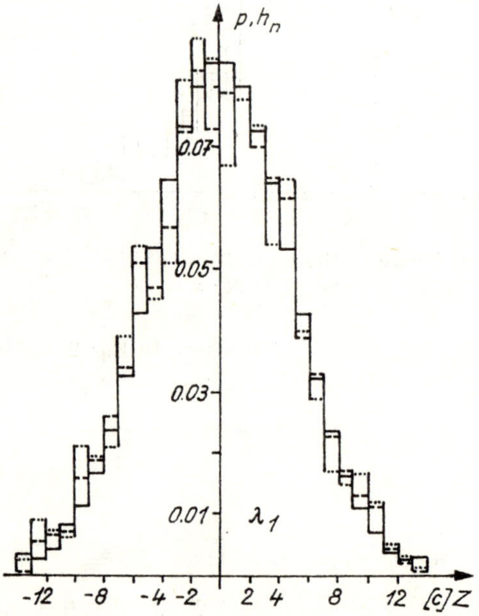

Fig. 2.1b. Comparison between theoretical probability and relative frequency of λ_1 ($q = 13$; $c = 0.025$; $s = 1$)

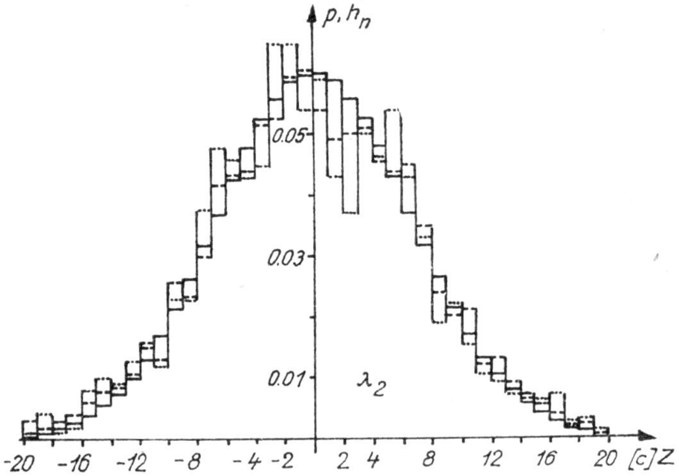

Fig. 2.1c. Comparison between theoretical probability and relative frequency of λ_2 ($q = 19$; $c = 0.025$; $s = 1$)

We now introduce the sum of the squares of the differences $h_n(z) - p(z)$ as a measure of the difference between $h_n(z)$ and $p(z)$:

$$F_n^2(q) \doteq \sum_{i=-q-1}^{q} \left(h_n((i + \tfrac{1}{2})\, c) - p((i + \tfrac{1}{2})\, c)\right)^2 .$$

Table 2.3a gives the simulation results for $N = 10$, and Table 2.3b for $N = 20$. We can conclude that $F_{n_1}(q) > F_{n_2}(q)$ for $n_1 < n_2$, and $F_{3000}(q)$ is much less than $F_{300}(q)$, for instance. Between the F-values for a Gaussian distribution of the ζ_i and the F-values for a uniform distribution we cannot find any

Table 2.3a

$F_n(q)\backslash$	n 300	1000	2000	3000	500	1000	q
for b_{11}	0.0634	0.0401	0.0258	0.0214	0.0401	0.0239	20
for b_{22}	0.0590	0.0312	0.0270	0.0224	0.0500	0.0337	12
for $A_{1\varepsilon}$	0.0650	0.0479	0.0298	0.0227	0.0477	0.0397	14
for $A_{2\varepsilon}$	0.0512	0.0376	0.0252	0.0205	0.0489	0.0384	20
	$s = 1$				$\sigma = 1/\sqrt{12}$		

Table 2.3b

$F_n(q)\backslash$	n 1000	2000	3000	q
for $A_{1\varepsilon}$	0.0347	0.0207	0.0197	14
for $A_{2\varepsilon}$	0.0303	0.0247	0.0180	20
	$s = 1$			

Table 2.4

$\bar{A}_{2\varepsilon}$ [c] \ $\bar{A}_{1\varepsilon}$ [c]	−9 to −7	−7 to −5	−5 to −3	−3 to −1	−1 to 1	1 to 3	3 to 5	5 to 7	7 to 9
11 to 9							**0.0001** 0 0	**0.0018** 0.0030 0.0030	**0.0009** 0.0010 0.0003
9 to 7							**0,0050** 0.0070 0.0060	**0.0084** 0.0150 0.0120	**0.0002** 0 0
7 to 5						**0.0054** 0.0030 0.0060	**0.0376** 0.0360 0.0427	**0.0047** 0.0020 0.0047	
5 to 3					**0.0022** 0.0020 0.0017	**0.0764** 0.0850 0.0883	**0.0379** 0.0300 0.0337	**0.0002** 0 0	
3 to 1				**0.0003** 0.0010 0.0007	**0.0690** 0.0600 0.0657	**0.1266** 0.1210 0.1210	**0.0036** 0.0050 0.0030		
1 to −1				**0.0256** 0.0280 0.0293	**0.1874** 0.1710 0.1710	**0.0256** 0.0220 0.0223			
−1 to −3			**0.0036** 0.0030 0.0050	**0.1266** 0.1150 0.1160	**0.0690** 0.0810 0.0717	**0.0003** 0 0			
−3 to −5		**0.0002** 0 0	**0.0379** 0.0460 0.0413	**0.0764** 0.0830 0.0777	**0.0022** 0.0030 0.0030				
−5 to −7	**0.0002** 0 0	**0.0047** 0.0020 0.0037	**0.0376** 0.0430 0.0410	**0.0054** 0.0040 0.0043					
−7 to −9	**0.0008** 0 0.0003	**0.0084** 0.0130 0.0127	**0.0050** 0.0080 0.0067						
−9 to −11	**0.0003** 0 0.0003	**0.0018** 0.0070 0.0050	**0.0001** 0 0						

$c = 0.05$

essential distinction. The F-values for $N = 20$ are less than the F-values for $N = 10$. This fact is in agreement with the convergence as $N \to \infty$. By means of simulation we also get useful distribution relations for the random vectors (b_{11}, b_{22}) and $(\bar{A}_{1\varepsilon}, \bar{A}_{2\varepsilon})$. The necessary assumptions are the same as in case of the investigated one-dimensional distributions. In Table 2.4 the theoretical probabilities $p(J_1, J_2)$ are compared with the relative frequencies $h_n(J_1, J_2)$. The theoretical probability is defined by

$$\mathsf{P}(\bar{\Lambda}_{1\varepsilon} \in J_1, \bar{\Lambda}_{2\varepsilon} \in J_2) = \int_{J_1} \int_{J_2} \bar{\varphi}_{12}(t_1, t_2) \, \mathrm{d}t_1 \, \mathrm{d}t_2 = p(J_1, J_2) ,$$

where $\bar{\varphi}_{12}(t_1, t_2)$ denotes the density function of the limit distribution of $(\bar{\Lambda}_{1\varepsilon}, \bar{\Lambda}_{2\varepsilon})$, namely

$$\bar{\varphi}_{12}(t_1, t_2) = \frac{1}{2\pi \sqrt{\det \sigma^{\Lambda}}} \exp\left(-\frac{1}{2 \det \sigma^{\Lambda}} \{\sigma_{22}^{\Lambda} t_1^2 - 2\sigma_{12}^{\Lambda} t_1 t_2 + \sigma_{11}^{\Lambda} t_2^2\}\right),$$

with

$$\sigma^{\Lambda} = \begin{pmatrix} \sigma_{11}^{\Lambda} & \sigma_{12}^{\Lambda} \\ \sigma_{21}^{\Lambda} & \sigma_{22}^{\Lambda} \end{pmatrix} = \frac{\sigma^2}{N} \begin{pmatrix} 0.4133 & 0.5533 \\ 0.5533 & 0.8133 \end{pmatrix}.$$

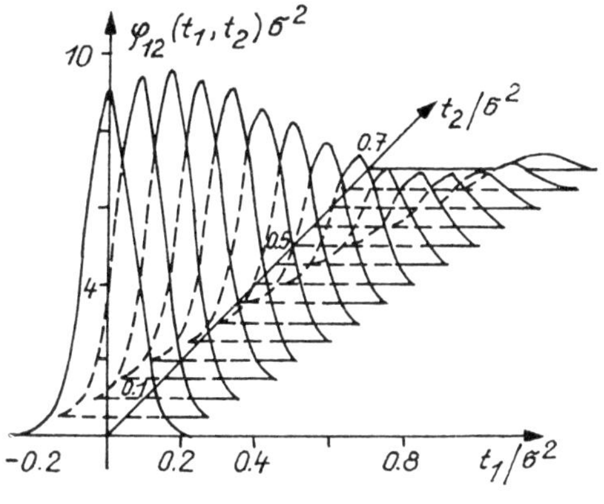

Fig. 2.2. Density function of the limit distribution of $(\bar{\Lambda}_{1\varepsilon}, \bar{\Lambda}_{2\varepsilon})$

The density function $\bar{\varphi}_{12}(\bar{t}_1, \bar{t}_2)\,\sigma^2$ with $\bar{t}_i = t_i/\sigma$, for $i = 1, 2$ and $N = 10$ is represented in Fig. 2.2. Because of $\bar{\varphi}_{12}(\bar{t}_1, \bar{t}_2) = \bar{\varphi}_{12}(-\bar{t}_1, -\bar{t}_2)$ we have only to draw the values for $\bar{t}_2 \geqq 0$. Furthermore, put

$$h_n(J_1, J_2) = \frac{1}{n} H_n(J_1, J_2) .$$

$H_n(J_1, J_2)$ gives the number of realizations of $(\bar{\Lambda}_{1\varepsilon}, \bar{\Lambda}_{2\varepsilon})$ for which $\bar{\Lambda}_{1\varepsilon}$ is contained in J_1 and $\bar{\Lambda}_{2\varepsilon}$ in J_2. The first bold number of Table 2.4 gives the calculated probability p with which the random vector $(\bar{\Lambda}_{1\varepsilon}, \bar{\Lambda}_{2\varepsilon})$ is contained in the considered square. The second number denotes the relative frequency h_{1000} for this square, and the third number h_{3000}. We notice a good agreement between the calculated results and the results obtained by simulation. As a measure of the coincidence between h_n and p we introduce the expression F_n defined by

$$F_n^2 \doteq \sum_i (h_n^i - p^i)^2 .$$

The sum is taken over all the squares listed in Table 2.4. We obtain for the ase considered, with parameters $s = 1$, $N = 10$ and $c = 0.05$:

n	300	1000	2000	3000
F_n	0.0635	0.0330	0.0290	0.0273

2.2.4. Applications

The method given in the last section is of importance for applications in which the eigenvalues and eigenfunctions are difficult to calculate for the differential equation problem and the method of Ritz must be used. The examples considered are clear from the point of view of applications, and we also note the eigenvalues and eigenfunctions can also be determined by the direct method.

Consider the random eigenvalue problem

$$(fu'')'' = \lambda(-g_1 u'' + gu) \,, \qquad u(0) = u(1) = u''(0) = u''(1) = 0 \,, \tag{2.61}$$

where $0 \leqq x \leqq 1$, and $f(x, \omega)$ and $g(x, \omega)$ are assumed to be random processes with the properties

$$f(x, \omega) \geqq c(\omega) > 0 \,, \qquad g(x, \omega) \geqq \bar{c}(\omega) > 0 \quad \text{a.s.}$$

Theorem 2.10 can be applied to this eigenvalue problem so that the existence of a discrete spectrum and the convergence of the Ritz approximate solutions are guaranteed.

The buckling problem of a simply supported bar follows from problem (2.61) with $g = 0$, $g_1 = 1$ and $f = EI$, where EI denotes the bending stiffness (E is the modulus of elasticity and I the moment of inertia of the cross-sectional area). Hence, the eigenvalue problem for the buckling problem of a bar can be written as

$$(EIu'')'' = -\lambda u'' \,, \qquad u(0) = u(1) = u''(0) = u''(1) = 0 \,, \tag{2.62}$$

with $0 \leqq x \leqq 1$.

Furthermore, the problem of the eigenvibrations of a simply supported bar leads to an eigenvalue problem of the form of (2.61):

$$(EIu'')'' = \lambda \varrho A u \,, \qquad u(0) = u(1) = u''(0) = u''(1) = 0 \,, \tag{2.63}$$

with $0 \leqq x \leqq 1$. Here A denotes the cross-sectional area of the bar, ϱ its mass per unit, and λ the square of the eigenfrequency of the vibrations.

In this section the method of Ritz is applied to this simple eigenvalue problem (2.61), and the essential properties of the eigenvalues and eigenfunctions are investigated. The random character of the coefficients f and g is induced by the random behaviour of E, I, A or ϱ, respectively.

Let the bar possess always a circular cross-sectional area, but a random radius $r(x, \omega)$. Hence, I and A are random processes with

$$I(x, \omega) = \tfrac{1}{4}\pi r^4(x, \omega)\,, \qquad A(x, \omega) = \pi r^2(x, \omega)\,.$$

In the same way, the modulus of elasticity E and the mass per unit length ϱ can be treated as random processes.

Now the random eigenvalue problem (2.61) is replaced by the equations of Ritz:

$$\big(A_0 + B(\omega)\big)\, U = \Lambda\big(C_0 + D(\omega)\big)\, U\,, \tag{2.64}$$

with the notations of Section 2.2.2. We put

$$a_{ij} = (\!(\langle f\rangle\, \varphi_i'', \varphi_j'')\!)\,, \qquad c_{ij} = (\!(g_1\varphi_i', \varphi_j')\!) + (\!(\langle g\rangle\, \varphi_i, \varphi_j)\!)\,,$$

$$b_{ij} = (\!(\bar f\varphi_i'', \varphi_j'')\!)\,, \qquad d_{ij} = (\!(\bar g\varphi_i, \varphi_j)\!)\,,$$

for $1 \leqq i, j \leqq n$, and $\bar f$ and $\bar g$ are defined by $\bar f \doteq f - \langle f\rangle$ and $\bar g \doteq g - \langle g\rangle$. The system of functions $\{\varphi_i\}$ is assumed to be a basis for the energetic space $\mathcal{H}_M$ associated with the operator $Mu = (fu'')''$ and the boundary conditions of (2.61).

The eigenvalues ${}^n\Lambda_h$ and the eigenfunctions ${}^nu_h = \sum_{i=1}^{n} {}^nu_{hi}\varphi_i$, calculated from the eigenvectors ${}^nU_h = ({}^nu_{hi})^{\mathsf T}_{1\leqq i\leqq n}$, approximate the eigenvalues λ_h and the eigenfunctions in the sense of Theorem 2.10. For abbreviation, in subsequent investigations, we assume

$$\langle f\rangle = f_0 = \text{const} > 0\,, \qquad \langle g\rangle = g_0 = \text{const} > 0 \quad \text{and} \quad g_1 = \text{const} \geqq 0\,.$$

First, we consider the averaged eigenvalue problem

$$\begin{aligned} &(f_0w'')'' = \mu(-g_1w'' + g_0w)\,,\\ &w(0) = w(1) = w''(0) = w''(1) = 0 \end{aligned} \tag{2.65}$$

associated with the eigenvalue problem (2.61). We calculate the eigenvalues and eigenfunctions for this averaged problem in order to obtain a comparison of the solutions of the averaged problem and of (2.64). Assuming $w = e^{sx}$, we have from (2.65)

$$f_0s^4 - \mu(-g_1s^2 + g_0) = 0\,;$$

and with $s^2 = t$ the solutions

$$t_{1/2} = -\frac{\mu g_1}{2f_0} \mp \sqrt{\left(\frac{\mu g_1}{2f_0}\right)^2 + \frac{\mu g_0}{f_0}}\,,$$

and $s_{1/2} = \pm\, i\sqrt{-t_1}$, $s_{3/4} = \pm\sqrt{t_2}$. The solution of (2.65) is

$$w(x) = c_1 \cos\left(\sqrt{-t_1}\, x\right) + c_2 \sin\left(\sqrt{-t_1}\, x\right) + c_3\, e^{\sqrt{t_2}\,x} + c_4\, e^{-\sqrt{t_2}\,x}$$

which satisfies the boundary conditions if

$$(t_1 - t_2)^2 \sin \sqrt{-t_1} \cosh \sqrt{t_2} = 0 .$$

Non-trivial solutions of the eigenvalue problem (2.65) exists only for $t_{1k} = -(k\pi)^2$, $k = 1, 2, \ldots$ This means that

$$\mu_k = \frac{f_0(k\pi)^4}{g_0 + g_1(k\pi)^2}, \qquad k = 1, 2, \ldots$$

are the eigenvalues. For the corresponding eigenfunctions we have

$$w_k(x) = c_k \sin(k\pi x), \qquad k = 1, 2, \ldots$$

If we assume that w_k is normalized by the condition

$$\int_0^1 (-g_1 w_k'' + g_0 w_k)\, w_k \,\mathrm{d}x = 1$$

then

$$c_k = \sqrt{\frac{2}{g_0 + g_1(k\pi)^2}} .$$

Hence, the eigenfunctions associated with the eigenvalues μ_k are

$$w_k(x) = \sqrt{\frac{2}{g_0 + g_1(k\pi)^2}} \sin(k\pi x),$$

for $k = 1, 2, \ldots$ The averaged problem (2.65) only has simple eigenvalues.

In the following we deal with the eigenvalue problem (2.64). The corresponding averaged problem is of the form

$$A_0\,{}^nU_0 = {}^n\Lambda_0 C_0\,{}^nU_0 . \tag{2.66}$$

We select a sequence of polynomials

$$\begin{aligned} \varphi_1(x) &= x - 2x^3 + x^4, \\ \varphi_2(x) &= 7x - 10x^3 + 3x^5, \\ \varphi_i(x) &= x^i(1-x)^3, \quad \text{for } i = 3, 4, \ldots, \end{aligned} \tag{2.67}$$

as a basis (cf. Michlin [2]), and have with the notations $c_{ij}^1 = (\!(\varphi_i', \varphi_j')\!)\, g_1$, $c_{ij}^2 = (\!(\varphi_i, \varphi_j)\!)\, g_0$

$$(a_{ij})_{1 \le i,j \le 4} = \begin{pmatrix} 4.800000 & 36.000000 & 0.171429 & 0.085714 \\ & 274.285714 & 1.285714 & 0.714286 \\ & & 0.057143 & 0.028571 \\ & & & 0.020779 \end{pmatrix} f_0,$$

$$(c_{ij}^1)_{1 \le i,j \le 4} = \begin{pmatrix} 0.485714 & 3.642857 & 0.019048 & 0.009524 \\ & 27.428571 & 0.142857 & 0.073593 \\ & & 0.001299 & 0.000649 \\ & & & 0.000400 \end{pmatrix} g_1,$$

$$(c_{ij}^2)_{1\leqq i,\, j\leqq 4} = \begin{pmatrix} 0.049206 & 0.369048 & 0.001948 & 0.000974 \\ & 2.770563 & 0.014610 & 0.007362 \\ & & 0.000083 & 0.000042 \\ & & & 0.000022 \end{pmatrix} g_0 \,.$$

Because of the symmetry of the matrices, the terms are only written for $j \geqq i$.

For $n = 2$, the eigenvalues ${}^2\Lambda_{10}$, ${}^2\Lambda_{20}$ of the eigenvalue problem (2.66) can be calculated easily by

$$\frac{1}{f_0}\,{}^2\Lambda_{1/20} = \frac{2.595918 g_1 + 0.223871 g_0 \mp \sqrt{2.456576 g_1^2 + 0.039163 g_0^2 + 0.620348 g_0 g_1}}{0.000266 g_0^2 + 0.104082 g_1^2 + 0.013173 g_0 g_1}\,.$$

The results are given in Table 2.5:

Table 2.5

		$g_1 = 0$	$g_0 = 0$
$h = 1$	${}^2\Lambda_{h0}$	$97.5483\ f_0/g_0$	$9.8824\ f_0/g_1$
	μ_h	$97.4091\ f_0/g_0$	$9.8696\ f_0/g_1$
	deviations	0.14%	0.13%
$h = 2$	${}^2\Lambda_{h0}$	$1584.0000\ f_0/g_0$	$40.0000\ f_0/g_1$
	μ_h	$1558.5454\ f_0/g_0$	$39.4784\ f_0/g_1$
	deviations	1.63%	1.32%

Further calculations for the eigenvalue problem (2.64) associated with the problem (2.61) are very extensive, so that we confine ourselves to the cases $g_1 = 0$ and $g_0 = 0$, since these are important for applications.

2.2.4.1. Bending vibrations of bars

This case corresponds to the special case $g_1 = 0$. We deal with the matrix eigenvalue problem (2.64) associated with the eigenvalue problem (2.63) where $C = C^2 = (c_{ij}^2)_{1\leqq i,\, j\leqq n}$. The eigenvalues of the averaged problem (2.66) are given in Table 2.6, and the corresponding eigenvectors in Table 2.7. We observe a very good correspondence between the eigenfunctions $w_k(x)$ and ${}^n u_{k0}(x)$ for $k \leqq n$, and also between the first and second derivatives. This is clear from

Table 2.6

${}^n\Lambda_{h0}g_0/f_0$	$n = 2$	$n = 3$	$n = 4$	μ_h
$h = 1$	97.5484	97.4091	97.4091	97.4091
$h = 2$	1584.0000	1584.0000	1558.6401	1558.5454
$h = 3$	—	8337.9414	8337.9414	7890.1363
$h = 4$	—	—	28151.1516	24936.7270

Table 2.7

${}^{n}u_{hi0}\sqrt{g_0}$	$n = 2$		$n = 3$			$n = 4$			
	$h = 1$	$h = 2$	$h = 1$	$h = 2$	$h = 3$	$h = 1$	$h = 2$	$h = 3$	$h = 4$
$i = 1$	4.5080	144.1874	4.4423	144.1874	−16.0108	4.4423	132.9890	−16.0195	364.6832
$i = 2$	0	−19.2250	0	−19.2250	− 0.0001	0	−17.7319	0.0011	−48.6246
$i = 3$	—	—	1.6606	− 0.0003	403.9649	1.6606	35.2891	403.9913	−1142.1741
$i = 4$	—	—	—	—	—	0	−70.5783	−0.0528	2284.4111

Table 2.8

$\lvert w_h - {}^{n}u_{h0}\rvert\sqrt{g_0} \leqq$	$h = 1$	$h = 2$	$\lvert w_h'' - {}^{n}u_{h0}''\rvert\sqrt{g_0} \leqq$	$h = 1$	$h = 2$
$n = 2$	0.0070	0.050	$n = 2$	0.550	9.00
$n = 3$	0.0001	0.050	$n = 3$	0.015	9.00
$n = 4$	0.0001	0.002	$n = 4$	0.015	0.56

Table 2.8 where

$$\max_{0\leq x\leq 1} |w_1''(x)| = 13.9577 \frac{1}{\sqrt{g_0}}, \qquad \max_{0\leq x\leq 1} |w_2''(x)| = 55.8309 \frac{1}{\sqrt{g_0}}.$$

In order to apply the limit theorem for the eigenvalues and eigenfunctions proved in Section 2.2.2, the random vector $(\bar{f}(x, \omega), \bar{g}(x, \omega))$ is assumed to be weakly correlated connected. We write $(\bar{f}, \bar{g}) = (\bar{f}_\varepsilon, \bar{g}_\varepsilon)$. Furthermore, let $(\bar{f}_\varepsilon, \bar{g}_\varepsilon)$ be a wide-sense stationary vector process such that the intensities a_{11} belonging to $\bar{f}_\varepsilon$, a_{12} belonging to $\bar{f}_\varepsilon, \bar{g}_\varepsilon$ and a_{22} belonging to $\bar{g}_\varepsilon$ are constants. Then, from Theorem 2.11 we have convergence in distribution

$$\lim_{\varepsilon\downarrow 0} \frac{1}{\sqrt{\varepsilon}} ({}^n\Lambda_{1\varepsilon} - {}^n\Lambda_{10}, {}^n\Lambda_{2\varepsilon} - {}^n\Lambda_{20}, \dots, {}^n\Lambda_{n\varepsilon} - {}^n\Lambda_{n0}) = ({}^n\xi_1, {}^n\xi_2, \dots, {}^n\xi_n)$$

for the eigenvalues ${}^n\Lambda_{i\varepsilon}$, $i = 1, 2, \dots, n$, of problem (2.64). The correlation relations of $({}^n\xi_1, {}^n\xi_2, \dots, {}^n\xi_n)$ are given by Theorem 2.12; namely,

$$\langle {}^n\xi_g {}^n\xi_h \rangle = a_{11} \langle\!\langle {}^n\bar{u}_{g0}^2, {}^n\bar{u}_{h0}^2 \rangle\!\rangle - a_{12} [{}^n\Lambda_{g0} \langle\!\langle {}^n\bar{u}_{h0}^2, {}^n u_{g0}^2 \rangle\!\rangle + {}^n\Lambda_{h0} \langle\!\langle {}^n\bar{u}_{g0}^2, {}^n u_{h0}^2 \rangle\!\rangle] + a_{22} {}^n\Lambda_{g0} {}^n\Lambda_{h0} \langle\!\langle {}^n u_{g0}^2, {}^n u_{h0}^2 \rangle\!\rangle,$$

where

$${}^n\bar{u}_{g0}(x) = \sum_{k=1}^{n} {}^n u_{gk0} \varphi_k''(x), \qquad {}^n u_{g0}(x) = \sum_{k=1}^{n} {}^n u_{gk0} \varphi_k(x).$$

The values which are necessary for the calculation can be obtained from Tables 2.6 and 2.7. Using the notation

$${}^nS_{gh}^1 \doteq \langle\!\langle {}^n\bar{u}_{g0}^2, {}^n\bar{u}_{h0}^2 \rangle\!\rangle, \qquad {}^nS_{gh}^{12} \doteq \langle\!\langle {}^n\bar{u}_{g0}^2, {}^n u_{h0}^2 \rangle\!\rangle, \qquad {}^nS_{gh}^2 \doteq \langle\!\langle {}^n u_{g0}^2, {}^n u_{h0}^2 \rangle\!\rangle,$$

Table 2.9 can be determined.

For small ε, the random vector $({}^n\Lambda_{1\varepsilon}, \dots, {}^n\Lambda_{n\varepsilon})$ is approximately Gaussian with mean $({}^n\Lambda_{10}, \dots, {}^n\Lambda_{n0})$ and correlation matrix

$$(\langle ({}^n\Lambda_{g\varepsilon} - {}^n\Lambda_{g0}) ({}^n\Lambda_{h\varepsilon} - {}^n\Lambda_{h0}) \rangle)_{1\leq g, h\leq n} \approx \varepsilon (\langle {}^n\xi_g\, {}^n\xi_h \rangle)_{1\leq g, h\leq n}.$$

We now calculate the intensities of the incoming weakly correlated processes. To simplify the situation we assume that the cross-sectional area of the bars is a circle with a random radius $r(x, \omega)$. Let $\bar{r}_\varepsilon(x, \omega) = r(x, \omega) - \langle r(x) \rangle$ be a weakly correlated process having a stationary Gaussian distribution with

$$\langle r_\varepsilon(x) \rangle \equiv r_0 \quad \text{and} \quad \langle \bar{r}_\varepsilon(x)\, \bar{r}_\varepsilon(y) \rangle = R_\varepsilon(y - x).$$

Furthermore, the mass per unit length $\varrho(x, \omega)$ is assumed to be a wide-sense stationary process with mean $\langle \varrho(x) \rangle = \varrho_0$, and $\bar{\varrho}_\varepsilon(x, \omega) \doteq \varrho(x, \omega) - \varrho_0$ is a weakly correlated process with correlation length ε. The processes $\bar{\varrho}_\varepsilon(x, \omega)$ and $\bar{r}_\varepsilon(x, \omega)$ are independent. Hence, the processes

$$\bar{f}_\varepsilon(x, \omega) = EI(x, \omega) - E\langle I(x) \rangle = \tfrac{1}{4} E\pi[r^4(x, \omega) - \langle r^4(x) \rangle],$$

$$\bar{g}_\varepsilon(x, \omega) = \varrho(x, \omega)\, A(x, \omega) - \langle \varrho(x)\, A(x) \rangle$$

$$= \pi[\varrho(x, \omega)\, r^2(x, \omega) - \varrho_0 \langle r^2(x) \rangle]$$

Table 2.9

h	g	n	${}^nS^1_{gh}g_0^2$	${}^nS^{12}_{hg}g_0^2$	${}^nS^{12}_{gh}g_0^2$	${}^nS^2_{gh}g_0^2$
1	1	2	13594	142,26	142,26	1.491813
		3	14229	146,09	146.09	1.499961
		4	14229	146.09	146.09	1.499961
2	1	2	140471	1388.19	97.55	0.972684
		3	134622	1381.04	94.43	0.968869
		4	151485	1553.99	97.41	0.999292
	2	2	3684626	2317.94	2317.94	1.494570
		3	3684624	2317.94	2317.94	1.494568
		4	3672550	2347.99	2347.99	1.501294
3	1	3	614586	6312.51	89.12	0.915192
		4	614587	6312.51	89.12	0.915218
	2	3	14337072	7924.15	1770.82	1.016229
		4	11833540	7567.22	1542.56	0.984709
	3	3	110816517	12480.42	12480.42	1.512442
		4	110823507	12483.35	12483.35	1.512445
4	1	4	1734785	17815.67	81.99	0.841835
	2	4	36175336	23355.96	1489.17	0.959155
	3	4	317074980	29581.44	10769.44	1.098581
	4	4	1472916532	44900.51	44900.51	1.566899

are also weakly correlated with correlation length ε. Calculations of the means yield

$$f_0 = \tfrac{1}{4}\, E\pi\langle r^4(x)\rangle = \tfrac{1}{4}\, E\pi \sum_{i=0}^{4} \binom{4}{i} \langle (r(x) - r_0)^i\rangle\, r_0^{4-i}$$

$$= \tfrac{1}{4}\, E\pi(r_0^4 + 6\sigma_r^2 r_0^2 + 3\sigma_r^4)\,,$$

$$g_0 = \pi\varrho_0\langle r^2(x)\rangle = \pi\varrho_0(\sigma_r^2 + r_0^2)\,,$$

where $R_\varepsilon(0) = \sigma_r^2$, and the correlation functions

$$\langle \bar f_\varepsilon(x)\, \bar f_\varepsilon(y)\rangle = \left(\frac{E\pi}{4}\right)^2 [\langle r^4(x)\, r^4(y)\rangle - \langle r^4\rangle^2]$$

$$= \left(\frac{E\pi}{4}\right)^2 [24R_\varepsilon^4(y - x) + 96r_0^2 R_\varepsilon^3(y - x)$$

$$+72(r_0^2+\sigma_r^2)^2 R_\varepsilon^2(y-x) + 16r_0^2(r_0^2+3\sigma_r^2)^2 R_\varepsilon(y-x)]\,,$$

$$\langle \bar f_\varepsilon(x)\, \bar g_\varepsilon(y)\rangle = \frac{E}{4}\pi^2[\langle r^4(x)\, r^2(y)\, \varrho(y)\rangle - \langle r^4\rangle\, \langle r^2\varrho\rangle]$$

$$= \pi^2 E\varrho_0[3(\sigma_r^2 + r_0^2)\, R_\varepsilon^2(y - x) + 2r_0^2(3\sigma_r^2 + r_0^2)\, R_\varepsilon(y - x)]\,,$$

$$\langle \bar g_\varepsilon(x)\, \bar g_\varepsilon(y)\rangle = \pi^2[\langle \varrho(x)\, \varrho(y)\, r^2(x)\, r^2(y)\rangle - \langle \varrho r^2(x)\rangle\, \langle \varrho r^2(y)\rangle]$$

$$= \pi^2[R_\varepsilon^\varrho(y - x)\, \{(\sigma^2 + r_0^2)^2 + 4r_0^2 R_\varepsilon(y - x) + 2R_\varepsilon^2(y - x)\}$$

$$+ 2\varrho_0^2\{2r_0^2 R_\varepsilon(y - x) + R_\varepsilon^2(y - x)\}]\,,$$

where $R_\varepsilon^\varrho(y - x) \doteq \langle (\varrho(x) - \varrho_0)\, (\varrho(y) - \varrho_0) \rangle$. The correlation functions for $r(x, \omega)$ and $\varrho(x, \omega)$

$$\frac{1}{\sigma_r^2} R_\varepsilon(x) = \frac{1}{\sigma_\varrho^2} R_\varepsilon^\varrho(x) = \begin{cases} 1 - \dfrac{|x|}{\varepsilon} & \text{for} \quad |x| \leqq \varepsilon\,, \\ 0 & \text{otherwise} \end{cases}$$

are choosen very simply. Then we have

$$\frac{1}{\varepsilon} \int_{-\varepsilon}^{\varepsilon} \left(1 - \frac{|x|}{\varepsilon}\right)^k \mathrm{d}x = \frac{2}{k+1}$$

and

$$\begin{aligned} a_{11} &= \lim_{\varepsilon \downarrow 0} \frac{1}{\varepsilon} \int_{-\varepsilon}^{\varepsilon} \langle \bar{f}_\varepsilon(x)\, \bar{f}_\varepsilon(x + y) \rangle\, \mathrm{d}y \\ &= \left(\frac{E\pi}{4}\right)^2 [57{,}6\sigma_r^6 + 288 r_0^2 \sigma_r^4 + 144 r_0^4 \sigma_r^2 + 16 r_0^6]\, \sigma_r^2\,, \\ a_{12} &= \lim_{\varepsilon \downarrow 0} \frac{1}{\varepsilon} \int_{-\varepsilon}^{\varepsilon} \langle \bar{f}_\varepsilon(x)\, \bar{g}_\varepsilon(x + y) \rangle\, \mathrm{d}y \\ &= 2E\varrho_0 \pi^2 [\sigma_r^4 + 4 r_0^2 \sigma_r^2 + r_0^4]\, \sigma_r^2\,, \\ a_{22} &= \lim_{\varepsilon \downarrow 0} \frac{1}{\varepsilon} \int_{-\varepsilon}^{\varepsilon} \langle \bar{g}_\varepsilon(x)\, \bar{g}_\varepsilon(x + y) \rangle\, \mathrm{d}y \\ &= \pi^2 \left[\sigma_\varrho^2 \left(2\sigma_r^4 + \frac{14}{3} \sigma_r^2 r_0^2 + r_0^4\right) + 4\varrho_0^2 \sigma_r^2 \left(r_0^2 + \frac{1}{3} \sigma_r^2\right)\right]. \end{aligned} \tag{2.68}$$

Put $\sigma_r = \beta r_0$ for further applications. Then for every $x \in [0, 1]$, the Gaussian radius $r(x, \omega)$ is contained in the interval

$$(-3\beta + 1)\, r_0 \leqq r(x, \omega) \leqq (3\beta + 1)\, r_0$$

with probability 0.9972. The inequality $r_0/2 \leqq r(x, \omega) \leqq 3/2 r_0$ follows from $\beta = 1/6$, i.e. the radius of the bar is contained between $r_0/2$ and $3r_0/2$ with given probability. Using (2.68) we have

$$\begin{aligned} a_{11} &= (\tfrac{1}{4} E\pi r_0^4)^2\, \beta^2 [57.6\beta^6 + 288\beta^4 + 144\beta^2 + 16]\,, \\ a_{12} &= 2E\varrho_0 (\pi r_0^3)^2\, \beta^2 [\beta^4 + 4\beta^2 + 1]\,, \\ a_{22} &= \pi^2 r_0^4 [\sigma_\varrho^2 (2\beta^4 + \tfrac{14}{3} \beta^2 + 1) + 4\varrho_0^2 \beta^2 (1 + \tfrac{1}{3} \beta^2)]\,. \end{aligned} \tag{2.69}$$

The Ritz approximate solutions ${}^n u_{g0}$ converge to w_g as $n \to \infty$ in the space $\mathscr{H}_M$; that is

$$\lim_{n \to \infty} {}^n u_{g0} = w_g \quad \text{in } \mathscr{H}_M\,.$$

It then follows that

$$\lim_{n\to\infty} {}^n u''_{g0} = w''_g \quad \text{in } \mathbf{L}_2(0, 1);$$

and we also have convergence in $\mathbf{C}(0, 1)$:

$$\lim_{n\to\infty} {}^n u_{g0} = w_g .$$

Hence we have

$$\lim_{n\to\infty} \langle\!\langle {}^n u_{g0}^2, {}^n u_{h0}^2 \rangle\!\rangle = \langle\!\langle w_g^2, w_h^2 \rangle\!\rangle ,$$

$$\lim_{n\to\infty} \langle\!\langle {}^n \bar{u}_{g0}^2, {}^n u_{h0}^2 \rangle\!\rangle = \langle\!\langle w_g''^2, w_h^2 \rangle\!\rangle .$$

The second equation can be proved using the inequalities

$$\begin{aligned} |\langle\!\langle {}^n \bar{u}_{g0}^2, {}^n u_{h0}^2 \rangle\!\rangle - \langle\!\langle w_g''^2, w_h^2 \rangle\!\rangle| &\leqq |\langle\!\langle {}^n \bar{u}_{g0}^2 - w_g''^2, w_h^2 \rangle\!\rangle + \langle\!\langle {}^n \bar{u}_{g0}^2, {}^n u_{h0}^2 - w_h^2 \rangle\!\rangle| \\ &\leqq \|{}^n \bar{u}_{g0} - w_g''\| \left(\|w_g'' w_h^2\| + \|{}^n \bar{u}_{g0} w_h^2\|\right) \\ &\quad + \max_{0 \leqq x \leqq 1} |{}^n u_{h0}^2 - w_h^2| \, \|{}^n \bar{u}_{g0}\| . \end{aligned}$$

Now using

$$\langle\!\langle w_g^2, w_h^2 \rangle\!\rangle = \frac{4}{g_0^2} \int_0^1 \sin^2 (g\pi x) \sin^2 (h\pi x) \, dx = \begin{cases} \dfrac{3}{2} g_0^{-2} , & \text{for } g = h , \\ g_0^{-2} , & \text{for } g \neq h , \end{cases}$$

$$\langle\!\langle w_g''^2, w_h^2 \rangle\!\rangle = \frac{4(g\pi)^4}{g_0^2} \int_0^1 \sin^2 (g\pi x) \sin^2 (h\pi x) \, dx$$

$$= \begin{cases} \dfrac{3}{2} (g\pi)^4 g_0^{-2} , & \text{for } g = h , \\ (g\pi)^4 g_0^{-2} , & \text{for } g \neq h , \end{cases}$$

the values given in Table 2.10 can be determined. The values given in Table 2.9 agree very well with the limit values for increasing positive integer n. We first consider the case of a random mass per unit $\varrho(x, \omega)$ and with constant

Table 2.10

g	h	$\langle\!\langle w_g^2, w_h^2 \rangle\!\rangle g_0^2$	$\langle\!\langle w_g''^2, w_h^2 \rangle\!\rangle g_0^2$
1	1	1.5	146.11
2	2	1.5	2337.82
3	3	1.5	11835.20
4	4	1.5	37405.09
1	2, 3, 4	1	97.41
2	1, 3, 4	1	1558.55
3	1, 2, 4	1	7890.14
4	1, 2, 3	1	24936.73

radius r_0. It then follows from (2.69) that

$$a_{11} = a_{12} = 0 \quad \text{and} \quad a_{22} = (\pi r_0^2 \sigma_\varrho)^2 ,$$

because $\beta = 0$. Hence, the random vector $({}^n\Lambda_{1\varepsilon}, \dots, {}^n\Lambda_{n\varepsilon})$ is approximately Gaussian with mean vector $({}^n\Lambda_{10}, \dots, {}^n\Lambda_{n0})$ and correlation relations

$$\langle({}^n\Lambda_{g\varepsilon} - {}^n\Lambda_{g0})\,({}^n\Lambda_{h\varepsilon} - {}^n\Lambda_{h0})\rangle \approx \varepsilon(\pi r_0^2\sigma_\varrho)^2\, {}^n\Lambda_{g0}\, {}^n\Lambda_{h0}\, {}^nS_{gh}^2 \doteq {}^n\sigma_{gh} .$$

Applying ${}^nS_{gh}^2 \approx S_{gh}^2 = \langle\!\langle w_g^2, w_h^2\rangle\!\rangle$ and ${}^n\Lambda_{g0} \approx \mu_g$ we have

$${}^n\sigma_{gg} \approx \sigma_{gg} = \frac{3}{2}\,\varepsilon\,\frac{\sigma_\varrho^2}{\varrho_0^2}\,\mu_g^2 ,$$

and

$$\frac{{}^n\sigma_{gh}}{\sqrt{{}^n\sigma_{gg}\,{}^n\sigma_{hh}}} \approx \frac{\sigma_{gh}}{\sqrt{\sigma_{gg}\sigma_{hh}}} = \frac{2}{3} ,$$

i.e. the variance of the g-th eigenvalue increases with the number of the eigenvalue and the correlation coefficient between two arbitrary eigenvalues has the constant value 2/3.

In the following, a bar is considered to have the length 1m and material constants

$$r_0 = 0.01\ \text{m} , \qquad \varrho_0 = 8000\ \text{kg/m}^3 , \qquad E = 212 \cdot 10^9\ \text{N/m}^2 \tag{2.70}$$

(the material of the bar is assumed to be steel). Then we have

$$f_0 = 1665.044\ \text{Nm}^2 , \qquad g_0 = 2.513\ \text{Ns}^2/\text{m}^2$$

and the calculated eigenvalues of the averaged problem are

$${}^4\Lambda_{10} = 6.4533 \cdot 10^4\ \text{s}^{-2} , \qquad {}^4\Lambda_{20} = 1.032599 \cdot 10^6\ \text{s}^{-2} ,$$

$${}^4\Lambda_{30} = 5.523886 \cdot 10^6\ \text{s}^{-2}.$$

The eigenfrequencies ${}^4\omega_{g0} \doteq \sqrt{{}^4\Lambda_{g0}}$ are

$${}^4\omega_{10} = 254.03\ \text{s}^{-1} , \qquad {}^4\omega_{20} = 1016.17\ \text{s}^{-1} , \qquad {}^4\omega_{30} = 2350.29\ \text{s}^{-1} .$$

Furthermore, the variances ${}^n\sigma_{gg}$ for $g = 1, 2, 3$ and $n = 4$, are given by

$${}^4\sigma_{11} = \varepsilon\sigma_\varrho^2 \cdot 97.61\ \text{m}^7/\text{N}^2\text{s}^8 , \qquad {}^4\sigma_{22} = \varepsilon\sigma_\varrho^2 \cdot 24990.49\ \text{m}^7/\text{N}^2\text{s}^8 ,$$

$${}^4\sigma_{33} = \varepsilon\sigma_\varrho^2 \cdot 715155.86\ \text{m}^7/\text{N}^2\text{s}^8 .$$

The Gaussian distribution of the eigenvalue ${}^n\Lambda_{g\varepsilon}$ implies that the probability of the event $\{\omega: {}^n\Lambda_{g\varepsilon} < 0\}$ is much less than 1, but different from zero. This contradiction to ${}^n\Lambda_{g\varepsilon} \geqq 0$ a.s. is obtained by the applying the limit theorem and the other approximations. Nevertheless, the Gaussian distribution is a good approximation for the distribution of the eigenvalues ${}^n\Lambda_{g\varepsilon}$. For $x > 0$ we have

$$\mathsf{P}({}^n\Lambda_{g\varepsilon} < x) \approx \frac{1}{\sqrt{2\pi\, {}^n\sigma_{gg}}} \int_0^x \exp\left(-\frac{1}{2\, {}^n\sigma_{gg}}(t - {}^n\Lambda_{g0})^2\right) \mathrm{d}t = \mathsf{P}({}^n\omega_{g\varepsilon} < \sqrt{x}) .$$

Now let $\varrho(x, \omega)$ have a uniform distribution an the interval $[\varrho_0 - \delta, \varrho_0 + \delta]$. Then it follows that $\sigma_\varrho^2 = \delta^2/3$, and the density function of ${}^n\omega_{g\varepsilon}$ is given by

$$ {}^nf_g(y) = \frac{2y}{\sqrt{2\pi\, {}^n\sigma_{gg}}} \exp\left(-\frac{1}{2\, {}^n\sigma_{gg}} (y^2 - {}^n\Lambda_{g0})^2\right) $$

which is plotted at $\varepsilon = 0.1$ m in Fig. 2.3. The distribution function of ${}^n\omega_{g\varepsilon}$,

$$ {}^nF_g(y) = \frac{1}{\sqrt{2\pi\, {}^n\sigma_{gg}}} \int\limits_{-\infty}^{y^2} \exp\left(-\frac{1}{2\, {}^n\sigma_{gg}} (t - {}^n\Lambda_{g0})^2\right) dt $$

is plotted at $\varepsilon = 0.1$ m, $n = 4$ in Fig. 2.4. We remark that the frequency for $g = 1$ is concentrated around the mean if this frequency is compared with the

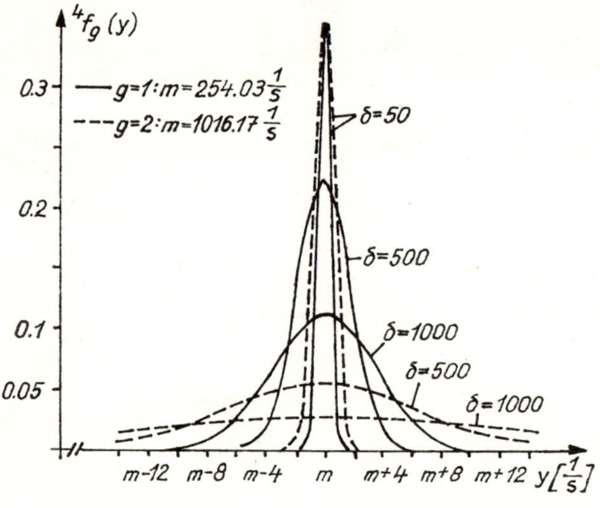

Fig. 2.3. Density function of the frequency ${}^4\omega_{g\varepsilon}$

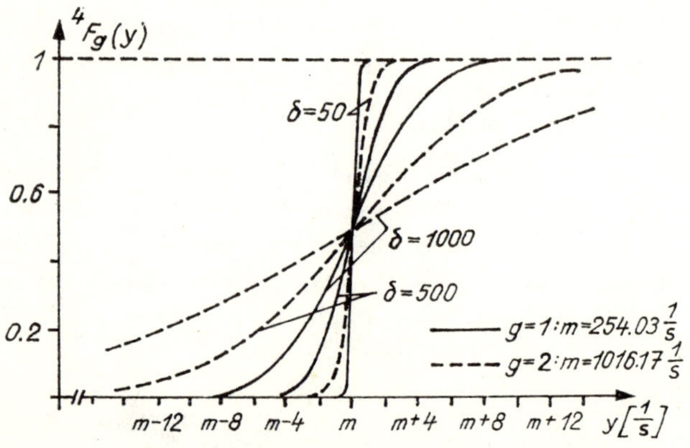

Fig. 2.4. Distribution function of the frequency ${}^4\omega_{g\varepsilon}$

frequency for $g = 2$. With an increasing variance σ_ϱ^2 the frequencies spread larger around the mean.

Now we consider the case of a random radius and a random mass per unit $\varrho(x, \omega)$ being independent of $r(x, \omega)$. Let $\sigma_r = \beta r_0$. Thus

$$^4\Lambda_{g0} \doteq \overline{^4\Lambda_{g0}} \frac{f_0}{g_0},$$

where $\overline{^4\Lambda_{10}} = 97.41$, $\overline{^4\Lambda_{20}} = 1558.64$ and

$$\frac{f_0}{g_0} = \frac{Er_0^2}{4\varrho_0} \frac{1 + 6\beta^2 + 3\beta^4}{1 + \beta^2}.$$

Furthermore, $^4\sigma_{gh}$ can be represented as

$$\begin{aligned} ^4\sigma_{gh} &= \langle (^4\Lambda_{g\varepsilon} - {}^4\Lambda_{g0}) (^4\Lambda_{h\varepsilon} - {}^4\Lambda_{h0}) \rangle \\ &\approx \varepsilon(a_{11}\, {}^4S_{gh}^1 - a_{12} [^4\Lambda_{g0}\, {}^4S_{hg}^{12} + {}^4\Lambda_{h0}\, {}^4S_{gh}^{12}] + a_{22}\, {}^4\Lambda_{g0}\, {}^4\Lambda_{h0}\, {}^4S_{gh}^2) \end{aligned}$$

where a_{ij} is determined by (2.69) and $^nS_{gh}^{\cdot\cdot}$ by Table 2.9. Setting $^nS_{gh}^{\cdot\cdot} \doteq \overline{^nS_{gh}^{\cdot\cdot}} g_0^{-2}$ the following formula obtains:

$$\begin{aligned} ^4\sigma_{gh} \approx \varepsilon \left(\frac{Er_0^2}{4\varrho_0}\right)^2 \Bigg[& \frac{\beta^2(57.6\,\beta^6 + 288\beta^4 + 144\beta^2 + 16)}{(1 + \beta^2)^2} \overline{^4S_{gh}^1} \\ & - \frac{8\beta^2(\beta^4 + 4\beta^2 + 1)(3\beta^4 + 6\beta^2 + 1)}{(1 + \beta^2)^3} \\ & \times (\overline{^4\Lambda_{g0}}\, \overline{^4S_{hg}^{12}} + \overline{^4\Lambda_{h0}}\, \overline{^4S_{gh}^{12}}) \\ & + \frac{(3\beta^4 + 6\beta^2 + 1)^2}{(1 + \beta^2)^4} \left(\frac{\sigma_\varrho^2}{\varrho_0^2}\left(2\beta^4 + \frac{14}{3}\beta^2 + 1\right)\right. \\ & \left. + 4\beta^2\left(1 + \frac{1}{3}\beta^2\right)\right) \overline{^4\Lambda_{g0}}\, \overline{^4\Lambda_{h0}}\, \overline{^4S_{gh}^2} \Bigg]. \end{aligned}$$

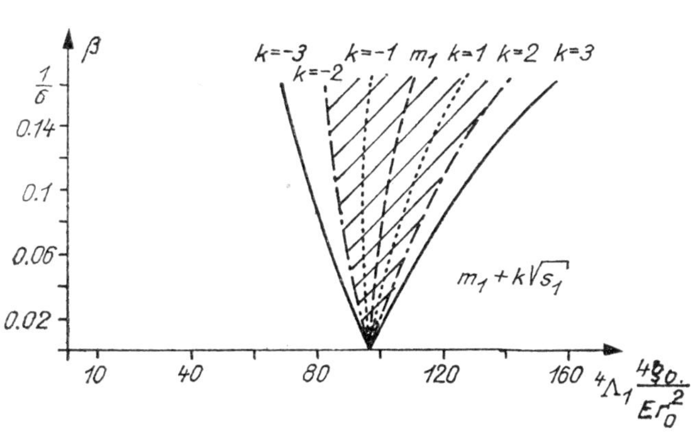

Fig. 2.5. Confidence intervals for the first eigenvalue

In Fig. 2.5 the intervals are plotted which satisfy the relations

$$\mathsf{P}\left(\left|{}^4\Lambda_{1\varepsilon}\frac{4\varrho_0}{Er_0^2}-m_1\right|\leqq ks_1\right)=\begin{cases}0.6826 & \text{for } k=1\,,\\ 0.9544 & \text{for } k=2\,,\\ 0.9972 & \text{for } k=3\,.\end{cases}$$

Fig. 2.5 is given for $\sigma_\varrho^2 = 0$ and different values of β. We have put

$$m_1 \doteq {}^4\Lambda_{10}\frac{4\varrho_0}{Er_0^2}, \qquad s_1 \doteq \sqrt{{}^4\sigma_{11}}\,\frac{4\varrho_0}{Er_0^2} \quad \text{and} \quad \varepsilon = 0.1\ \text{m}\,.$$

For instance, we have the inequality

$$86.5\frac{Er_0^2}{4\varrho_0} \leqq {}^4\Lambda_1(\omega) \leqq 118\frac{Er_0^2}{4\varrho_0}$$

with probability 0.9544 for $\beta = 0.1$. The values of ${}^4\Lambda_{1\varepsilon}\,4\varrho_0/Er_0^2$ are contained in the hatched domain of Fig. 2.5 with probability 0.9544 for a fixed value of β, respectively. The same results can also obtained for $g = 2, 3, \ldots$ where the corresponding intervals are, however, much larger.

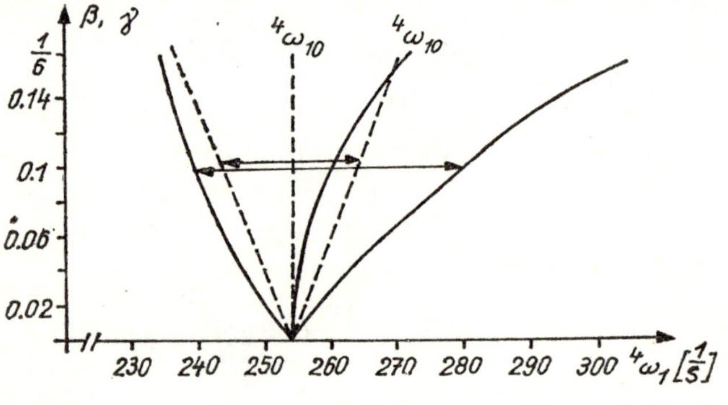

Fig. 2.6. Comparison of the influence of $r(x, \omega)$ and $\varrho(x, \omega)$ on the first frequency

We will now consider the case of random influence of the mass per unit and radius $r(x, \omega)$ on the eigenfrequencies of bars. Let $\varrho(x, \omega)$ and $r(x, \omega)$ be Gaussian processes with means $\langle\varrho(x)\rangle = \varrho_0$, $\langle r(x)\rangle = r_0$ and variances $\sigma_\varrho = \gamma\varrho_0$, $\sigma_r = \beta r_0$, respectively. The values γ and β denote measures for the deviation of the considered random variables from the means. In Fig. 2.6 the intervals are plotted containing the first frequency ${}^4\omega_{1\varepsilon}$ if ϱ is random (hatched line) and if r is random (bold line). We have used the same values for r_0, ϱ_0 and E as given in (2.70). We put $\varepsilon = 0.1$ m. For this example we can observe that the geometry influence more strongly the first eigenfrequency than the mass per unit. From

$$0.7r_0 \leqq r(x, \omega) \leqq 1.3r_0\,, \quad \text{with probability} \quad 0.9972\,,$$

we obtain the inequality

$$239\ \text{s}^{-1} \leqq {}^4\omega_{1\varepsilon} \leqq 270\ \text{s}^{-1}\,, \quad \text{with probability } 0.9544;$$

and from

$$0.7\varrho_0 \leqq \varrho(x, \omega) \leqq 1.3\varrho_0\,, \quad \text{with probability } 0.9972\,,$$

the inequality

$$244\ \mathrm{s}^{-1} \leqq {}^4\omega_{1\varepsilon} \leqq 264\ \mathrm{s}^{-1}\,, \quad \text{with probability } 0.9544\,.$$

2.2.4.2. Buckling problems

The matrix eigenvalue problem (2.64) associated with the eigenvalue problem (2.62) is considered where

$$a_{ij} = f_0\langle\!\langle \varphi_i'''', \varphi_j \rangle\!\rangle = f_0\langle\!\langle \varphi_i'', \varphi_j'' \rangle\!\rangle, \qquad b_{ij} = \langle\!\langle (\bar{f}\varphi_i'')'', \varphi_j \rangle\!\rangle = \langle\!\langle \bar{f}\varphi_i'', \varphi_j'' \rangle\!\rangle\,,$$
$$c_{ij} = c_{ij}^1 = -\langle\!\langle \varphi_i'', \varphi_j \rangle\!\rangle = \langle\!\langle \varphi_i', \varphi_j' \rangle\!\rangle\,, \qquad d_{ij} = 0\,.$$

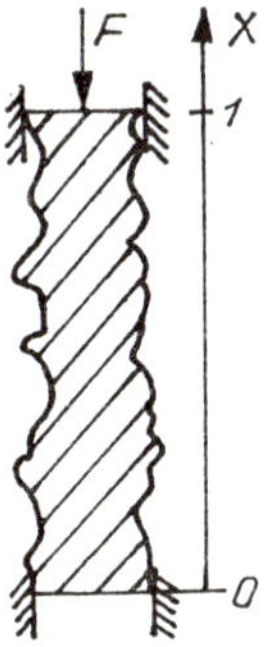

Fig. 2.7. A loaded bar with random surface

Because of the boundary conditions $u(0) = u(1) = u''(0) = u''(1) = 0$ the functions φ_i, $i = 1, 2, \ldots$, from (2.67) can be taken as a basis. We put $f(x, \omega) = EI$ and $\bar{f}(x, \omega) \doteq f(x, \omega) - \langle f(x) \rangle$. Let $f(x, \omega)$ be a wide-sense stationary process with $\langle f(x) \rangle = f_0$. Hence, we consider the simply supported buckling problem for a bar with a random surface of the bar or a random modulus of elasticity (cf. Fig. 2.7). This problem corresponds to the case $g_0 = 0, g_1 = 1$. The values for a_{ij} and $c_{ij} = c_{ij}^1$ were given in Section 2.2.4. The eigenvalues of the averaged problem ${}^n\Lambda_{g0} = {}^nM_g f_0$ are given by Table 2.11, and the corresponding eigenvectors ${}^nU_{g0} = ({}^nu_{gi0})_{1 \leqq i \leqq n}^{\mathrm{T}}$ in Table 2.12. A good correspondence between ${}^nu_{g0}(x) = \sum_{i=1}^{n} {}^nu_{gi0}\varphi_i(x)$ and $w_g(x) = \sqrt{2}/g\pi \sin(g\pi x)$ is obtained in this case for $g = 1, 2, 3, 4$. This correspondence is improved for increasing n.

Table 2.11

${}^nM_g$	$n = 2$	$n = 3$	$n = 4$	$n = \infty$
$g = 1$	9.8824	9.8696	9.8696	9.8696
$g = 2$	40.0000	40.0000	39.4806	39.4784
$g = 3$	—	92.5919	92.5919	88.8264
$g = 4$	—	—	172.2178	157.9137

Table 2.12

$^{n}u_{gi0}$	$n = 2$		$n = 3$		
	$g = 1$	$g = 2$	$g = 1$	$g = 2$	$g = 3$
$i = 1$	1.4349	22.9129	1.4140	22.9129	1.6872
$i = 2$	0	−3.0551	0	−3.0551	0
$i = 3$	—	—	0.5284	0	−42.5697

$^{4}u_{gi0}$	$g = 1$	$g = 2$	$g = 3$	$g = 4$
$i = 1$	1.4140	21.1711	1.6872	28.5068
$i = 2$	0	− 2.8228	0	− 3.8009
$i = 3$	0.5284	5.5999	−42.5697	−89.3423
$i = 4$	0	−11.1998	− 0.0001	178.6846

Let $\bar{f}(x, \omega) = \bar{f}_\varepsilon(x, \omega)$ denote a weakly correlated process. Then Theorem 2.12 can be applied to this matrix eigenvalue problem. We have the convergence in distribution

$$\lim_{\varepsilon \downarrow 0} \frac{1}{\sqrt{\varepsilon}} ({}^{n}\Lambda_{1\varepsilon} - {}^{n}\Lambda_{10}, \dots, {}^{n}\Lambda_{n\varepsilon} - {}^{n}\Lambda_{n0}) = ({}^{n}\xi_1, \dots, {}^{n}\xi_n)$$

for the eigenvalues ${}^{n}\Lambda_{g\varepsilon}(\omega)$ of the random matrix eigenvalue problem as in the example above. The Gaussian random vector $({}^{n}\xi_1, \dots, {}^{n}\xi_n)$ possesses the first moments $\langle {}^{n}\xi_g \rangle = 0$ and the second moments

$$\langle {}^{n}\xi_g \, {}^{n}\xi_h \rangle = a \, {}^{n}S_{gh} ,$$

where ${}^{n}S_{gh}$ is defined by

$${}^{n}S_{gh} \doteq \langle\!\langle {}^{n}\bar{u}_{g0}^{2}, {}^{n}\bar{u}_{h0}^{2} \rangle\!\rangle , \quad \text{with} \quad {}^{n}\bar{u}_{g0}(x) \doteq \sum_{i=1} {}^{n}u_{gi0}\varphi_i''(x) .$$

These values of ${}^{n}S_{gh}$ are calculated in Table 2.13 from the eigenvectors ${}^{n}U_{g0}$ given by Table 2.12.

Table 2.13

g	h	$^{2}S_{gh}$	$^{3}S_{gh}$	$^{4}S_{gh}$
1	1	139.52	146.08	146.08
1	2	359.36	344.45	388.65
2	2	2349.65	2349.65	2355.99
1	3	—	691.53	691.53
2	3	—	4020.47	3330.44
3	3	—	13666.50	13666.50
1	4	—	—	1075.91
2	4	—	—	5608.93
3	4	—	—	21532.53
4	4	—	—	55103.39

We have the following approximations:

$$\langle {}^n\Lambda_{g\varepsilon} \rangle \approx {}^n\Lambda_{g0}$$

and

$$ {}^n\sigma_{gh} = \langle ({}^n\Lambda_{g\varepsilon} - {}^n\Lambda_{g0}) ({}^n\Lambda_{h\varepsilon} - {}^n\Lambda_{h0}) \rangle \approx \varepsilon a \, {}^nS_{gh} ,$$

for $g, h = 1, 2, \ldots, n$.

To determine the intensity a of the process $\bar{f}_\varepsilon$ we assume bars whose cross-sectional area is circular with radius $r(x, \omega)$, as in the example of the eigenfrequencies. Let $\langle r(x) \rangle = r_0$. $\bar{r}_\varepsilon(x, \omega) \doteq r(x, \omega) - r_0$ is assumed to be a weakly correlated stationary process. Then the process

$$\bar{f}_\varepsilon(x, \omega) = \frac{E\pi}{4} \left(r^4(x, \omega) - \langle r^4(x) \rangle \right)$$

is also weakly correlated. Let the modulus of elasticity E be constant. It is also possible to consider a random modulus of elasticity. The intensity a of $\bar{f}_\varepsilon$ is given by

$$a = (\tfrac{1}{4} E\pi r_0^4)^2 \, [57.6\beta^6 + 288\beta^4 + 144\beta^2 + 16] \, \beta^2$$

where $\sigma = \beta r_0$ and $\sigma^2 = \langle \bar{r}_\varepsilon^2 \rangle$ if the same conditions for $\bar{r}_\varepsilon(x, \omega)$ are satisfied as in case of bending vibrations (cf. (2.69)).

Setting

$$\overline{{}^n\Lambda_{g\varepsilon}} = \frac{4}{E\pi r_0^4} \, {}^n\Lambda_{g\varepsilon} ,$$

we have

$$\overline{{}^n\sigma_{gh}} = \langle (\overline{{}^n\Lambda_{g\varepsilon}} - \overline{{}^n\Lambda_{g0}}) (\overline{{}^n\Lambda_{h\varepsilon}} - \overline{{}^n\Lambda_{h0}}) \rangle = \left(\frac{4}{E\pi r_0^4} \right)^2 {}^n\sigma_{gh} ;$$

and finally

$$\langle \overline{{}^n\Lambda_{g\varepsilon}} \rangle \approx (1 + 6\beta^2 + 3\beta^4) \, {}^nM_g ,$$

$$\overline{{}^n\sigma_{gh}} \approx \varepsilon\beta^2 (57.6\beta^6 + 288\beta^4 + 144\beta^2 + 16) \, {}^nS_{gh} .$$

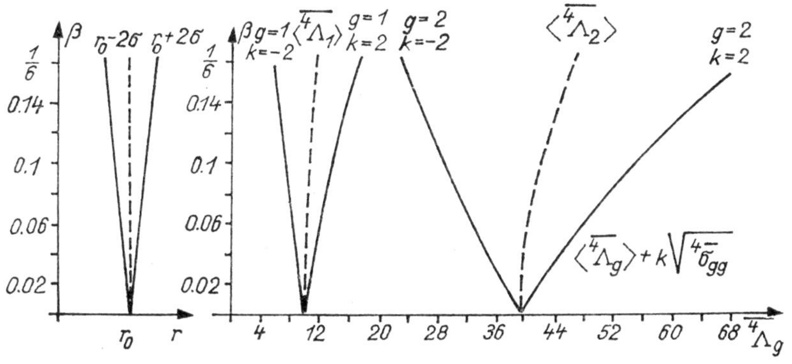

Fig. 2.8. Confidence intervals for the eigenvalues of buckling problem

Fig. 2.8 shows the domain in which $r(x, \omega)$, and $\overline{{}^4\Lambda_{1\varepsilon}}(\omega)$ and $\overline{{}^4\Lambda_{2\varepsilon}}(\omega)$ are contained as functions of β. The variables belong to the drawn intervals with probability 0.9544, respectively. We have put $\varepsilon = 0.1$. The large domain of $\overline{{}^4\Lambda_{2\varepsilon}}(\omega)$ is remarkable if it is compared with the domain of $\overline{{}^4\Lambda_{1\varepsilon}}(\omega)$. We obtain

$$\frac{\sqrt{{}^4\sigma_{22}}}{\sqrt{{}^4\sigma_{11}}} = 4.0160\,,$$

so that the domain of $\overline{{}^4\Lambda_{2\varepsilon}}(\omega)$ is about the four times of the domain of $\overline{{}^4\Lambda_{1\varepsilon}}(\omega)$.

The density function of $\overline{{}^n\Lambda_{g\varepsilon}}$ can be written as

$${}^n\varphi_{g\beta}(x) = \frac{1}{\sqrt{2\pi\, {}^n\sigma_{gg}}} \exp\left(-\frac{1}{2\, {}^n\sigma_{gg}} (x - \langle \overline{{}^n\Lambda_{g\varepsilon}} \rangle)^2\right);$$

and the density function of ${}^n\Lambda_{g\varepsilon}$ as

$${}^nf_{g\beta}(x) = \frac{4}{E\pi r_0^4}\, {}^n\varphi_{g\beta}\left(\frac{4x}{E\pi r_0^4}\right).$$

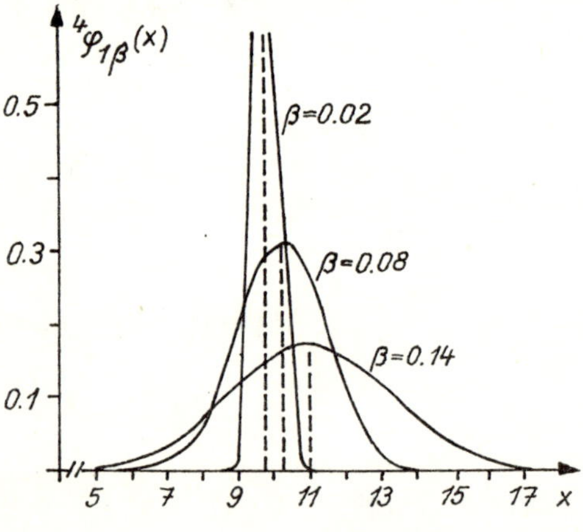

Fig. 2.9. Density function of $\overline{{}^4\Lambda_{1\varepsilon}}$ for various values of β

The density function of $\overline{{}^4\Lambda_{g\varepsilon}}$ is given for $g = 1$ and various values of β in Fig. 2.9, where $\varepsilon = 0.1$. If β increases then the mean of ${}^4\Lambda_{1\varepsilon}$ moves to the right-hand side and the variance increases.

Hence, from the mathematical point of view we have obtained maximum information for the buckling forces of this problem. A number of technological applications are possible.

We now consider a bar with a random surface as described above. The probability that a bar fails if it is loaded up to a force F can be given by

$$\mathsf{P}({}^4\Lambda_{1\varepsilon} \leqq F) = \mathsf{P}(\overline{{}^4\Lambda_{1\varepsilon}} \leqq \overline{F}) = \int_{-\infty}^{\overline{F}} {}^4\varphi_{1\beta}(x)\, dx\,,$$

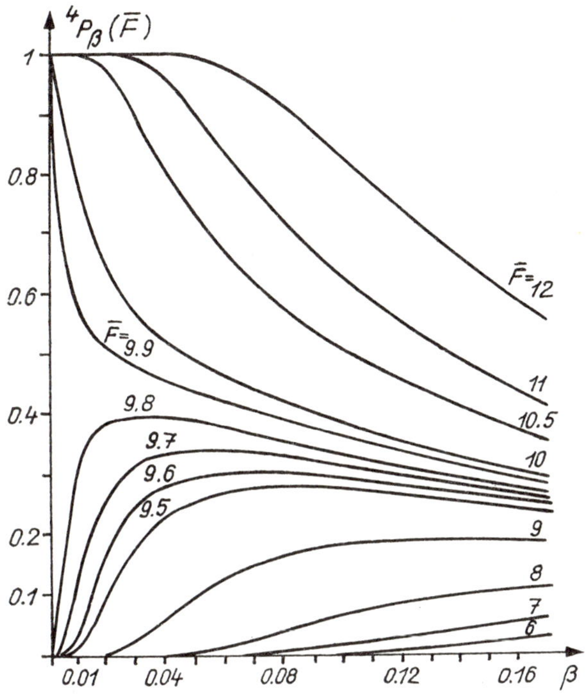

Fig. 2.10. The probability ${}^4p_\beta(\bar{F})$ as a function of β

where $\bar{F} = 4F/E\pi r_0^4$. In Fig. 2.10 the probability ${}^4p_\beta(\bar{F}) \doteq \mathsf{P}(\overline{{}^4\Lambda_{1\varepsilon}} \leqq \bar{F})$ is given as a function of the "degree of the randomness" β, for $6 \leqq \bar{F} \leqq 11$ and $\varepsilon = 0.1$. From this figure the probability ${}^4p_\beta$ of the bar failing if it is loaded up to a force corresponding to $\bar{F}$ can be determined. Using the definition of ${}^4p_\beta(\bar{F})$, we have

$$\lim_{\beta\to 0} {}^4p_\beta(\bar{F}) = \begin{cases} 0 & \text{for } \bar{F} < 9.8696\,, \\ 1 & \text{for } \bar{F} \geqq 9.8696\,, \end{cases}$$

i.e. for a small β the bar fails with probability 1 for $\bar{F} > 9.8696$, and with probability 0 for $\bar{F} < 9.8696$ (see also the non-random problem with $r(x) \equiv r_0$). The connection between F and $\bar{F}$ is given by $F = 1665.04\,\bar{F}$; and the inequality

$$9990.26N \leqq F \leqq 18315.49N$$

follows from $6 \leqq \bar{F} \leqq 11$ if the material constants are taken from (2.70). The probability ${}^4p_\beta(\bar{F})$ is plotted in Fig. 2.11 as a function of $\bar{F}$ for various values of β, i.e. in Fig. 2.11 the distribution function of $\overline{{}^4\Lambda_{1\varepsilon}}$ is plotted. By means of this figure Fig. 2.10 can be better interpreted.

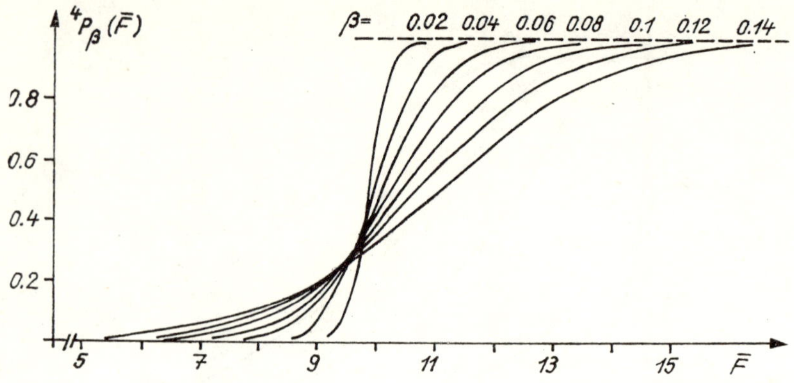

Fig. 2.11. The probability $^4p_\beta(\bar F)$ as a function of $\bar F$

In the following we deal with the determination of the limit distribution laws of the eigenvectors. With the aid of Theorem 2.12 for

$$^n u_{g\varepsilon}(x, \omega) = \sum_{i=1}^{n} {}^n u_{gi\varepsilon}(\omega)\, \varphi_i(x) = {}^n U_{g\varepsilon}^{\mathsf T}\, \Phi_n\,,$$

where

$$^n U_{g\varepsilon} = ({}^n u_{gi\varepsilon})_{1\leq i\leq n}^{\mathsf T}\,, \qquad \Phi_n = (\varphi_i)_{1\leq i\leq n}^{\mathsf T}\,,$$

we have

$$\langle {}^n u_{g\varepsilon}(x)\rangle \approx {}^n u_{g0}(x)$$

and

$$\begin{aligned} {}^n K_{gh}(x, y) &= \langle \left({}^n u_{g\varepsilon}(x) - {}^n u_{g0}(x)\right) \left({}^n u_{h\varepsilon}(y) - {}^n u_{h0}(y)\right)\rangle \\ &= \Phi_n^{\mathsf T}(x)\, \langle ({}^n U_{g\varepsilon} - {}^n U_{g0})\, ({}^n U_{h\varepsilon} - {}^n U_{h0})^{\mathsf T}\rangle\, \Phi_n(y) \\ &\approx \varepsilon \Phi_n^{\mathsf T}(x)\, \langle L_g L_h^{\mathsf T}\rangle\, \Phi_n(y)\,. \end{aligned}$$

In our case we have

$$\lim_{n\to\infty} {}^n u_{g\varepsilon}(x, \omega) = u_{g\varepsilon}(x, \omega) \quad \text{in} \quad \mathbf{C}(0, 1) \text{ a.s.};$$

$u_{g\varepsilon}(x, \omega)$ denotes the eigenfunction associated with the eigenvalue $\lambda_{g\varepsilon}$ of the eigenvalue problem (2.62). The process $^n u_{g\varepsilon}(x, \omega)$ is approximatively a Gaussian process.

Using Theorem 2.12 with $\eta_i(x) = \varphi_i''(x)$ and $\psi_i(x) = 0$ we have convergence in distribution

$$\lim_{\varepsilon\downarrow 0} \frac{1}{\sqrt{\varepsilon}} \left({}^n u_{1\varepsilon}(x) - {}^n u_{10}(x), {}^n u_{2\varepsilon}(x) - {}^n u_{20}(x)\right) = \left({}^n l_1(x), {}^n l_2(x)\right).$$

If we assume that $\bar{f}_\varepsilon(x, \omega)$ is a wide-sense stationary process the Gaussian vector process $({}^nl_1, {}^nl_2)$ possesses the first two moments given by

$$\langle {}^nl_g(x)\rangle = 0 ,$$

$$\langle {}^nl_g(x)\, {}^nl_h(y)\rangle$$

$$= a \sum_{\substack{i=1\\ i\neq g}}^{n} \sum_{\substack{j=1\\ j\neq h}}^{n} \frac{1}{({}^n\Lambda_{i0} - {}^n\Lambda_{g0})\,({}^n\Lambda_{j0} - {}^n\Lambda_{h0})} \mathbb{C}({}^n\bar{u}_{g0}\, {}^n\bar{u}_{h0}, {}^n\bar{u}_{i0}\, {}^n\bar{u}_{j0}) \cdot {}^nu_{i0}(x)\, {}^nu_{j0}(y)$$

where

$${}^n\bar{u}_{g0}(x) = \sum_{i=1}^{n} {}^nu_{gi0}\varphi_i''(x)$$

and

$${}^nu_{g0}(x) = \sum_{i=1}^{n} {}^nu_{gi0}\varphi_i(x) .$$

Hence, we have

$$\langle {}^nu_{g\varepsilon}(x)\rangle \approx {}^nu_{g0}(x)$$

and

$${}^nK_{gh}(x, y) \approx \varepsilon\langle {}^nl_g(x)\, {}^nl_h(y)\rangle .$$

Fig. 2.12 is given for the example of the buckling problem with a random radius $r(x, \omega)$. This figure shows the confidence domains of ${}^4u_1(x, \omega)$ and ${}^4u_2(x, \omega)$ setting $\beta = 0.14$, i.e. for each $x \in [0, 1]$ the domain is plotted in which ${}^4u_g(x, \omega)$ is contained with probability 0.9972. It yields the inequality

$$v_g^-(x) \leqq {}^4u_g(x, \omega) \leqq v_g^+(x)$$

for each $x \in [0, 1]$, with probability 0.9972 for $\varepsilon = 0.1$ where

$$v_g^{\mp}(x) \doteq {}^4u_{g0}(x) \mp 3\sqrt{{}^4K_{gg}(x, x)} .$$

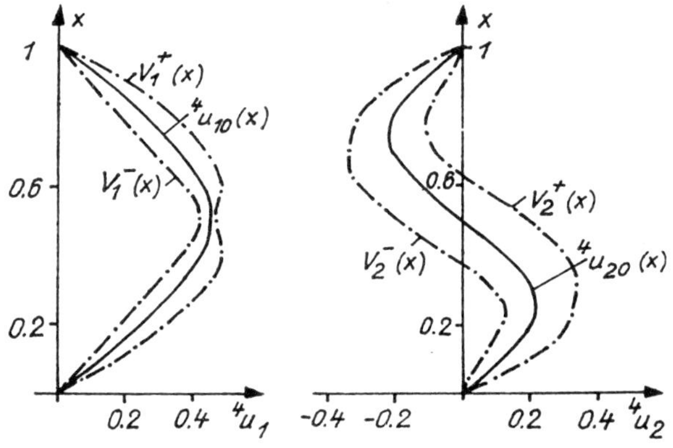

Fig. 2.12. Confidence domains of eigenfunctions for $\beta = 0.14$

The confidence domain for $g = 2$ is larger than that for $g = 1$. This fact was also observed for the eigenvalues (see Fig. 2.8). The maximum for ${}^4u_1(x, \omega)$ is at $x = 0.5$ relatively stable since the variation ${}^4K_{11}(x, x)$ has a minimum at $x = 0.5$ and the zero for ${}^4u_2(x, \omega)$ is relatively instable (see also Fig. 2.13). Fig. 2.13 is plotted for values of x in the interval $[0, 0.5]$ since $\langle {}^4l_g^2(x) \rangle$ is symmetric relative to $x = 0.5$. Fig. 2.12 is represented for $\beta = 0.14$ so that for a smaller value of β the confidence domain is still essential smaller. We observe

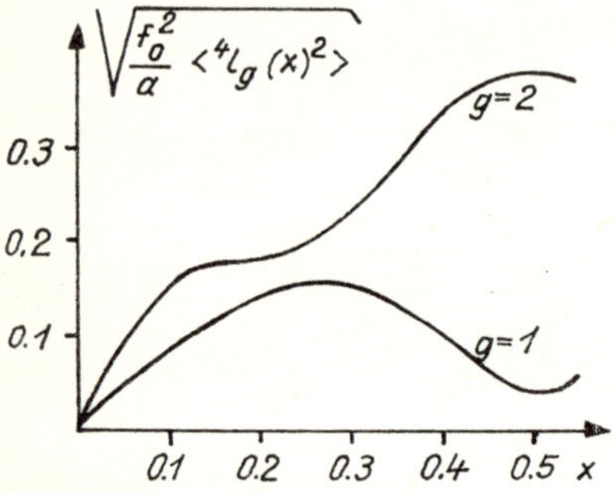

Fig. 2.13. The dispersion of ${}^4l_g(x)$ for $\beta = 0.14$

Table 2.14

g	i	j	$⟪{}^4\bar{u}_{g0}^2, {}^4\bar{u}_{i0}\, {}^4\bar{u}_{j0}⟫$
1	2	3	−0.0001
1	2	4	−375.4964
1	3	4	0.0004
2	1	3	453.4543
2	1	4	−0.0011
2	3	4	0.0029

that the typical form of the eigenfunctions ${}^4u_g(x, \omega)$ for $g = 1$ and $g = 2$ is preserved. The values necessary for the calculation of the dispersion of ${}^4l_g(x)$ are contained in Table 2.14 and ${}^4S_{gh} = ⟪{}^4\bar{u}_{g0}^2, {}^4\bar{u}_{h0}^2⟫$ in Table 2.13.

Similar considerations can be applied to the eigenvalues and eigenfunctions of partial differential operators. In this case additional difficulties do not appear except in the numerical calculations.

2.3. Limit distributions of eigenvalues and eigenfunctions of random differential operators

This section is concerned with eigenvalue problems for random differential operators. We will prove limit theorems similar to these in the case of random matrices. In proving these limit theorems, the perturbation expansions of the solutions of eigenvalue problems in terms of the coefficients of the differential operator are used. Section 2.3.1 deals with these perturbation expansions. In the following sections the distribution laws of the eigenvalues and eigenfunctions are deduced and some typical examples are investigated.

Random eigenvalue problems of the form

$$M(\omega)u = \lambda N(\omega)\, u \tag{2.71}$$

are considered in the interval $[a, b]$ with the boundary conditions

$$U_i[u] = 0\,, \qquad i = 1, 2, \ldots, 2m\,, \tag{2.72}$$

where M and N denote the random differential operators

$$M(\omega)\, u = \sum_{i=0}^{m} (-1)^i \,[f_i(x, \omega)\, u^{(i)}]^{(i)} ,$$

$$N(\omega)\, u = \sum_{i=0}^{m'} (-1)^i \,[g_i(x, \omega)\, u^{(i)}]^{(i)} ,$$

with $m > m'$. The conditions of Theorem 2.10 imply a discrete spectrum of the eigenvalue problem (2.71), (2.72) almost surely. Let

$$\lambda_1(\omega) \leqq \lambda_2(\omega) \leqq \dots \leqq \lambda_n(\omega) \leqq \dots$$

be the eigenvalues where,

$$\lim_{n\to\infty} \lambda_n(\omega) = \infty \quad \text{a.s.}$$

The eigenfunctions associated with $\lambda_n(\omega)$ are denoted by $u_n(x, \omega)$. These eigenfunctions are complete in $\mathscr{H}_M$ and $\mathscr{H}_N$.

Define

$$f_i(x, \omega) \doteq f_{0i}(x) + f_{1i}(x, \omega) \quad \text{for} \quad i = 0, 1, \dots, m ,$$

$$g_j(x, \omega) \doteq g_{0j}(x) + g_{1j}(x, \omega) \quad \text{for} \quad j = 0, 1, \dots, m'$$

where

$$f_{0i}(x) \doteq \langle f_i(x) \rangle , \qquad g_{0j}(x) \doteq \langle g_j(x) \rangle ;$$

$$f_{1i}(x, \omega) \doteq f_i(x, \omega) - f_{0i}(x) , \qquad g_{1j}(x, \omega) \doteq g_j(x, \omega) - g_{0j}(x) ,$$

and

$$M_k u \doteq \sum_{i=0}^{m} (-1)^i \,[f_{ki}(x)\, u^{(i)}]^{(i)} , \quad \text{for} \quad k = 0, 1 ,$$

$$N_k u \doteq \sum_{i=0}^{m'} (-1)^i \,[g_{ki}(x)\, u^{(i)}]^{(i)} , \quad \text{for} \quad k = 0, 1 .$$

Hence, (2.71) can be written as

$$M_0 u + M_1(\omega)\, u = \lambda\big(N_0 u + N_1(\omega)\, u\big) . \tag{2.73}$$

The eigenvalue problem

$$M_0 w = \mu N_0 w \tag{2.74}$$

with the boundary conditions (2.72) is called the averaged eigenvalue problem for the problem (2.73), (2.72).

Assume that the equations

$$\langle M(\omega)\, u \rangle = \langle M(\omega) \rangle\, u = M_0 u ,$$

$$\langle N(\omega)\, u \rangle = \langle N(\omega) \rangle\, u = N_0 u$$

are true for $u \in \mathrm{D}(M)$ or $u \in \mathrm{D}(N)$, respectively. Then the symmetry of M_0 and N_0 follows from the symmetry of $M(\omega)$ and $N(\omega)$. The averaged operators M_0, N_0 are positive if the operators $M(\omega)$, $N(\omega)$ possess this property almost surely. The spectrum of the averaged eigenvalue problem is assumed to be discrete.

Thus

$$\mu_1 \leqq \mu_2 \leqq \dots \leqq \mu_n \leqq \dots$$

are the eigenvalues. The eigenfunctions are denoted by $w_n(x)$. In many cases the existence of a discrete spectrum of the averaged problem follows directly from the random eigenvalue problem (2.71), (2.72) (see also Section 2.2.1).

In the following we assume that the averaged problem (2.74), (2.72) for the random problem (2.71), (2.72) has a discrete spectrum. Furthermore, the operators $M_1(\omega)$ and $N_1(\omega)$ are considered as random perturbations of M_0 and N_0, respectively. Assume that the eigenvalues and eigenfunctions are expanded in homogeneous terms of the coefficients of the operators $M_1(\omega)$ and $N_1(\omega)$. A discrete spectrum of the averaged eigenvalue problem also induces a discrete spectrum for the random problem for "small" perturbations $M_1(\omega)$ and $N_1(\omega)$.

In Section 2.3.1 the perturbations are assumed to be non-random and the corresponding expansions for the solution are deduced.

2.3.1. Perturbation results

Consider the eigenvalue problem

$$M_0 u + M_1 u = \lambda(N_0 u + N_1 u) \tag{2.75}$$

in the interval $[a, b]$ with the boundary conditions (2.72). The operators

$$M_k u = \sum_{i=0}^{m} (-1)^i [f_{ki}(x) u^{(i)}]^{(i)}$$

and

$$N_k u = \sum_{i=0}^{m'} (-1)^i [g_{ki}(x) u^{(i)}]^{(i)}$$

for $k = 0, 1$ with $m > m'$ are assumed to be symmetric and positive. Furthermore, the equation

$$\sum_{i=0}^{m} \sum_{t=0}^{i-1} (-1)^{i+t} \big((p f_{ki}(x) + q g_{ki}(x)) u^{(i)}\big)^{(i-t-1)} v^{(t)}\big|_a^b = 0 \tag{2.76}$$

holds for $k = 0, 1$; $p, q = 0, 1$ and for all functions u, v which satisfy the boundary conditions (2.72) where $g_{ki} \equiv 0$ for $k > m'$. Then the equations

$$\langle\!\langle M_k u, v\rangle\!\rangle = \sum_{i=0}^{m} \langle\!\langle f_{ki} u^{(i)}, v^{(i)}\rangle\!\rangle, \qquad \langle\!\langle N_k u, v\rangle\!\rangle = \sum_{i=0}^{m'} \langle\!\langle g_{ki} u^{(i)}, v^{(i)}\rangle\!\rangle, \tag{2.76'}$$

for $k = 0, 1$ are obtained from (2.76) for the functions u, v above.

Assume that the unperturbed eigenvalue problem for (2.75), (2.72)

$$M_0 w = \mu N_0 w \tag{2.77}$$

with the boundary conditions (2.72) possesses a discrete spectrum. Let the eigenvalue μ be simple. $w(x)$ denotes the eigenfunction associated with μ. We write $\lambda_0 = \mu$ and $u_0(x) = w(x)$.

Expansions of the form

$$\lambda = \sum_{k=0}^{\infty} \lambda_k , \qquad u(x) = \sum_{k=0}^{\infty} u_k(x) \tag{2.78}$$

are assumed for the eigenvalue λ and the eigenfunction $u(x)$ of the problem (2.75), (2.72). The quantities λ_k and $u_k(x)$ denote the terms of λ and $u(x)$ which are homogeneous of the k-th order in the perturbation coefficients $f_{1i}(x)$, $i = 0, 1, \dots, m$, $g_{1j}(x)$, $j = 0, 1, \dots, m'$, respectively.

We state an important theorem for our further investigations concerning the expansions of the eigenvalues and eigenfunctions. The domain of convergence is estimated. The application of such an expansion theorem is often used. For our further considerations only the convergence is important. We state that these expansions are a tool for the proof of the probabilistic limit theorems to be shown. Similar expansions for the eigenvalues and eigenelements of perturbed operators can be found in papers of RELLICH [1, 2, 3, 4].

Theorem 2.13. *Let the eigenvalue μ of the unperturbed problem* (2.77) *be simple, and let $w(x)$ be the eigenfunction associated with μ, where $\mu = \lambda_0$, $w(x) = u_0(x)$. Furthermore, let the order of M_1 be less than m, i.e. $f_{1m}(x) \equiv 0$, and $f_{0m}(x) \neq 0$. The conditions and notations above are used. We take for abbreviation $Y_{ij} \doteq (y_i, y_{i+1}, \dots, y_j)$, $Y_j = Y_{1j}$. Then the homogeneous terms λ_k, $u_k(x)$ of λ and $u(x)$, respectively, are of the form*

$$\lambda_k = \sum_{i_1, \dots, i_k = 0}^{2m-1} \int_a^b \dots \int_a^b F_{i_1 \dots i_k}(Y_k)\, h_{i_1}(y_1) \, .. \, h_{i_k}(y_k)\, \mathrm{d}y_1 \dots \mathrm{d}y_k , \tag{2.79}$$

$$u_k(x) = \sum_{i_1, \dots, i_k = 0}^{2m-1} \int_a^b \dots \int_a^b H_{i_1 \dots i_k}(x; Y_k)\, h_{i_1}(y_1) \dots h_{i_k}(y_k)\, \mathrm{d}y_1 \dots \mathrm{d}y_k , \tag{2.80}$$

for $k \geqq 1$, where

$$F_{i_1 \dots i_k}(Y_k) \in \mathbf{C}\big((a, b)^k\big) ,$$

$$\frac{\partial^l}{\partial x^l} H_{i_1 \dots i_k}(x; Y_k) = H^{(l)}_{i_1 \dots i_k}(x; Y_k) \in \mathbf{C}\big((a, b)^{k+1}\big) ,$$

for $0 \leqq l \leqq m - 1$, and $H^{(m)}_{i_1 \dots i_k}(x; Y_k)$ is bounded. The notations

$$h_i(y) \doteq \begin{cases} \bar{f}_{1i}(y) & \text{for } 0 \leqq i \leqq m - 1 , \\ g_{1i-m}(y) & \text{for } m \leqq i \leqq 2m - 1 , \end{cases}$$

with

$$\bar{f}_{1i}(y) \doteq f_{1i}(y) - \mu g_{1i}(y) , \qquad g_{1j}(y) \equiv 0 \quad \text{for } j > m'$$

are used. The expansion of $u(x)$ is determined so that $u(x)$ is normalized by $\langle\!\langle N_0 u + N_1 u, u \rangle\!\rangle = 1$. The estimate

$$|F_{i_1 \dots i_k}(Y_k)| , \qquad |H^{(j)}_{i_1 \dots i_k}(x; Y_k)| \leqq \frac{1}{2p} \left| \binom{1/2}{k} \right| (4x_1)^k ,$$

for

$$k \geqq 1 , \quad j = 0, 1, \dots, m , \quad 0 \leqq i_1, \dots, i_k \leqq 2m - 1 , \quad x, y_1, \dots, y_k \in [a, b]$$

can be found where

$$p \doteq \max\{rd, 2gd(1 + dr^2)\},$$

$$x_1 \doteq \max\{p(1 + r^2), r^2(p(1 + g)^2 + \tfrac{1}{2} r)\}.$$

The values d, g, and r are chosen so that

$$|u_0^{(j)}(x)|, \qquad |N_0 u_0(x)| \leqq r, \quad \textit{for } x \in [a, b], \quad 0 \leqq j \leqq m,$$

$$\left|\frac{\partial^{i+j} G(x, y)}{\partial x^i\, \partial y^j}\right|, \qquad |N_0 G(x, y)| \leqq g, \quad \textit{for } x, y \in [a, b], \quad 0 \leqq i + j \leqq 2m - 1,$$

and $d = \max\{b - a, 1\}$ *are satisfied. The inequalities*

$$|\lambda - \sum_{k=0}^{n} \lambda_k| \leqq \frac{1}{2p}(c\bar{\eta})^{n+1},$$

$$|u^{(j)}(x) - \sum_{k=0}^{n} u_k^{(j)}(x)| \leqq \frac{1}{2p}(c\bar{\eta})^{n+1}, \quad \textit{for } j = 0, 1, \ldots, m, \quad x \in [a, b],$$

show that the expansions (2.78) *converge for* $\bar{\eta} < 1/c$. *Put*

$$c \doteq 8m(b - a)\, x_1, \qquad \eta(1 + \mu) \leqq \bar{\eta},$$

where

$$|f_{1i}(x)| \leqq \eta, \qquad |g_{1i}(x)| \leqq \eta, \quad \textit{for all } x \in [a, b].$$

Particularly, the expressions

$$F_i(y) = \begin{cases} (u_0^{(i)}(y))^2, & \textit{for } 0 \leqq i \leqq m - 1, \\ 0, & \textit{for } m \leqq i \leqq 2m - 1, \end{cases}$$

and

$$H_i(x; y) = \begin{cases} -\dfrac{\partial^i G(x, y)}{\partial y^i}\, u_0^{(i)}(y), & \textit{for } 0 \leqq i \leqq m - 1, \\ -\tfrac{1}{2}\, (u_0^{(i-m)}(y))^2\, u_0(x), & \textit{for } m \leqq i \leqq 2m - 1, \end{cases}$$

are deduced. $G(x, y)$ *denotes the Green's function of the operator* $M_0 - \mu N_0$ *and the boundary conditions* (2.72), *where* $G(x, y)$ *is normalized by*

$$\int_a^b G(x, y)\, N_0 u_0(x)\, \mathrm{d}x = 0.$$

The elements λ, $u(x)$ *from* (2.78), *with the terms of the expansions from* (2.79) *and* (2.80), *determine a solution of the generalized problem belonging to* (2.75), (2.72), *i.e. the equation*

$$\sum_{i=0}^{m} (\!((f_{0i} + f_{1i})\, u^{(i)}, v^{(i)})\!) = \lambda \sum_{i=0}^{m'} (\!((g_{0i} + g_{1i})\, u^{(i)}, v^{(i)})\!)$$

is satisfied for all admissible functions $v(x)$, *and* $u(x)$ *satisfies the required boundary conditions* (*see Section* 2.2.1). *A function is said to be admissible if it has m continuous derivatives and satisfies the required boundary conditions of* (2.72).

Proof. First, we determine formally the terms λ_k and $u_k(x)$ by substituting the expansions (2.78) in the eigenvalue problem (2.75). Then the elements λ_k,

$u_k(x)$ follow successively from the equations

$$M_0u_k - \lambda_0N_0u_k = -(M_1 - \lambda_0N_1)\,u_{k-1} + \sum_{s=1}^{k-1} \lambda_s(N_0u_{k-s} + N_1u_{k-s-1}) + \lambda_kN_0u_0, \qquad (2.81)$$

$$U_i[u_k] = 0\,, \qquad i = 1, 2, \dots, 2m\,.$$

For $k = 1$ the sum on the right-hand side of (2.81) is omitted. As a necessary and sufficient condition for the solution we have

$$\lambda_k = ⟪(M_1 - \lambda_0N_1)\,u_{k-1}, u_0⟫ - \sum_{s=1}^{k-1} \lambda_s⟪N_0u_{k-s} + N_1u_{k-s-1}, u_0⟫ \qquad (2.82)$$

because $⟪N_0u_0, u_0⟫ = 1$.

If $G(x, y)$ denotes the generalized Green's function of the differential operator $M_0 - \lambda_0N_0$ and the boundary conditions (2.72), then the operator

$$Tv \doteq \int_a^b G(x, y)\,v(y)\,\mathrm{d}y$$

satisfies the poperties

$$⟪Tu, v⟫ = ⟪u, Tv⟫\,,$$

$$(M_0 - \lambda_0N_0)\,Tu = T(M_0 - \lambda_0N_0)\,u = u - ⟪u, u_0⟫\,u_0\,,$$

$$TN_0u_0 = 0\,,$$

for all functions u and v which have $2m$ continuous derivatives and satisfy the boundary conditions (2.72). Applying T to (2.81) the equation

$$u_k(x) = -T(M_1 - \lambda_0N_1)\,u_{k-1} + \sum_{s=1}^{k-1} \lambda_sT(N_0u_{k-s} + N_1u_{k-s-1}) + a_ku_0 \qquad (2.83)$$

is obtained. The constant a_k is determined by the normalization condition $⟪N_0u + N_1u, u⟫ = 1$. This condition can be written in the form

$$⟪N_0u + N_1u, u⟫ = \sum_{k=0}^{\infty} \Big[\sum_{s=0}^{k} ⟪N_0u_s, u_{k-s}⟫ + \sum_{s=1}^{k} ⟪N_1u_{s-1}, u_{k-s}⟫\Big] = 1$$

and from this the equations

$$⟪N_0u_0, u_0⟫ = 1\,,$$

and

$$⟪N_0u_0, u_k⟫ = -\tfrac{1}{2}\Big[\sum_{s=1}^{k-1} ⟪N_0u_s, u_{k-s}⟫ + \sum_{s=1}^{k} ⟪N_1u_{s-1}, u_{k-s}⟫\Big]\,, \quad \text{for } k \geqq 1\,,$$

follow. Using (2.83) we obtain

$$a_k = -⟪-T(M_1 - \lambda_0N_1)\,u_{k-1} + \sum_{s=1}^{k-1} \lambda_sT(N_0u_{k-s} + N_1u_{k-s-1}), N_0u_0⟫ - \tfrac{1}{2}\Big[\sum_{s=1}^{k-1} ⟪N_0u_s, u_{k-s}⟫ + \sum_{s=1}^{k} ⟪N_1u_{s-1}, u_{k-s}⟫\Big]\,. \qquad (2.84)$$

Hence, the terms λ_k and $u_k(x)$, for $k = 1, 2, \ldots$, can be determined successively from (2.82) and (2.83) in connection with (2.84).

First, λ_1 and $u_1(x)$ are determined. Equation (2.82) for $k = 1$ yields

$$\lambda_1 = (\!(\bar{M}_1 u_0, u_0)\!) = \sum_{i=0}^{m-1} (\!(\bar{f}_{1i} u_0^{(i)}, u_0^{(i)})\!) = \sum_{i=0}^{2m-1} \int_a^b F_i(y)\, h_i(y)\, dy ,$$

where

$$\bar{M}_1 \doteq M_1 - \lambda_0 N_1 ,$$

$$\bar{f}_{1i} \doteq f_{1i} - \lambda_0 g_{1i} , \qquad g_{1i} \equiv 0 \quad \text{for } i > m' \quad \text{and}$$

$$h_i(y) \doteq \begin{cases} \bar{f}_{1i}(y) & \text{for } 0 \leqq i \leqq m-1 , \\ g_{1i-m}(y) & \text{for } m \leqq i \leqq 2m-1 . \end{cases}$$

Thus

$$F_i(y) = \begin{cases} (u_0^{(i)}(y))^2 & \text{for } 0 \leqq i \leqq m-1 , \\ 0 & \text{for } m \leqq i \leqq 2m-1 . \end{cases}$$

The equation

$$u_1(x) = \bar{u}_1(x) - [(\!(\bar{u}_1, N_0 u_0)\!) + \tfrac{1}{2} (\!(N_1 u_0, u_0)\!)]\, u_0(x)$$

is obtained from (2.83) for $k = 1$ where

$$\bar{u}_1(x) = -T\bar{M}_1 u_0$$

and if we use (2.84) for $k = 1$, we have

$$a_1 = -(\!(\bar{u}_1, N_0 u_0)\!) - \tfrac{1}{2} (\!(N_1 u_0, u_0)\!) .$$

We find that $u_1(x)$ can be written in the form

$$u_1(x) = \sum_{i=0}^{2m-1} \int_a^b H_i(x; y)\, h_i(y)\, dy , \tag{2.85}$$

where

$$H_i(x; y) \doteq \begin{cases} -G_{0i}(x, y)\, u_0^{(i)}(y) , & \text{for } 0 \leqq i \leqq m-1 , \\ -\frac{1}{2} (u_0^{(i-m)}(y))^2\, u_0(x) , & \text{for } m \leqq i \leqq 2m-1 . \end{cases}$$

The term $H_i(x;y)$ has the same derivatives with respect to x as $G(x, y)$. We put

$$G_{ij}(x, y) \doteq \frac{\partial^{i+j} G(x, y)}{\partial x^i\, \partial y^j} ,$$

and we note that $G_{ij}(x, y)$ is continuous for $i + j \leqq 2m - 2$.

Suppose that (2.79) and (2.80) are derived for 1, 2, ... , $k - 1$. We establish these equations for k. According to (2.82) we obtain

$$\lambda_k = \sum_{i=0}^{m-1} \langle\!\langle \bar{f}_{1i} u_{k-1}^{(i)}, u_0^{(i)} \rangle\!\rangle - \sum_{s=1}^{k-1} \lambda_s [\langle\!\langle u_{k-s}, N_0 u_0 \rangle\!\rangle + \sum_{i=0}^{m'} \langle\!\langle g_{1i} u_{k-s-1}^{(i)}, u_0^{(i)} \rangle\!\rangle] = \sum_{i_1, \dots, i_k = 0}^{2m-1} \int_a^b \dots \int_a^b h_{i_1}(y_1) \dots h_{i_k}(y_k)$$

$$\times \left[\begin{Bmatrix} H_{i_1 \dots i_{k-1}}^{(i_k)}(y_k;\, Y_{k-1})\, u_0^{(i_k)}(y_k) \\ 0 \end{Bmatrix} - \sum_{s=1}^{k-1} F_{i_1 \dots i_s}(Y_s) \left[\langle\!\langle H_{i_{s+1} \dots i_k}(\cdot\,; Y_{s+1,\,k}), N_0 u_0 \rangle\!\rangle + \begin{Bmatrix} 0 \\ H_{i_{s+1} \dots i_{k-1}}^{(i_k - m)}(y_k;\, Y_{s+1,\,k-1})\, u_0^{(i_k - m)}(y_k) \end{Bmatrix} \right] \right] \mathrm{d}y_1 \dots \mathrm{d}y_k \tag{2.86}$$

where the first relation is used for $0 \leqq i_k \leqq m - 1$ and the second relation for $m \leqq i_k \leqq 2m - 1$. From (2.86) we have

$$F_{i_1 \dots i_k}(y_1, \dots, y_k)$$
$$= \begin{Bmatrix} H_{i_1 \dots i_{k-1}}^{(i_k)}(y_k;\, Y_{k-1})\, u_0^{(i_k)}(y_k) \\ 0 \end{Bmatrix} - \sum_{s=1}^{k-1} F_{i_1 \dots i_s}(Y_s) \left[\langle\!\langle H_{i_{s+1} \dots i_k}(\cdot\,; Y_{s+1,\,k}), N_0 u_0 \rangle\!\rangle + \begin{Bmatrix} 0 \\ H_{i_{s+1} \dots i_{k-1}}^{(i_k - m)}(y_k;\, Y_{s+1,\,k-1})\, u_0^{(i_k - m)}(y_k) \end{Bmatrix} \right]. \tag{2.87}$$

for $k \geqq 1$, where

$$H_{i_{s+1} \dots i_{k-1}}^{(j)}(y;\, Y_{s+1,\,k-1})|_{s=k-1} \doteq u_0^{(j)}(y) \,.$$

The corresponding term for $u_k(x)$ is determined by (2.83) and (2.84). Let $\bar{u}_k(x)$ denote the expression

$$\bar{u}_k(x) \doteq -T \bar{M}_1 u_{k-1} + \sum_{s=1}^{k-1} \lambda_s T(N_0 u_{k-s} + N_1 u_{k-s-1}) \,.$$

Then we have

$$\bar{u}_k(x) = \sum_{i_1, \dots, i_k = 0}^{2m-1} \int_a^b \dots \int_a^b h_{i_1}(y_1) \dots h_{i_k}(y_k)$$

$$\times \left[\begin{Bmatrix} -H_{i_1 \dots i_{k-1}}^{(i_k)}(y_k;\, Y_{k-1})\, G_{0 i_k}(x, y_k) \\ 0 \end{Bmatrix} + \sum_{s=1}^{k-1} F_{i_1 \dots i_s}(Y_s) \left[T N_0 H_{i_{s+1} \dots i_k}(\cdot\,; Y_{s+1,\,k}) + \begin{Bmatrix} 0 \\ G_{0(i_k - m)}(x, y_k)\, H_{i_{s+1} \dots i_{k-1}}^{(i_k - m)}(y_k;\, Y_{s+1,\,k-1}) \end{Bmatrix} \right] \right]$$
$$\times\, \mathrm{d}y_1 \dots \mathrm{d}y_k \,,$$

and define

$$\overline{H}_{i_1\ldots i_k}(x;\,Y_k) \doteq \begin{Bmatrix} -H^{(i_k)}_{i_1\ldots i_{k-1}}(y_k;\,Y_{k-1})\,G_{0i_k}(x,y_k) + \sum\limits_{s=1}^{k-1} F_{i_1\ldots i_s}(Y_s)\left[TN_0 H_{i_{s+1}\ldots i_k}(\cdot\,;\,Y_{s+1,\,k}) + \begin{Bmatrix} 0 \\ G_{0(i_k-m)}(x,y_k)\,H^{(i_k-m)}_{i_{s+1}\ldots i_{k-1}}(y_k;\,Y_{s+1,\,k-1}) \end{Bmatrix}\right] \\ 0 \end{Bmatrix}. \tag{2.88}$$

It is easy to see that

$$a_k = -\sum_{i_1,\,\ldots,\,i_k=0}^{2m-1} \int_a^b \ldots \int_a^b h_{i_1}(y_1)\ldots h_{i_k}(y_k)$$
$$\times\left[\langle\!\langle \overline{H}_{i_1\ldots i_k}(\cdot\,;\,Y_k),\,N_0 u_0\rangle\!\rangle + \tfrac{1}{2}\sum_{s=1}^{k-1}\langle\!\langle N_0 H_{i_1\ldots i_s}(\cdot\,;\,Y_s),\,H_{i_{s+1}\ldots i_k}(\cdot\,;\,Y_{s+1,\,k})\rangle\!\rangle + \begin{Bmatrix} 0 \\ \tfrac{1}{2}\sum\limits_{s=1}^{k} H^{(i_k-m)}_{i_1\ldots i_{s-1}}(y_k;\,Y_{s-1})\,H^{(i_k-m)}_{i_s\ldots i_{k-1}}(y_k;\,Y_{s,\,k-1}) \end{Bmatrix}\right] \mathrm{d}y_1\ldots\mathrm{d}y_k\,,$$

and this implies

$$u_k(x) = \sum_{i_1,\,\ldots,\,i_k=0}^{2m-1}\int_a^b \ldots \int_a^b h_{i_1}(y_1)\ldots h_{i_k}(y_k)\,H_{i_1\ldots i_k}(x;\,Y_k)\,\mathrm{d}y_1\ldots\mathrm{d}y_k\,,$$

where

$$H_{i_1\ldots i_k}(x;\,Y_k) = \overline{H}_{i_1\ldots i_k}(x;\,Y_k)$$
$$-\left[\langle\!\langle \overline{H}_{i_1\ldots i_k}(\cdot\,;\,Y_k),\,N_0 u_0\rangle\!\rangle + \tfrac{1}{2}\sum_{s=1}^{k-1}\langle\!\langle N_0 H_{i_1\ldots i_s}(\cdot\,;\,Y_s),\,H_{i_{s+1}\ldots i_k}(\cdot\,;\,Y_{s+1,\,k})\rangle\!\rangle + \begin{Bmatrix} 0 \\ \tfrac{1}{2}\sum\limits_{s=1}^{k} H^{(i_k-m)}_{i_1\ldots i_{s-1}}(y_k;\,Y_{s-1})\,H^{(i_k-m)}_{i_s\ldots i_{k-1}}(y_k;\,Y_{s,\,k-1}) \end{Bmatrix}\right] u_0(x)\,, \tag{2.89}$$

for $k \geqq 1$. $\overline{H}_{i_1\ldots i_k}(x;\,Y_k)$ is written in (2.88). We can easily verify the properties concerning the derivatives stated in Theorem 2.13, and we obtain

$$u_k^{(j)}(x) = \sum_{i_1,\,\ldots,\,i_k=0}^{2m-1}\int_a^b \ldots \int_a^b H^{(j)}_{i_1\ldots i_k}(x;\,Y_k)\,h_{i_1}(y_1)\ldots h_{i_k}(y_k)\,\mathrm{d}y_1\ldots\mathrm{d}y_k \tag{2.90}$$

for $0 \leqq j \leqq m$.

We now consider the estimate of the convergence of the series (2.78). Let

$$|u_0^{(i)}(x)| \leqq r\,, \qquad \text{for} \quad 0 \leqq i \leqq m-1\,, \quad a \leqq x \leqq b\,,$$
$$|N_0 u_0(x)| \leqq r\,, \qquad \text{for} \quad a \leqq x \leqq b\,,$$

and

$$\left.\begin{array}{ll} |G_{ij}(x, y)| \leqq g, & \text{for} \quad 0 \leqq i + j \leqq 2m-1\,, \\ |N_0 G_{0i}(x, y)| \leqq g\,, & \text{for} \quad 0 \leqq i \leqq m\,, \end{array}\right. \quad a \leqq x, y \leqq b\,.$$

By means of (2.87) and (2.89), for $k = 1$, we obtain

$$|F_i(y)| \leqq r^2\,,$$
$$|N_0 H_i(x; y)|\,, \qquad |H_i^{(j)}(x; y)| \leqq \max\,\{gr, r^{3/2}\}\,, \quad \text{for} \quad 0 \leqq j \leqq m\,.$$

Furthermore, assuming

$$|F_{i_1 \ldots i_{k-1}}(Y_{k-1})| \leqq a_{k-1}\,,$$
$$|N_0 H_{i_1 \ldots i_{k-1}}(x;\, Y_{k-1})| \leqq b_{k-1}\,, \tag{2.91}$$

and

$$|H^{(j)}_{i_1 \ldots i_{k-1}}(x;\, Y_{k-1})| \leqq b_{k-1}\,, \quad \text{for} \quad j = 0, 1, \ldots, m\,,$$

it follows, using (2.87)

$$|F_{i_1 \ldots i_k}(Y_k)|$$
$$\leqq r\Big(b_{k-1} + \sum_{s=1}^{k-2} a_s((b-a)\, b_{k-s} + b_{k-s-1}) + a_{k-1}(b_1(b-a) + r)\Big)$$

for $k \geqq 2$, and by means of

$$a_k \doteq \max_{\substack{0 \leqq i_1, \ldots, i_k \leqq m-1 \\ y_1, \ldots, y_k \in [a, b]}} [|F_{i_1 \ldots i_k}(Y_k)|] \quad \text{for } k \geqq 2$$

we have

$$a_k \leqq r\Big(b_{k-1} + \sum_{s=1}^{k-2} a_s((b-a)\, b_{k-s} + b_{k-s-1}) + a_{k-1}(b_1(b-a) + r)\Big)\,, \quad \text{for} \quad k \geqq 2\,, \tag{2.92}$$

and $a_1 = r^2$.

Similarly, substituting the assumptions (2.91) and

$$b_k \doteq \max_{\substack{0 \leqq i_1, \ldots, i_k \leqq m-1 \\ y_1, \ldots, y_k \in [a, b], \\ j = 0, 1, \ldots, m}} \{|H^{(j)}_{i_1 \ldots i_k}(x;\, Y_k)|, |N_0 H_{i_1 \ldots i_k}(x;\, Y_k)|\}\,, \quad \text{for} \quad k \geqq 2\,,$$

in (2.89) the inequality

$$b_k \leqq g\big(1 + (b-a)\, r^2\big)\, \Big(b_{k-1} + \sum_{s=1}^{k-2} a_s((b-a)\, b_{k-s} + b_{k-s-1})$$
$$+ a_{k-1}((b-a)\, b_1 + r)\Big)$$
$$+ \tfrac{1}{2} r \sum_{s=1}^{k-2} b_s\big((b-a)\, b_{k-s} + b_{k-s-1}\big) + b_{k-1}\big(r + \tfrac{1}{2}(b-a)\, b_1\big)\, r$$

is verified, and with the aid of $d \doteq \max\{1, b-a\}$ we have

$$b_k \leqq gd(1+dr^2)\big(b_{k-1} + \sum_{s=1}^{k-2} a_s(b_{k-s}+b_{k-s-1}) + a_{k-1}(b_1+r)\big)$$
$$+\tfrac{1}{2}rd\sum_{s=1}^{k-2} b_s(b_{k-s}+b_{k-s-1}) + rdb_{k-1}(r+\tfrac{1}{2}b_1)\,, \qquad (2.93)$$

for $k \geqq 2$. We now put

$$\xi_s \doteq a_s + a_{s-1}\,, \qquad \zeta_s \doteq b_s + b_{s-1} \qquad \text{for} \quad s \geqq 2\,,$$

and the inequality (2.92) can be written as

$$\xi_k \leqq rd\big(\zeta_{k-1}(1+a_1) + \xi_{k-1}(r+b_1) + \sum_{s=2}^{k-2}\xi_s\zeta_{k-s}\big)\,,$$

and also for the inequality (2.93)

$$\zeta_k \leqq gd(1+dr^2)\,[\zeta_{k-1}(1+a_1) + \xi_{k-1}(r+b_1) + \sum_{s=2}^{k-2}\xi_s\zeta_{k-s}]$$
$$+\tfrac{1}{2}rd\sum_{s=2}^{k-2}\zeta_s\zeta_{k-s} + dr(r+b_1)\,\zeta_{k-1}\,,$$

in each case for $k \geqq 3$. The substitution of

$$\xi_1 \doteq 1 + a_1\,, \qquad \zeta_1 \doteq r + b_1$$

leads to the inequalities

$$\xi_k \leqq rd\sum_{s=1}^{k-1}\xi_s\zeta_{k-s}\,,$$
$$\zeta_k \leqq gd(1+dr^2)\sum_{s=1}^{k-1}\xi_s\zeta_{k-s} + \tfrac{1}{2}rd\sum_{s=1}^{k-1}\zeta_s\zeta_{k-s}\,, \qquad (2.94)$$

for $k \geqq 3$. These inequalities (2.94) are also true for $k = 2$. This follows from the estimate

$$\xi_2 = a_1 + a_2 \leqq a_1 + rb_1 + rda_1(b_1+r) \leqq rd\,\xi_1\zeta_1\,,$$
$$\zeta_2 = b_1 + b_2 \leqq b_1 + gd(1+dr^2)\big(b_1 + a_1(b_1+r)\big) + b_1rd(r+\tfrac{1}{2}b_1)$$
$$\leqq gd(1+dr^2)\,(\xi_1\zeta_1 - r) + \tfrac{1}{2}rd(\zeta_1^2 - r^2) + b_1$$
$$\leqq gd(1+dr^2)\,\zeta_1\xi_1 + \tfrac{1}{2}rd\,\zeta_1^2\,.$$

Using

$$p \doteq \max\{rd, 2gd(1+dr^2)\}$$

and

$$p\zeta_s \doteq \bar{\zeta}_s\,, \qquad p\xi_s \doteq \bar{\xi}_s\,, \quad \text{for} \quad s \geqq 1\,,$$

we finally get the inequalities

$$\bar{\xi}_k \leqq \sum_{s=1}^{k-1} \bar{\xi}_s \bar{\zeta}_{k-s} , \qquad \bar{\zeta}_k \leqq \tfrac{1}{2} \sum_{s=1}^{k-1} \bar{\xi}_s \bar{\zeta}_{k-s} + \tfrac{1}{2} \sum_{s=1}^{k-1} \bar{\zeta}_s \bar{\zeta}_{k-s} . \tag{2.95}$$

The sequences (x_k) and (y_k) are defined by

$$x_1 = y_1 \doteq \max \{\bar{\xi}_1, \bar{\zeta}_1\} ,$$

$$\left. \begin{aligned} x_k &\doteq \sum_{s=1}^{k-1} x_s y_{k-s} , \\ y_k &\doteq \tfrac{1}{2} \sum_{s=1}^{k-1} x_s y_{k-s} + \tfrac{1}{2} \sum_{s=1}^{k-1} y_s y_{k-s} \end{aligned} \right\} \quad \text{for} \quad k \geqq 2 ,$$

and then we can easily conclude that

$$x_k = y_k , \quad \text{for} \quad k \geqq 1 ,$$

and

$$\bar{\xi}_k \leqq x_k , \qquad \bar{\zeta}_k \leqq x_k , \quad \text{for} \quad k \geqq 1 . \tag{2.96}$$

For the function

$$f(z) \doteq \sum_{k=1}^{\infty} x_k z^k$$

we have

$$f^2(z) = \sum_{k=2}^{\infty} \Big(\sum_{s=1}^{k-1} x_s x_{k-s} \Big) z^k = \sum_{k=2}^{\infty} x_k z^k = f(z) - x_1 z ;$$

and from the above

$$f(z) = \tfrac{1}{2} (1 - \sqrt{1 - 4x_1 z}) = \tfrac{1}{2} \sum_{k=1}^{\infty} \left| \binom{1/2}{k} \right| (4x_1 z)^k .$$

Thus, x_k can be written explicitly as

$$x_k = \tfrac{1}{2} \left| \binom{1/2}{k} \right| (4x_1)^k ,$$

and so

$$a_k \leqq x_k/p , \qquad b_k \leqq x_k/p \quad \text{for} \quad k \geqq 1 .$$

Finally, the inequalities

$$|F_{i_1 \dots i_k}(Y_k)| \leqq \frac{1}{2p} \left| \binom{1/2}{k} \right| (4x_1)^k ,$$

and

$$\left. \begin{aligned} &|H^{(j)}_{i_1 \dots i_k}(x;\, Y_k)| \\ &|N_0 H_{i_1 \dots i_k}(x;\, Y_k)| \end{aligned} \right\} \leqq \frac{1}{2p} \left| \binom{1/2}{k} \right| (4x_1)^k \tag{2.97}$$

follow, where

$$0 \leqq i_1, \dots , i_k \leqq 2m - 1 , \qquad k \geqq 1 , \qquad j = 0, 1, \dots , m .$$

Let

$$|h_i(x)| \leqq \bar{\eta} .$$

Then, by means of the inequalities (2.97) from (2.79), (2.80) we can estimate λ_k and $u_k^{(j)}(x)$, and obtain

$$\left.\begin{array}{l} |\lambda_k| \\ |u_k^{(j)}(x)| \end{array}\right\} \leqq \frac{1}{2p} \left|\binom{1/2}{k}\right| (8m(b-a)\,\bar{\eta} x_1)^k ,$$

for $x \in [a, b]$, $j = 0, 1, \dots, m$. This implies that the series (2.78) and its derivatives for

$$\bar{\eta} < 1/c , \quad \text{with} \quad c \doteq 8m(b-a)\, x_1 ,$$

converge for $x \in [a, b]$. The estimate

$$|\lambda - \sum_{k=0}^{n} \lambda_k| \leqq \sum_{k=n+1}^{\infty} |\lambda_k|$$

$$\leqq \frac{1}{2p} (c\bar{\eta})^{n+1} \sum_{k=n+1}^{\infty} \left|\binom{1/2}{k}\right| (c\bar{\eta})^{k-n-1} \leqq \frac{1}{2p} (c\bar{\eta})^{n+1}$$

shows this convergence of λ; and we can easily write a corresponding estimate for $u^{(j)}(x)$, where $0 \leqq j \leqq m$.

The pair $(\lambda, u(x))$ is called a solution of the generalized problem for the eigenvalue problem (2.75) if the equation

$$\sum_{i=0}^{m} \langle\!\langle (f_{0i} + f_{1i})\, u^{(i)}, v^{(i)} \rangle\!\rangle = \lambda \sum_{i=0}^{m'} \langle\!\langle (g_{0i} + g_{1i})\, u^{(i)}, v^{(i)} \rangle\!\rangle \quad \text{a. s.}$$

is satisfied for all admissible functions v. From

$$u^{(i)}(x) = \sum_{k=0}^{\infty} u_k^{(i)}(x) ,$$

for $0 \leqq i \leqq m$, and the uniform convergence of these series it follows that the required boundary conditions are satisfied and the equation

$$\sum_{k=0}^{\infty} \langle\!\langle (M_0 + M_1)\, u_k, v \rangle\!\rangle = \sum_{k=0}^{\infty} \lambda_k \sum_{k=0}^{\infty} \langle\!\langle (N_0 + N_1)\, u_k, v \rangle\!\rangle$$

is obtained. Hence, we have

$$\langle\!\langle M_0 u_0, v \rangle\!\rangle + \sum_{k=1}^{\infty} [\langle\!\langle M_0 u_k, v \rangle\!\rangle + \langle\!\langle M_1 u_{k-1}, v \rangle\!\rangle]$$

$$= \lambda_0 \langle\!\langle N_0 u_0, v \rangle\!\rangle + \sum_{k=1}^{\infty} [\lambda_k \langle\!\langle N_0 u_0, v \rangle\!\rangle + \sum_{s=0}^{k-1} \lambda_s \langle\!\langle N_0 u_{k-s} + N_1 u_{k-s-1}, v \rangle\!\rangle] ,$$

and because of (2.81) this completes the proof of Theorem 2.13. ◀

2.3.2. A limit theorem for eigenvalues and eigenfunctions

For the deduction of the limit theorem for the eigenvalues and eigenfunctions of the eigenvalue problem (2.75), with the boundary conditions (2.72), we use the notation

$$E_{gr}(\omega) \doteq \begin{cases} \lambda_g(\omega)\,, & \text{for} \quad r = 0\,, \\ u_g(x_r, \omega)\,, & \text{for} \quad r \neq 0\,, \end{cases} \qquad g = 1, 2, \dots\,,$$

where x_r, for $r = 1, 2, \dots$, are given numbers belonging to the interval $[a, b]$. Furthermore, we put

$$E_{gr,k}(\omega) \doteq \begin{cases} \lambda_{gk}(\omega)\,, & \text{for} \quad r = 0\,, \\ u_{gk}(x_r, \omega)\,, & \text{for} \quad r \neq 0\,. \end{cases}$$

In this formula $\lambda_{gk}(\omega)$ denotes the term in the expansion of $\lambda_g(\omega)$ that is homogeneous of the k-th order (see (2.79)), and $u_{gk}(x, \omega)$ denotes the corresponding term of $u_g(x, \omega)$ (see (2.80)). Hence, with the above notation we obtain

$$E_{gr}(\omega) = \sum_{k=0}^{\infty} E_{gr,k}(\omega)\,,$$

and the series converges a.s. The summands can be written as

$$E_{gr,k}(\omega) = \sum_{i_1,\dots,i_k=0}^{2m-1} \int_a^b \dots \int_a^b h_{i_1}(y_1, \omega) \dots h_{i_k}(y_k, \omega)\, K^g_{i_1\dots i_k, r}(y_1, \dots, y_k)\, \mathrm{d}y_1 \dots \mathrm{d}y_k\,,$$

where

$$K^g_{i_1\dots i_k, r}(y_1, \dots, y_k) = \begin{cases} F^g_{i_1\dots i_k}(y_1, \dots, y_k)\,, & \text{for} \quad r = 0\,, \\ H^g_{i_1\dots i_k}(x_r; y_1, \dots, y_k)\,, & \text{for} \quad r \neq 0\,. \end{cases}$$

The functions $F^g_{\dots}$ and $H^g_{\dots}$ are the corresponding expressions of (2.79) and (2.80) for the g-th eigenvalue $\lambda_g(\omega)$ and the g-th eigenfunction $u_g(x, \omega)$, respectively.

Now, with the aid of above notation we will state the following limit theorem.

Theorem 2.14. *Assume the coefficient processes*

$$\big(f_{10\varepsilon}(x, \omega), \dots, f_{1m-1\varepsilon}(x, \omega),\ g_{10\varepsilon}(x, \omega), \dots, g_{1m'\varepsilon}(x, \omega)\big)$$

of the eigenvalue problem (2.75) *with the boundary conditions* (2.72) *to be a weakly correlated connected vector process with the intensities* $a^{11}_{ij}(x)$ *between the processes* $f_{1i\varepsilon}$ *and* $f_{1j\varepsilon}$, $a^{12}_{ij}(x)$ *between* $f_{1i\varepsilon}$ *and* $g_{1j\varepsilon}$, *and* $a^{22}_{ij}(x)$ *between* $g_{1i\varepsilon}$ *and* $g_{1j\varepsilon}$. *Furthermore, let these processes satisfy the condition of "smallness" of Theorem* 2.13, *so that the expansions derived there can be applied. For this the eigenvalues of the averaged problem* (2.77) *to be dealt with are assumed to be simple. The eigenvalues of the averaged problem are denoted by* μ_g *and their eigenfunctions by* $w_g(x)$. *The generalized Green's function* $G^g(x, y)$ *associated with the differential operator* $M_0 - \mu_g N_0$ *and the boundary conditions* (2.72) *is normalized by*

$$\int_a^b G^g(x, y)\, N_0 w_g(x)\, \mathrm{d}x = 0\,.$$

Then the random vector

$$\frac{1}{\sqrt{\varepsilon}}\,(E_{g_1r_1}-E_{g_1r_1,0},\,E_{g_2r_2}-E_{g_2r_2,0},\,\dots,\,E_{g_sr_s}-E_{g_sr_s,0})$$

converges in distribution to a Gaussian random vector

$$(\xi_{g_1r_1},\,\xi_{g_2r_2},\,\dots,\,\xi_{g_sr_s})$$

as $\varepsilon \downarrow 0$, *and this limit vector has the first two moments which are given by*

$$\langle \xi_{g_ir_i}\rangle = 0\,,$$

$$\langle \xi_{gr}\xi_{ht}\rangle = \sum_{i,j=0}^{m+m'} \int_a^b K_{i,r}^g(y)\,K_{j,t}^h(y)\,b_{ij}^{gh}(y)\,\mathrm{d}y\,.$$

The functions $b_{ij}^{gh}(y)$ *are defined by*

$$b_{ij}^{gh} \doteq \begin{cases} a_{ij}^{11}-\mu_h a_{ij}^{12}-\mu_g a_{ji}^{12}+\mu_g\mu_h a_{ij}^{22}, & \text{for } \; 0\leqq i,j\leqq m-1\,,\\ a_{ij-m}^{12}-\mu_g a_{ij-m}^{22}\,, & \text{for } \; 0\leqq i\leqq m-1\,,\\ & \phantom{\text{for }}\; m\leqq j\leqq m+m'\,,\\ a_{i-mj-m}^{22} & \text{for } \; m\leqq i,\,j\leqq m+m'\,, \end{cases}$$

where $a_{ij}^{12}=0$ *for* $j>m'$, *and* $a_{ij}^{22}=0$ *for* $i>m'$ *or* $j>m'$, *and the functions* $K_{i,r}^g(y)$ *are given by*

$$K_{i,0}^g(y) \doteq \begin{cases} \left(w_g^{(i)}(y)\right)^2, & \text{for } \; 0\leqq i\leqq m-1\,,\\ 0 & \text{for } \; m\leqq i\leqq m+m'\,, \end{cases}$$

$$K_{i,r}^g(y) \doteq \begin{cases} -\dfrac{\partial^i G^g(x_r,y)}{\partial y^i}\,w_g^{(i)}(y)\,, & \text{for } \; 0\leqq i\leqq m-1\,,\\[2mm] -\frac{1}{2}\left(w_g^{(i-m)}(y)\right)^2 w_g(x_r)\,, & \text{for } \; m\leqq i\leqq m+m'\,. \end{cases}$$

In addition to convergence in distribution, the first moments converge as $\varepsilon\downarrow 0$. *Thus*

$$\lim_{\varepsilon\downarrow 0}\,\langle E_{g_ir_i}\rangle = E_{g_ir_i,0}\,,$$

$$\lim_{\varepsilon\downarrow 0}\frac{1}{\varepsilon}\,\langle (E_{g_ir_i}-E_{g_ir_i,0})\,(E_{g_jr_j}-E_{g_jr_j,0})\rangle = \langle \xi_{g_ir_i}\xi_{g_jr_j}\rangle\,,$$

for $i,j\in\{1,2,\dots,s\}$.

Proof. 1. First, we prove convergence of the k-th moments

$$\frac{1}{\sqrt{\varepsilon^k}}\,\langle E_{i_1}^{(n)}E_{i_2}^{(n)}\dots E_{i_k}^{(n)}\rangle\,,$$

where $i_j\in\{1,2,\dots,s\}$, $j=1,2,\dots,k$ and $E_i^{(n)}$ is given by

$$E_i^{(n)} \doteq \sum_{j=1}^{n} E_{g_ir_i,\,j}\,.$$

We show that the moment above converges to the corresponding moment of $(\xi_{g_1r_1}, \dots, \xi_{g_sr_s})$,

$$\lim_{\varepsilon\downarrow 0} \frac{1}{\sqrt{\varepsilon^k}} \langle \prod_{p=1}^{k} E_{i_p}^{(n)} \rangle = \langle \prod_{p=1}^{k} \xi_{g_{i_p}r_{i_p}} \rangle .$$

For this we deal with the order of the term

$$P_t(\varepsilon) = \langle \sum_{j_1,\dots,j_t=0}^{2m-1} \int_a^b \dots \int_a^b h_{j_1}(y_1) \dots h_{j_t}(y_t) L_{j_1\dots j_t}(y_1, \dots, y_t) \, dy_1 \dots dy_t \rangle$$

as $\varepsilon \downarrow 0$, for $t \geqq 2$ if $L_{i_1\dots i_t}(y_1, \dots, y_t)$ is assumed to be bounded in the interval $[a, b]^t$ for $0 \leqq j_1, \dots, j_t \leqq 2m-1$,

$$|L_{j_1\dots j_t}(y_1, \dots, y_t)| \leqq c .$$

Hence, terms of the form

$$Q(\varepsilon) = \int_a^b \dots \int_a^b |\langle h_{j_1}(y_1) \dots h_{j_t}(y_t) \rangle| \, dy_1 \dots dy_t$$

must be considered. Because of the assumptions of this theorem, the random vector

$$\big(h_{0\varepsilon}(x, \omega), \dots, h_{m+m'\varepsilon}(x, \omega)\big)$$

is also weakly correlated connected. Using Theorem 2.8 we obtain

$$Q(\varepsilon) = \begin{cases} O(\varepsilon^{t/2}) , & \text{for } t \text{ even,} \\ o(\varepsilon^{t/2}) , & \text{for } t \text{ odd;} \end{cases}$$

and then

$$|P_t(\varepsilon)| \leqq \begin{cases} O(\varepsilon^{t/2}) , & \text{for } t \text{ even,} \\ o(\varepsilon^{t/2}) , & \text{for } t \text{ odd.} \end{cases}$$

This result leads to

$$\frac{1}{\sqrt{\varepsilon^k}} \langle \prod_{p=1}^{k} E_{i_p}^{(n)} \rangle = \frac{1}{\sqrt{\varepsilon^k}} \langle \prod_{p=1}^{k} \big(\sum_{j=1}^{n} E_{g_{i_p}r_{i_p},j} \big) \rangle$$

$$= \begin{cases} \dfrac{1}{\sqrt{\varepsilon^k}} \langle \prod_{p=1}^{k} E_{g_{i_p}r_{i_p},1} \rangle + o(\sqrt{\varepsilon}) , & \text{for } k \text{ even,} \\ \dfrac{1}{\sqrt{\varepsilon^k}} o(\sqrt{\varepsilon^k}) , & \text{for } k \text{ odd,} \end{cases}$$

and it follows that

$$\lim_{\varepsilon\downarrow 0} \frac{1}{\sqrt{\varepsilon^k}} \langle \sum_{p=1}^{k} E_{i_p}^{(n)} \rangle = \begin{cases} \lim\limits_{\varepsilon\downarrow 0} \dfrac{1}{\sqrt{\varepsilon^k}} \langle \prod_{p=1}^{k} E_{g_{i_p}r_{i_p},1} \rangle , & \text{for } k \text{ even,} \\ 0 , & \text{for } k \text{ odd.} \end{cases}$$

By the use of

$$E_{gr,1} = \sum_{i=0}^{m+m'} \int_a^b h_{i\varepsilon}(y) \, K^g_{i,r}(y) \, dy ,$$

and Theorem 2.8, we have

$$\lim_{\varepsilon\downarrow 0} \frac{1}{\sqrt{\varepsilon^k}} \langle \prod_{p=1}^{k} E_{g_{i_p} r_{i_p}, 1} \rangle = \langle \prod_{p=1}^{k} \xi_{g_{i_p} r_{i_p}} \rangle$$

if the random vector $\{\xi_{g_1 r_1}, \ldots, \xi_{g_s r_s}\}$ denotes a zero mean Gaussian vector possessing the second moments

$$\langle \xi_{gr} \xi_{ht} \rangle = \sum_{i,j=0}^{m+m'} \int_a^b K^g_{i,r}(y)\, K^h_{j,t}(y)\, b^{gh}_{ij}(y)\, dy\,.$$

The intensities b^{gh}_{ij} are defined by

$$b^{gh}_{ij}(x) \doteq \lim_{\varepsilon\downarrow 0} \frac{1}{\varepsilon} \int_{-\varepsilon}^{\varepsilon} \langle h^g_{i\varepsilon}(x)\, h^h_{j\varepsilon}(x+y) \rangle\, dy\,,$$

and these values depend on the number of the eigenvalue through the processes $h^g_{i\varepsilon}$. We have put

$$h^g_{i\varepsilon} \doteq \begin{cases} f_{1i\varepsilon} - \mu_g g_{1i\varepsilon}\,, & \text{for } 0 \leqq i \leqq m-1\,, \\ g_{1i-m\varepsilon}\,, & \text{for } m \leqq i \leqq m+m'\,, \end{cases}$$

and obtain

$$b^{gh}_{ij} = \begin{cases} a^{11}_{ij} - \mu_h a^{12}_{ij} - \mu_g a^{12}_{ji} + \mu_g \mu_h a^{22}_{ij}\,, & \text{for } 0 \leqq i, j \leqq m-1\,, \\ a^{12}_{ij-m} - \mu_g a^{22}_{ij-m}\,, & \text{for } 0 \leqq i \leqq m-1\,, \\ & \quad m \leqq j \leqq m+m'\,, \\ a^{22}_{i-mj-m}\,, & \text{for } m \leqq i, j \leqq m+m'\,, \end{cases}$$

where $a^{12}_{ij} = 0$ for $j > m'$, and $a^{22}_{ij} = 0$ for $i > m'$ or $j > m'$. Using the results of Theorem 2.13 the statement of Theorem 2.14 for $\langle \xi_{gr} \xi_{ht} \rangle$ follows. Hence, the proof is complete for the vector process

$$\frac{1}{\sqrt{\varepsilon}} \Big(\sum_{j=1}^{n} E_{g_1 r_1, j}\,, \sum_{j=1}^{n} E_{g_2 r_2, j}\,, \ldots, \sum_{j=1}^{n} E_{g_s r_s, j} \Big);$$

but we must still show that the proved statement is also true as n tends to infinity.

2. To complete the proof of this theorem, uniform convergence in mean square (in ε) of $(\sum_{j=0}^{n} E_{gr,j} - E_{gr})/\sqrt{\varepsilon}$ to zero as $n \to \infty$ has to be established, i.e. we now show that

$$\lim_{n\to\infty} \frac{1}{\varepsilon} \langle (\sum_{j=0}^{n} E_{gr,j} - E_{gr})^2 \rangle = 0 \quad \text{uniformly in } \varepsilon\,. \tag{2.98}$$

This convergence (2.98) implies uniform convergence (in ε) in probability

$$\text{P-}\lim_{n\to\infty} \frac{1}{\sqrt{\varepsilon}} (E^{(n)}_{i_1}, E^{(n)}_{i_2}, \ldots, E^{(n)}_{i_k})$$

$$= \frac{1}{\sqrt{\varepsilon}} (E_{g_{i_1} r_{i_1}} - E_{g_{i_1} r_{i_1}, 0}, \ldots, E_{g_{i_k} r_{i_k}} - E_{g_{i_k} r_{i_k}, 0})$$

if we apply Chebyshev's inequality. Furthermore, from this we obtain uniform convergence (in ε) of the joint distribution functions

$$\lim_{n\to\infty} F_\varepsilon^{(n)}(t_1, \ldots, t_k) = F_\varepsilon(t_1, \ldots, t_k) \tag{2.99}$$

for all $(t_1, \ldots, t_k) \in R^k$ and $n \geqq 1$, where

$$F_\varepsilon^{(n)}(t_1, \ldots, t_k) \doteq \mathsf{P}\left(\frac{1}{\sqrt{\varepsilon}} E_{i_1}^{(n)} < t_1, \ldots, \frac{1}{\sqrt{\varepsilon}} E_{ik}^{(n)} < t_k\right)$$

and

$$F_\varepsilon(t_1, \ldots, t_k)$$
$$\doteq \mathsf{P}\left(\frac{1}{\sqrt{\varepsilon}}(E_{g_{i_1} r_{i_1}} - E_{g_{i_1} r_{i_1}, 0}) < t_1, \ldots, \frac{1}{\sqrt{\varepsilon}}(E_{g_{ik} r_{ik}} - E_{g_{ik} r_{ik}, 0}) < t_k\right).$$

We showed in the first part of this proof that

$$\lim_{\varepsilon\downarrow 0} F_\varepsilon^{(n)}(t_1, \ldots, t_k) = \Phi(t_1, \ldots, t_k) \tag{2.100}$$

for $n \geqq 1$, where $\Phi(t_1, \ldots, t_k)$ denotes the distribution function of $(\xi_{g_1 r_1}, \ldots, \xi_{g_k r_k})$, i.e.,

$$\Phi(t_1, \ldots, t_k) \doteq \mathsf{P}(\xi_{g_1 r_1} < t_1, \ldots, \xi_{g_k r_k} < t_k)\,.$$

Hence, because of (2.99) the inequality

$$F_\varepsilon^{(n)}(t_1, \ldots, t_k) - \eta \leqq F_\varepsilon(t_1, \ldots, t_k) \leqq F_\varepsilon^{(n)}(t_1, \ldots, t_k) + \eta$$

holds independently of ε for $n \geqq n_0(\eta)$ and by means of (2.100) we have

$$\Phi(t_1, \ldots, t_k) - \eta \leqq \overline{\lim_{\varepsilon\downarrow 0}}\, F_\varepsilon(t_1, \ldots, t_k) \leqq \Phi(t_1, \ldots, t_k) + \eta$$

for each $\eta > 0$. The above implies that

$$\lim_{\varepsilon\downarrow 0} F_\varepsilon(t_1, \ldots, t_k) = \Phi(t_1, \ldots, t_k)\,,$$

and the proof of Theorem 2.14 is complete if the convergence (2.98) is shown.

We now prove (2.98), which can be written as

$$\lim_{n\to\infty} \frac{1}{\varepsilon} \langle (\sum_{j=n+1}^{\infty} E_{gr,j})^2 \rangle = \lim_{n\to\infty} \frac{1}{\varepsilon} \sum_{s,t=n+1}^{\infty} \langle E_{gr,s} E_{gr,t} \rangle = 0\,. \tag{2.101}$$

From Theorem 2.13 we have

$$\frac{1}{\varepsilon} \langle E_{gr,s} E_{gr,t} \rangle \leqq \frac{1}{4p^2} \left|\binom{1/2}{s}\right| \left|\binom{1/2}{t}\right| (4x_1)^{s+t}$$
$$\times \sum_{\substack{i_1, \ldots, i_s = 0 \\ j_1, \ldots, j_t = 0}}^{2m-1} \frac{1}{\varepsilon} \int_a^b \ldots \int_a^b |\langle h_{i_1}(y_1) \ldots h_{i_s}(y_s)\, h_{j_1}(z_1) \ldots h_{j_t}(z_t) \rangle| \times \mathrm{d}y_1 \ldots \mathrm{d}y_s\, \mathrm{d}z_1 \ldots \mathrm{d}z_t$$

and we must consider an integral of the form

$$\int_a^b \dots \int_a^b |\langle h_{i_1}(y_1) \dots h_{i_k}(y_k)\rangle| \, dy_1 \dots dy_k \, .$$

For $k^2 \geqq (b-a)/2\varepsilon$ we have the inequality

$$\int_a^b \dots \int_a^b |\langle h_{i_1}(y_1) \dots h_{i_k}(y_k)\rangle| \, dy_1 \dots dy_k \leqq (b-a)^k \bar{\eta}^k$$
$$\leqq (b-a)^{k-1} 2\varepsilon k^2 \bar{\eta}^k$$

and for $k^2 < (b-a)/2\varepsilon$ the inequality

$$\int_a^b \dots \int_a^b |\langle h_{i_1}(y_1) \dots h_{i_k}(y_k)\rangle| \, dy_1 \dots dy_k \leqq \bar{\eta}^k((b-a)^k - V_k) \, .$$

In the above V_k denotes the volume of points $(y_1, \dots, y_k) \in [a, b]^k$ for which $\{\{y_1\}, \dots, \{y_k\}\}$ is the decomposition into ε-adjoining subsets. We compute

$$V_k = \int_a^b dy_1 \int_{|y_1-y_2|>\varepsilon} dy_2 \dots \int_{\substack{|y_1-y_k|>\varepsilon \\ |y_2-y_k|>\varepsilon \\ \vdots \\ |y_{k-1}-y_k|>\varepsilon}} dy_k \geqq (b-a) \prod_{i=1}^{k-1} (b-a-2\varepsilon i)$$
$$\geqq (b-a)(b-a-2\varepsilon k)^{k-1} \geqq (b-a)^k \left(1 - \frac{2\varepsilon k^2}{b-a}\right),$$

and obtain

$$\int_a^b \dots \int_a^b |\langle h_{i_1}(y_1) \dots h_{i_k}(y_k)\rangle| \, dy_1 \dots dy_k \leqq 2(b-a)^{k-1} \varepsilon k^2 \bar{\eta}^k$$

for all $k \geqq 2$. Thus

$$\frac{1}{\varepsilon} \langle E_{gr,s} E_{gr,t}\rangle$$
$$\leqq \frac{1}{2p^2} \left|\binom{1/2}{s}\right| \left|\binom{1/2}{t}\right| (4x_1)^{s+t} (2m\bar{\eta})^{s+t} (b-a)^{s+t-1} (s+t)^2 \, .$$

Using (2.101) this inequality leads to the uniform convergence in square mean (in ε) of (2.98).

3. From the above proof convergence of the second moments can also be deduced. Put $E_i \doteq E_{g_i r_i} - E_{g_i r_i, 0}$. Then a simple calculation yields

$$\frac{1}{\varepsilon} \langle E_1 E_2\rangle = \frac{1}{\varepsilon} [\langle (E_1 - E_1^{(n)}) (E_2 - E_2^{(n)})\rangle + \langle E_1^{(n)} (E_2 - E_2^{(n)})\rangle$$
$$+ \langle (E_1 - E_1^{(n)}) E_2^{(n)}\rangle + \langle E_1^{(n)} E_2^{(n)}\rangle]$$
$$\leqq \left[\frac{1}{\varepsilon} \langle (E_1 - E_1^{(n)})^2\rangle \frac{1}{\varepsilon} \langle (E_2 - E_2^{(n)})^2\rangle\right]^{1/2}$$
$$+ \left[\frac{1}{\varepsilon} \langle E_1^{(n)^2}\rangle \frac{1}{\varepsilon} \langle (E_2 - E_2^{(n)})^2\rangle\right]^{1/2}$$
$$+ \left[\frac{1}{\varepsilon} \langle E_2^{(n)^2}\rangle \frac{1}{\varepsilon} \langle (E_1 - E_1^{(n)})^2\rangle\right]^{1/2} + \frac{1}{\varepsilon} \langle E_1^{(n)} E_2^{(n)}\rangle \, ;$$

from which it follows that

$$\left| \overline{\lim_{\varepsilon \downarrow 0}} \frac{1}{\varepsilon} \langle E_1 E_2 \rangle - \langle \xi_{g_1 r_1} \xi_{g_2 r_2} \rangle \right| \leqq s(n) ,$$

because of (2.98), where $\lim_{n \to \infty} s(n) = 0$. Hence, we have

$$\lim_{\varepsilon \downarrow 0} \frac{1}{\varepsilon} \langle (E_{g_1 r_1} - E_{g_1 r_1, 0}) (E_{g_2 r_2} - E_{g_2 r_2, 0}) \rangle = \langle \xi_{g_1 r_1} \xi_{g_2 r_2} \rangle .$$

The convergence of higher moments of $(E_{gr} - E_{gr,0})/\sqrt{\varepsilon}$ can be shown by more detailed estimations than in the second part of this proof. Therefore the theorem is proved. ▶

In the following we will deal with some important special cases of Theorem 2.14. The notation introduced above are used.

(a) *Convergence of eigenvalues*

We first have

$$\lim_{\varepsilon \downarrow 0} \frac{1}{\sqrt{\varepsilon}} (\lambda_g - \mu_g) = \xi_{g0} \quad \text{in distribution}$$

for $r = 0$, where

$$\langle \xi_{g0} \rangle = 0 ,$$

$$\langle \xi_{g0}^2 \rangle = \sum_{i,j=0}^{m-1} \int_a^b b_{ij}^{gg}(y) \left(w_g^{(i)}(y)\, w_g^{(j)}(y)\right)^2 dy .$$

The functions $b_{ij}^{gg}(y)$ are determined by

$$b_{ij}^{gg}(y) = \lim_{\varepsilon \downarrow 0} \frac{1}{\varepsilon} \int_{-\varepsilon}^{\varepsilon} \langle \left(f_{1i\varepsilon}(y) - \mu_g g_{1i\varepsilon}(y)\right) \left(f_{1j\varepsilon}(y+z) - \mu_g g_{1j\varepsilon}(y+z)\right) \rangle \, dz$$

and $g_{1i\varepsilon}(y) \equiv 0$ for $i > m'$ is set. In case of independent processes $f_{10\varepsilon}(x, \omega), \dots,$ $f_{1m-1\varepsilon}(x, \omega)$, $g_{10\varepsilon}(x, \omega), \dots, g_{1m'\varepsilon}(x, \omega)$, the functions $b_{ij}^{gg}(y)$ can be simplified by the relation

$$b_{ij}^{gg}(y) = \delta_{ij}\left(a_{ii}^{11}(y) + \mu_g^2 a_{ii}^{22}(y)\right) ,$$

and the second moment of ξ_{g0} is given by

$$\langle \xi_{g0}^2 \rangle = \sum_{i=0}^{m-1} \int_a^b b_{ii}^{gg}(y) \left(w_g^{(i)}(y)\right)^4 dy .$$

For s eigenvalues we have

$$\lim_{\varepsilon \downarrow 0} \frac{1}{\sqrt{\varepsilon}} (\lambda_{g_1} - \mu_{g_1}, \dots, \lambda_{g_s} - \mu_{g_s}) = (\xi_{g_1 0}, \dots, \xi_{g_s 0}\rangle \quad \text{in distribution.}$$

The Gaussian vector $(\xi_{g_1 0}, \dots, \xi_{g_s 0})$ has moments

$$\langle \xi_{g0} \rangle = 0 ,$$
$$\langle \xi_{g0} \xi_{h0} \rangle = \sum_{i,j=0}^{m-1} \int_a^b b_{ij}^{gh}(y) \left(w_g^{(i)}(y) \, w_h^{(j)}(y) \right)^2 \mathrm{d}y . \tag{2.102}$$

As an application to the special eigenvalue problem

$$M_0 \bar{u} + M_1 \bar{u} = \bar{\lambda} N_0 \bar{u}$$

with the order of M_1 not greater than m', we have

$$\langle \bar{\xi}_{g0} \bar{\xi}_{h0} \rangle = \sum_{i,j=0}^{m-1} \int_a^b a_{ij}^{11}(y) \left(w_g^{(i)}(y) \, w_h^{(j)}(y) \right)^2 \mathrm{d}y ;$$

and for the eigenvalue problem

$$M_0 \bar{\bar{u}} = \bar{\bar{\lambda}} (N_0 \bar{\bar{u}} + M_1 \bar{\bar{u}})$$

the correlation relations

$$\langle \bar{\bar{\xi}}_{g0} \bar{\bar{\xi}}_{h0} \rangle = \mu_g \mu_h \sum_{i,j=0}^{m-1} \int_a^b a_{ij}^{11}(y) \left(w_g^{(i)}(y) \, w_h^{(j)}(y) \right)^2 \mathrm{d}y .$$

Hence, we have

$$\langle \bar{\bar{\xi}}_{g0} \bar{\bar{\xi}}_{h0} \rangle = \mu_g \mu_h \langle \bar{\xi}_{g0} \bar{\xi}_{h0} \rangle .$$

For $g = h$ it follows that the variance $\langle \bar{\bar{\xi}}_{g0}^2 \rangle$ of the second eigenvalue problem is equal to the variance $\langle \bar{\xi}_{g0}^2 \rangle$ of the first problem multiplied by μ_g^2. We compare the correlation between ξ_{g0} and ξ_{h0} will the same perturbation of N_0 and M_0. Then $\langle \xi_{g0} \xi_{h0} \rangle$ resulting from a perturbation of N_0 gives the result multiplied with $\mu_g \mu_h$ fold for the case of M_0 being perturbed.

Furthermore, the variance of the eigenvalues ξ_{g0} tends to infinity if g increases, i.e. small perturbations of the operators M_0 or N_0 effect a „melt'' of the eigenvalues if the number of the eigenvalue increases. In Section 2.3.3 further statements of this kind are given.

(b) Convergence of eigenfunctions

From Theorem 2.14 the convergence in distribution

$$\lim_{\varepsilon \downarrow 0} \frac{1}{\sqrt{\varepsilon}} \left(u_g(x, \omega) - w_g(x) \right) = \xi_g(x, \omega)$$

follows where $\xi_g(x, \omega)$ denotes a mean zero Gaussian process with correlation function

$$\begin{aligned} \langle \xi_g(x) \, \xi_g(y) \rangle = {} & \sum_{i,j=0}^{m-1} \int_a^b b_{ij}^{gg}(z) \, K_i^g(x, z) \, K_j^g(y, z) \, \mathrm{d}z \\ & + \sum_{i=0}^{m-1} \sum_{j=0}^{m'} \int_a^b [b_{ij+m}^{gg}(z) \, K_i^g(x, z) \, k_j^g(y, z) \\ & \qquad + b_{j+mi}^{gg}(z) \, k_j^g(x, z) \, K_i^g(y, z)] \, \mathrm{d}z \\ & + \sum_{i,j=0}^{m'} \int_a^b b_{i+mj+m}^{gg}(z) \, k_i^g(x, z) \, k_j^g(y, z) \, \mathrm{d}z . \end{aligned}$$

In the above formula $k_i^g(x, z)$ and $K_i^g(x, z)$ are defined by

$$k_i^g(x, z) \doteq -\tfrac{1}{2}\left(w_g^{(i)}(z)\right)^2 w_g(x)\,, \quad \text{for} \quad 0 \leqq i \leqq m'\,,$$

$$K_i^g(x, z) \doteq -G_{0i}^g(x, z)\, w_g^{(i)}(z)\,, \quad \text{for} \quad 0 \leqq i \leqq m-1\,,$$

where $G_{0i}^g(x, y) \doteq \partial^i G^g(x, y)/\partial y^i$. In particular, for $N_1 = 0$ we have

$$\langle \xi_g(x)\, \xi_g(y) \rangle = \sum_{i,j=0}^{m-1} \int_a^b a_{ij}^{11}(z)\, G_{0i}(x, z)\, G_{0j}(y, z)\, w_g^{(i)}(z)\, w_g^{(j)}(z)\, \mathrm{d}z;$$

and for $M_1 = 0$,

$$\begin{aligned}\langle \xi_g(x)\, \xi_g(y) \rangle = \sum_{i,j=0}^{m'} \Big[\mu_g^2 \int_a^b & a_{ij}^{22}(z)\, G_{0i}^g(x, z)\, G_{0j}^g(y, z)\, w_g^{(i)}(z)\, w_g^{(j)}(z)\, \mathrm{d}z \\ & - \tfrac{1}{2}\mu_g \int_a^b a_{ij}^{22}(z)\, \{G_{0i}^g(x, z)\, w_g^{(i)}(z)\left(w_g^{(j)}(z)\right)^2 w_g(y) \\ & + G_{0j}^g(y, z)\, w_g^{(j)}(z)\left(w_g^{(i)}(z)\right)^2 w_g(x)\}\, \mathrm{d}z \\ & + \tfrac{1}{4}\int_a^b a_{ij}^{22}(z)\, [w_g^{(i)}(z)\, w_g^{(j)}(z)]^2\, \mathrm{d}z\, w_g(x)\, w_g(y)\Big]\,.\end{aligned}$$

As a generalization we have convergence in distribution

$$\lim_{\varepsilon \downarrow 0} \frac{1}{\sqrt{\varepsilon}}\left(u_{g_1}(x) - w_{g_1}(x), \ldots, u_{g_s}(x) - w_{g_s}(x)\right) = \left(\xi_{g_1}(x), \ldots, \xi_{g_s}(x)\right)$$

for s eigenfunctions. The mean zero Gaussian vector process $\left(\xi_{g_1}(x), \ldots, \xi_{g_s}(x)\right)$ is determined by the second moments

$$\begin{aligned}\langle \xi_g(x)\, \xi_h(y) \rangle = & \sum_{i,j=0}^{m-1} \int_a^b b_{ij}^{gh}(z)\, K_i^g(x, z)\, K_j^h(y, z)\, \mathrm{d}z \\ & + \sum_{i=0}^{m-1} \sum_{j=0}^{m'} \int_a^b [b_{ij+m}^{gh}(z)\, K_i^g(x, z)\, k_j^h(y, z) \\ & \qquad + b_{j+mi}^{gh}(z)\, k_j^g(x, z)\, K_i^h(y, z)]\, \mathrm{d}z \\ & + \sum_{i,j=0}^{m'} \int_a^b b_{i+mj+m}^{gh}(z)\, k_i^g(x, z)\, k_j^h(y, z)\, \mathrm{d}z\,,\end{aligned} \tag{2.103}$$

for $g, h \in \{g_1, \ldots, g_s\}$.

Assuming

$$\|u^{(s)}\|_{\mathbf{L}_2} \leqq \|u\|_{\mathscr{H}_{M_0}}\,, \quad \text{for} \quad s = 0, 1, \ldots, t\,, \qquad u \in \mathscr{H}_{M_0}\,,$$

we have

$$G_{0s}^g(x, y) = \sum_{\substack{k=1 \\ k \neq g}}^{\infty} \frac{w_k(x)\, w_k^{(s)}(y)}{\mu_k - \mu_g} \tag{2.104}$$

for $s = 0, 1, \ldots, t$, where the series converges in $\mathbf{L}_2(a, b)$. Then the second moments of the limit processes can be computed by

$$\langle \xi_g(x)\, \xi_h(y) \rangle = \sum_{i,j=0}^{m-1} \sum_{\substack{p=1\\ p\neq g}}^{\infty} \sum_{\substack{q=1\\ q\neq h}}^{\infty} \frac{w_p(x)\, w_q(y)}{(\mu_p - \mu_g)\,(\mu_q - \mu_h)} \langle\!\langle b_{ij}^{gh} w_g^{(i)} w_h^{(j)},\, w_p^{(i)} w_q^{(j)} \rangle\!\rangle$$

$$+ \frac{1}{2} \sum_{i=0}^{m-1} \sum_{j=0}^{m'} \Bigg[\sum_{\substack{p=0\\ p\neq g}}^{\infty} \frac{w_p(x)\, w_h(y)}{\mu_p - \mu_g} \langle\!\langle b_{ij+m}^{gh} (w_h^{(j)})^2,\, w_g^{(i)} w_p^{(i)} \rangle\!\rangle$$

$$+ \sum_{\substack{p=1\\ p\neq h}}^{\infty} \frac{w_g(x)\, w_p(y)}{\mu_p - \mu_h} \langle\!\langle b_{j+mi}^{gh} (w_g^{(j)})^2,\, w_h^{(i)} w_p^{(i)} \rangle\!\rangle \Bigg]$$

$$+ \frac{1}{4} \sum_{i,j=0}^{m'} \langle\!\langle b_{i+mj+m}^{gh} (w_g^{(i)})^2,\, (w_h^{(j)})^2 \rangle\!\rangle\, w_g(x)\, w_h(y)\,.$$

This equation often permits an advantageous numerical analysis if the eigenvalues μ_g and the eigenfunctions $w_g(x)$ of the averaged problem are known.

Equation (2,104) can be proved by Theorem 2.10, applying this theorem to the averaged eigenvalue problem $M_0 w = \mu N_0 w$ with the boundary conditions (2.72). Using $G^g(x, .) \in \mathcal{H}_{M_0}$ we have

$$G^g(x, y) = \sum_{k=1}^{\infty} \langle\!\langle N_0 G^g(x, .),\, w_k \rangle\!\rangle\, w_k(y)\,,$$

and this series converges in $\mathcal{H}_{M_0}$. From $(M_0 - \mu_g N_0)\, u = \tilde{\mu} N_0 u$ and the boundary conditions (2.72) it follows that

$$u(x) = \tilde{\mu} \int_a^b G^g(x, y)\, N_0 u(y)\, dy;$$

and then

$$\langle\!\langle N_0 G^g(x, .),\, w_k \rangle\!\rangle = \frac{w_k(x)}{\mu_k - \mu_g}, \quad \text{for } g \neq k\,,$$

and

$$\langle\!\langle N_0 G^g(x, .),\, w_g \rangle\!\rangle = 0\,,$$

using the normalization condition of G^g. Hence, the equation

$$G^g(x, y) = \sum_{\substack{k=1\\ k\neq g}}^{\infty} \frac{w_k(x)\, w_k(y)}{\mu_k - \mu_g}$$

obtains, and because of the assumption, we obtain (2.104).

(c) Convergence for eigenvalues and eigenfunctions

Convergence in distribution

$$\lim_{\varepsilon \downarrow 0} \frac{1}{\sqrt{\varepsilon}} \left(\lambda_g - \mu_g,\, u_h(x) - w_h(x)\right) = \left(\xi_{g0},\, \xi_h(x)\right)$$

follows from Theorem 2.14. The mean zero Gaussian vector $(\xi_{g0}, \xi_h(x))$ has second moments

$$\langle \xi_{g0}\xi_h(x)\rangle = \sum_{i,j=0}^{m-1} \int_a^b b_{ij}^{gh}(z)\, K_j^h(x, z)\, \big(w_g^{(i)}(z)\big)^2\, dz$$

$$+ \sum_{i=0}^{m-1} \sum_{j=0}^{m'} \int_a^b b_{ij+m}^{gh}(z)\, k_j^h(x, z)\, \big(w_g^{(i)}(z)\big)^2\, dz \qquad (2.105)$$

because $K_{i,0}^g(y) = 0$ for $i \geqq m$, furthermore, $\langle \xi_{g0}^2\rangle$ from (a) and $\langle \xi_h(x)\, \xi_h(y)\rangle$ from (*b*). Here the notation of (*b*) is used. The generalization leads to

$$\lim_{\varepsilon\downarrow 0} \frac{1}{\sqrt{\varepsilon}} \big(\lambda_{g_1} - \mu_{g_1}, \dots, \lambda_{g_s} - \mu_{g_s}, u_{h_1}(x) - w_{h_1}(x), \dots, u_{h_t}(x) - w_{h_t}(x)\big)$$

$$= \big(\xi_{g_10}, \dots, \xi_{g_s0}, \xi_{h_1}(x), \dots, \xi_{h_t}(x)\big) \text{ in distribution;}$$

and the moments $\langle \xi_{g0}\xi_{h0}\rangle$ can be computed by (2.102), $\langle \xi_g(x)\, \xi_h(y)\rangle$ by (2.103) and finally $\langle \xi_{g0}\xi_h(x)\rangle$ by (2.105).

In Section 2.3.3 we will deal with a special class of eigenvalue problems which permit immediate calculations, and therefore leads to concrete results. The results obtained are discussed in this section.

Consider the connection between the results of the method of Ritz (see Section 2.2.2) and those of the perturbation method. We will restrict our attention to the eigenvalues since the results for the eigenfunctions are essentially the same, but these results are more complicated to obtain.

We now consider the averaged problem for the eigenvalue problem (2.75), (2.72):

$$M_0 w = \mu N_0 w\,,$$

with the boundary conditions (2.72) in the case, where the operators M_0, N_0 are symmetric, M_0 positive definite in $\mathbf{L}_2(a, b)$, and N_0 is positive in $\mathbf{L}_2(a, b)$. We then get the statements of Theorem 2.10. The corresponding Ritz eigenvalue problem

$$\sum_{k=1}^{n} {}^n u_k [(\!(M_0\varphi_k, \varphi_j)\!) - {}^n\mu (\!(N_0\varphi_k, \varphi_j)\!)] = 0\,, \quad \text{for} \quad j = 1, 2, \dots, n\,,$$

leads to the eigenvalues ${}^n\mu_g$, $g = 1, 2, \dots, n$, ${}^n\mu_1 \leqq {}^n\mu_2 \leqq \dots \leqq {}^n\mu_n$, and to the associated approximate eigenfunctions

$${}^n u_{g0}(x) = \sum_{k=1}^{n} {}^n u_{gk0}\varphi_k(x)\,, \quad \text{for} \quad g = 1, 2, \dots, n\,,$$

where $({}^n u_{gk0})_{1\leqq k\leqq n}$ denotes the solution of the Ritz eigenvalue problem for the eigenvalue ${}^n\mu_g$. We then have

$$\lim_{n\to\infty} {}^n\mu_g = \mu_g\,;$$

and for a simple eigenvalue μ_g

$$\lim_{n\to\infty} {}^n u_{g0}(x) = w_g(x) \quad \text{in } \mathscr{H}_{M_0}\,.$$

Assume now that the random eigenvalue problem

$$(M_0 + M_1(\omega))\, u = \lambda(N_0 + N_1(\omega))\, u\,,$$

with the boundary conditions (2.72), satisfies almost surely the conditions of Theorem 2.10. The Ritz approximations of the eigenvalues $\lambda_g(\omega)$ are denoted by ${}^n\lambda_g(\omega)$ and the eigenfunctions $u_g(x, \omega)$ by ${}^nu_g(x, \omega)$. Then we have

$$\lim_{\varepsilon \downarrow 0} \frac{1}{\varepsilon} \langle ({}^n\lambda_g - {}^n\mu_g)\,({}^n\lambda_h - {}^n\mu_h)\rangle = \sum_{i,j=0}^{m} \langle\!\langle b_{ij}^{gh}({}^nu_{g0}^{(i)})^2, ({}^nu_{h0}^{(j)})^2 \rangle\!\rangle \tag{2.106}$$

(cf. (2.46)), and the conditions (2.76) for the boundary conditions. For the proof of (2.106) the condition $f_{1m}(x, \omega) \equiv 0$ is not necessary, but this condition is required for the derivation of the results utilizing perturbation theory.

In general, from the convergence $\lim_{n\to\infty} {}^nu_{g0} = w_g$ in $\mathcal{H}_{M_0}$ we cannot make a statement concerning the convergence of

$$\langle\!\langle b_{ij}^{gh}({}^nu_{g0}^{(m)})^2, ({}^nu_{g0}^{(m)})^2 \rangle\!\rangle$$

as $n \to \infty$ in (2.106). However, we obtain for the eigenvalue problems which we are interested in

$$\lim_{n\to\infty} \sum_{i,j=0}^{m-1} \langle\!\langle b_{ij}^{gh}({}^nu_{g0}^{(i)})^2, ({}^nu_{h0}^{(j)})^2 \rangle\!\rangle = \sum_{i,j=0}^{m-1} \langle\!\langle b_{ij}^{gh}(u_{g0}^{(i)})^2, (u_{h0}^{(j)})^2 \rangle\!\rangle\,; \tag{2.107}$$

and then

$$\lim_{n\to\infty} \langle {}^n\xi_g\, {}^n\xi_k \rangle = \langle \xi_g \xi_h \rangle\,,$$

where $\langle \xi_g \xi_h \rangle$ is given in Theorem 2.14 and $\langle {}^n\xi_g\, {}^n\xi_h \rangle$ in Theorem 2.11. In order to prove the equation (2.107) it is sufficient to show that the convergence of ${}^nu_{g0}$ to w_g implies the convergence of the i-th derivative in $\mathbf{L}_4(a, b)$; that is

$$\lim_{n\to\infty} {}^nu_{g0}^{(i)} = w_g^{(i)} \quad \text{in} \quad \mathbf{L}_4(a, b)\,, \quad \text{for} \quad i = 0, 1, \dots, m-1\,. \tag{2.108}$$

This convergence holds for all examples given in Section 2.2.1. For the first example the convergence in (2.108) can be determined from the fact that the uniform convergence of all derivatives of an order less than m follows from the convergence in $\mathring{W}_2^m$. For the eigenvalue problem (2.34), we obtain the uniform convergence of functions from the convergence in $\mathcal{H}_{M_0}$; and for the eigenvalue problem (2.36) the uniform convergence of functions and its first derivatives follows from convergence in $\mathcal{H}_{M_0}$. In Section 2.2.4.1 we have dealt with these convergence problems; and these results were compared with concrete numerical calculations.

2.3.3. Applications to special eigenvalue problems

Consider the random eigenvalue problem

$$-(f_{10}u')' + (f_{00} + f_{01})\, u = \lambda(g_{00} + g_{01})\, u\,, \tag{2.109}$$

for $0 \leqq x \leqq l$, with the boundary conditions

$$u(0) = u(l) = 0 .$$

To simplify the numerical calculations, let the non-random functions $f_{10}(x)$, $f_{00}(x)$ and $g_{00}(x)$ be constants, say

$$f_{10} = c , \qquad f_{00} = 0 , \qquad g_{00} = 1 .$$

If $f_{00} \neq 0$ or $g_{00} \neq 1$ we can introduce a new eigenvalue parameter $\bar{\lambda} = \lambda g_{00} - f_{00}$. Let $(f_{01}(x, \omega), g_{01}(x, \omega))$ be a weakly correlated connected vector process with correlation length ε on the interval $[0, l]$ with intensities a_{11} for f_{01}, a_{12} between f_{01} and g_{01}, and a_{22} for g_{01}.

The averaged problem corresponding to (2.109) can be written as

$$-cw'' = \mu w , \qquad w(0) = w(l) = 0 ,$$

for $0 \leqq x \leqq l$. This eigenvalue problem possesses the simple eigenvalues $\mu_k = c(k\pi/l)^2$; and the associated eigenfunctions $w_k(x) = \sqrt{2/l} \sin(\pi k x/l)$ for $k = 1, 2, \ldots$ The generalized Green's function associated with the operator $-cw'' - \mu_g w$ and the boundary conditions $w(0) = w(l) = 0$ is given by

$$G^g(x, y)$$
$$= \begin{cases} \dfrac{1}{g\pi c} \left\{ \sin\left(\dfrac{g\pi y}{l}\right) \left[\dfrac{l}{2g\pi} \sin\left(\dfrac{g\pi x}{l}\right) - x \cos\left(\dfrac{g\pi x}{l}\right)\right] \right. \\ \qquad \left. + (l - y) \cos\left(\dfrac{g\pi y}{l}\right) \sin\left(\dfrac{g\pi x}{l}\right)\right\} \quad \text{for} \quad 0 \leqq x < y \leqq l , \\ \dfrac{1}{g\pi c} \left\{ \sin\left(\dfrac{g\pi x}{l}\right) \left[\dfrac{l}{2g\pi} \sin\left(\dfrac{g\pi y}{l}\right) - y \cos\left(\dfrac{g\pi y}{l}\right)\right] \right. \\ \qquad \left. + (l - x) \cos\left(\dfrac{g\pi x}{l}\right) \sin\left(\dfrac{g\pi y}{l}\right)\right\} ; \quad \text{for } 0 \leqq y < x \leqq l \end{cases}$$

and we have $\int\limits_0^l G^g(x, y)\, w_k(x)\, dx = 0$. On the other hand, $G^g(x, y)$ admits the representation

$$G^g(x, y) = \sum_{\substack{i=1 \\ i \neq g}}^{\infty} \frac{w_i(x)\, w_i(y)}{\mu_i - \mu_g} ,$$

and this series converges uniformly in $[0, l] \times [0, l]$.

Using Theorem 2.14 the random vector

$$\frac{1}{\sqrt{\varepsilon}} (E_{g_1 r_1} - E_{g_1 r_1, 0}, E_{g_2 r_2} - E_{g_2 r_2, 0}, \ldots , E_{g_s r_s} - E_{g_s r_s, 0})$$

converges in distribution to the Gaussian random vector $(\xi_{g_1r_1}, \ldots, \xi_{g_sr_s})$. This Gaussian vector has the moments

$$\langle \xi_{gr} \rangle = 0 ,$$

$$\langle \xi_{gr} \xi_{ht} \rangle = \sum_{i,j=0}^{1} \int_0^l K_{i,r}^g(y)\, K_{j,t}^h(y)\, b_{ij}^{gh}(y)\, dy , \qquad (2.110)$$

where

$$K_{i,0}^g(y) = \delta_{i0} w_g^2(y) ,$$

$$K_{0,r}^g(y) = -G^g(x_r, y)\, w_g(y) ,$$

$$K_{1,r}^g(y) = -\tfrac{1}{2} w_g^2(y)\, w_g(x_r) ,$$

and

$$b_{00}^{gh}(y) = a_{11}(y) - (\mu_g + \mu_h)\, a_{12}(y) + \mu_g \mu_h a_{22}(y) ,$$

$$b_{01}^{gh}(y) = a_{12}(y) - \mu_g a_{22}(y) , \qquad b_{10}^{gh}(y) = b_{01}^{hg}(y) ,$$

$$b_{11}^{gh}(y) = a_{22}(y) .$$

In particular, we obtain from these results:

(a) *for the eigenvalues*

Equation (2.110), for $r = t = 0$, leads to

$$\sigma_{gh} \doteq \langle \xi_{g0} \xi_{h0} \rangle = \int_0^l w_g^2(y)\, w_h^2(y)\, [a_{11}(y) - (\mu_g + \mu_h)\, a_{12}(y) + \mu_g \mu_h a_{22}(y)]\, dy \qquad (2.111)$$

for $g, h = 1, 2, \ldots$ We have

$$\lim_{\varepsilon \downarrow 0} F_\varepsilon^{gh}(p, q) = \Phi^{gh}(p, q)$$

where $F_\varepsilon^{gh}(p, q)$ denotes the distribution function of the vector $(\lambda_g - \mu_g, \lambda_h - \mu_h)/\sqrt{\varepsilon}$ and $\Phi^{gh}(p, q)$ denotes the distribution function of (ξ_{g0}, ξ_{h0}). $\Phi^{gh}(p, q)$ is the distribution function of a mean zero Gaussian vector with the second moments given by (2.111). Hence, for small ε the random vector (λ_g, λ_h) can be assumed to be a Gaussian random vector with mean vector (μ_g, μ_h) and correlation matrix

$$\varepsilon(\langle \xi_{i0} \xi_{j0} \rangle)_{i,j \in \{g,h\}} .$$

In particular, for small ε the eigenvalue λ_g is normally distributed with

$$\langle \lambda_g \rangle \approx \mu_g$$

and

$$\langle (\lambda_g - \mu_g)^2 \rangle \approx \varepsilon \langle \xi_{g0}^2 \rangle = \varepsilon \int_0^l w_g^4(y)\, [a_{11}(y) - 2\mu_g a_{12}(y) + \mu_g^2 a_{22}(y)]\, dy ,$$

i.e. $\lambda_g \sim \mathrm{N}(\mu_g, \varepsilon \sigma_{gg})$. Stability problems for technological systems expressed by the random eigenvalue problem (2.109) can now be solved since, for instance, the probability $\mathsf{P}(\lambda_1 < p)$ can be computed (cf. also Section 2.2.4).

For further considerations let $(f_{01}(x, \omega), g_{01}(x, \omega))$ be a wide-sense stationary connected random vector so that the intensities a_{ij} are constant. In this case we have

$$\langle \xi_{g0} \xi_{h0} \rangle = [a_{11} - (\mu_g + \mu_h)\, a_{12} + \mu_g \mu_h a_{22}] \left[1 + \frac{1}{2} \delta_{gh}\right] \frac{1}{l} .$$

The variance of the limit distribution of an eigenvalue

$$\langle \xi_{g0}^2 \rangle = \frac{3}{2l} [a_{11} - 2\mu_g a_{12} + \mu_g^2 a_{22}]$$

depends on the number of the eigenvalue, if $g_{01}(x, \omega) \not\equiv 0$. This property describes a "melt" of the eigenvalues if the number of the eigenvalues increases.

For $g_{01}(x, \omega) \equiv 0$ we have

$$\langle \xi_{g0}^2 \rangle = \frac{3}{2l} a_{11} ;$$

i.e. the variance of the limit distribution turns out to be independent of the number of the eigenvalues. In Fig. 2.14 the density functions $f_g^i(p)$ are plotted for the random variables ξ_{g0}. The function $f_g^1(p)$ denotes the density of the random variable ξ_{g0} in the case, where $g_{01}(x, \omega) \equiv 0$; and $f_g^2(p)$ the density in the case, where $f_{01}(x, \omega) \equiv 0$. In this figure we have put $a_{11} = 1$ and $a_{22} = 1$. It can be seen that $f_g^1(p) = f_g^2(p)$ follows from $\mu_g^2 a_{22} = a_{11}$. Consequently, a perturbation of the differential operator N_0 has a greater influence on the eigenvalues than a perturbation of the operator M_0. With an increasing length l the variance $\langle \xi_{g0}^2 \rangle$ converges to zero, so that the eigenvalues are more stable relative to perturbations if the length is large.

The correlation coefficient can be computed as

$$\varrho_{gh} = \frac{2}{3} \frac{a_{11} - (\mu_g + \mu_h)\, a_{12} + \mu_g \mu_h a_{22}}{\sqrt{(a_{11} - 2\mu_g a_{12} + \mu_g^2 a_{22})\,(a_{11} - 2\mu_h a_{12} + \mu_h^2 a_{22})}} ;$$

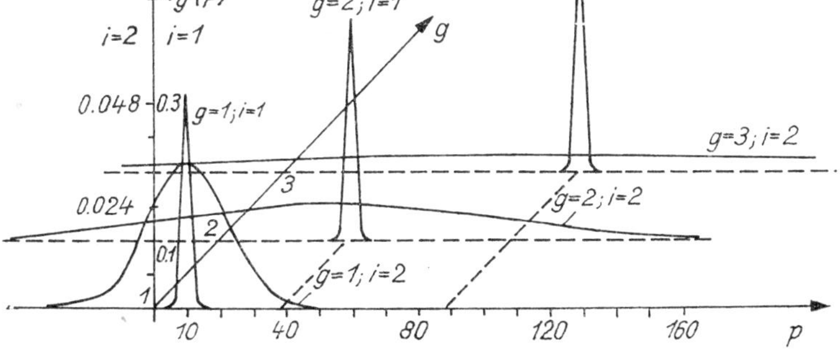

Fig. 2.14. Density functions of ξ_{g0}

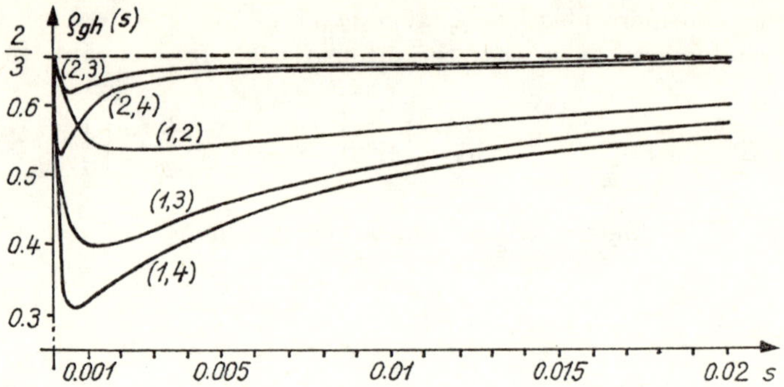

Fig. 2.15. The correlation coefficient of various eigenvalues

and for independent processes $f_{01}(x, \omega)$ and $g_{01}(x, \omega)$, and for $a_{22} = sa_{11}$ we have

$$\varrho_{gh}(s) = \frac{2}{3} \frac{1 + \mu_g \mu_h s}{\sqrt{(1 + \mu_g^2 s)(1 + \mu_h^2 s)}} .$$

This correlation coefficient is represented for various g and h in Fig. 2.15. We see that $\varrho_{gh}(s)$ possesses a minimum for $s_0 = 1/\mu_g \mu_h$, and the minimum is given by

$$\varrho_{gh}(s_0) = \frac{4}{3} \frac{\sqrt{\mu_g \mu_h}}{\mu_g + \mu_h} .$$

Hence, a minimum of the correlation is found for $a_{22} = a_{11}/\mu_g \mu_h$. Furthermore, we have, for a fixed g,

$$\lim_{h \to \infty} \varrho_{gh}(s) = \frac{2}{3} \frac{\mu_g \sqrt{s}}{\sqrt{1 + \mu_g^2 s}} .$$

These correlation relations hold also for the eigenvalue if ε is small, because the limit relation

$$\lim_{\varepsilon \downarrow 0} \frac{\langle (\lambda_g - \mu_g)(\lambda_h - \mu_h) \rangle}{\sqrt{\langle (\lambda_g - \mu_g)^2 \rangle \langle (\lambda_h - \mu_h)^2 \rangle}} = \frac{\langle \xi_{g0} \xi_{h0} \rangle}{\sqrt{\langle \xi_{g0}^2 \rangle \langle \xi_{h0}^2 \rangle}}$$

follows from

$$\lim_{\varepsilon \downarrow 0} \frac{1}{\varepsilon} \langle (\lambda_g - \mu_g)(\lambda_h - \mu_h) \rangle = \langle \xi_{g0} \xi_{h0} \rangle .$$

(*b*) *for the eigenfunctions*

Using (2.110) we obtain the relation

$$K_{gh}(x, y) \doteq \langle \xi_g(x) \xi_h(y) \rangle$$

$$= \int_0^l G^g(x,z) G^h(y,z) w_g(z) w_h(z) b_{00}^{gh}(z) \,\mathrm{d}z + \tfrac{1}{2} \int_0^l G^g(x,z) w_g(z) w_h^2(z) b_{10}^{gh}(z) \,\mathrm{d}z w_h(y)$$

$$+ \tfrac{1}{2} \int_0^l G^h(y,z) w_h(z) w_g^2(z) b_{01}^{gh}(z) \,\mathrm{d}z\, w_g(x) + \tfrac{1}{4} \int_0^l w_g^2(z) w_h^2(z) b_{11}^{gh}(z) \,\mathrm{d}z\, w_g(x) w_h(y) ,$$

for $x_r = x$ and $x_t = y$. For further considerations, let $(f_{01}(x, \omega), g_{01}(x, \omega))$ be a wide-sense stationary vector process. Then the intensities b_{ij}^{gh} are constant. The Green's function $G^g(x, y)$ was given in the beginning of this section, so that the function $K_{gh}(x, y)$ can be calculated. However, we compute the correlation function $K_{gh}(x, y)$ by means of the relation

$$K_{gh}(x, y) = \lim_{\varepsilon \downarrow 0} \frac{1}{\varepsilon} \langle u_{g1}(x)\, u_{h1}(y) \rangle ,$$

where the term $u_{p1}(x, \omega)$ is given by the boundary value problem

$$cu_{p1}'' + \mu_p u_{p1} = \left(h_0^p(x, \omega) - \langle\!\langle h_0^p, w_p^2 \rangle\!\rangle\right) w_p(x) , \qquad u_{p1}(0) = u_{p1}(l) = 0 ,$$

which follows from (2.81) and (2.82) for $k = 1$. In this boundary value problem we have put

$$h_0^p(x, \omega) \doteq f_{01}(x, \omega) - \mu_p g_{01}(x, \omega) .$$

The solution is given by

$$\begin{aligned} u_{p1}(x, \omega) = {} & -\frac{\sqrt{l}}{\sqrt{2c\mu_p}} \int_0^x h_0^p(t, \omega)\, w_p(t)\, w_p(t - x)\, dt \\ & + \frac{1}{2\mu_p} \left(x w_p'(x) - \frac{1}{2} w_p(x) \right) \int_0^l h_0^p(t, \omega)\, w_p^2(t)\, dt \\ & - \frac{1}{2\mu_p} w_p(x) \int_0^l h_0^p(t, \omega)\, (l - t)\, w_p(t)\, w_p'(t)\, dt \\ & - \frac{1}{2} w_p(x) \int_0^l g_{10}(t, \omega)\, w_p^2(t)\, dt , \end{aligned} \tag{2.112}$$

if u_{p1} is normalized by $\langle\!\langle u_{p1}, w_p \rangle\!\rangle = -\frac{1}{2} \langle\!\langle g_{01} w_p, w_p \rangle\!\rangle$ with the aid of (2.84). By means of Theorem 2.5 we obtain

$$\begin{aligned} & K_{gh}(x, y) \\ & = b_{00}^{gh} \left[\frac{l}{2c\sqrt{\mu_g \mu_h}} \int_0^{\min\{x, y\}} w_g(t)\, w_h(t)\, w_g(t - x)\, w_h(t - y)\, dt \right. \\ & \qquad \left. + D_{gh}(x, y) + D_{hg}(y, x) \right] \\ & \qquad + b_{01}^{gh} S_{gh}(x)\, w_h(y) + b_{10}^{gh} S_{hg}(y)\, w_g(x) \\ & \qquad + \frac{1}{4} b_{11}^{gh} w_g(x)\, w_h(y) \int_0^l w_g^2(t)\, w_h^2(t)\, dt , \end{aligned} \tag{2.113}$$

where

$$
\begin{aligned}
&D_{gh}(x, y) \\
&\doteq \frac{1}{4\mu_g\mu_h}\Bigg[\sqrt{\frac{2l\mu_g}{c}}\Bigg\{-\left(yw_g'(y) - \frac{1}{2}w_h(y)\right)\int_0^x w_g(t)\,w_h^2(t)\,w_g(t-x)\,dt \\
&\qquad + w_h(y)\int_0^x (1-t)\,w_g(t)\,w_h(t)\,w_h'(t)\,w_g(t-x)\,dt\Bigg\} \\
&\qquad + \frac{1}{2}\left(xw_g'(x) - \frac{1}{2}w_g(x)\right)\left(yw_h'(y) - \frac{1}{2}w_h(y)\right)\int_0^l w_g^2(t)\,w_h^2(t)\,dt \\
&\qquad - \left(xw_g'(x) - \frac{1}{2}w_g(x)\right)w_h(y)\int_0^l (l-t)\,w_g^2(t)\,w_h(t)\,w_h'(t)\,dt \\
&\qquad + \frac{1}{2}w_g(x)\,w_h(y)\int_0^l (l-t)^2\,w_g(t)\,w_h(t)\,w_g'(t)\,w_h'(t)\,dt\Bigg],
\end{aligned}
$$

and

$$
\begin{aligned}
S_{gh}(x) = \frac{1}{4\mu_g}\Bigg[&\sqrt{\frac{2l\mu_g}{c}}\int_0^x w_g(t)\,w_h^2(t)\,w_g(t-x)\,dt \\
&- \left(xw_g'(x) - \frac{1}{2}w_g(x)\right)\int_0^l w_g^2(t)\,w_h^2(t)\,dt \\
&+ w_g(x)\int_0^l (l-t)\,w_g(t)\,w_h^2(t)\,w_g'(t)\,dt\Bigg].
\end{aligned}
$$

Straightforward calculations show that for $y \geqq x$

$$
\int_0^x w_g(t)\,w_h(t)\,w_g(t-x)\,w_h(t-y)\,dt
$$

$$
= \begin{cases}
\dfrac{1}{\overline{\mu}_{gh}}\dfrac{1}{4l}\sqrt{\dfrac{\overline{\mu}_g}{\overline{\mu}_h}}\Bigg\{\overline{w}_g'(\overline{x})\,\overline{w}_h'(\overline{y})\left[2\left(1-\dfrac{\overline{\mu}_h}{\overline{\mu}_g}\right)\overline{x} - \dfrac{1}{\overline{\mu}_h}\overline{w}_h(\overline{x})\,\overline{w}_h'(\overline{x})\right] \\
\qquad + \overline{w}_h'(\overline{y})\,\overline{w}_g(\overline{x})\left(2\dfrac{\overline{\mu}_h}{\overline{\mu}_g} - \overline{w}_h^2(\overline{x})\right) \\
\qquad + \overline{w}_h(\overline{x})\,\overline{w}_h(\overline{y})\,[\overline{w}_g(\overline{x})\,\overline{w}_h'(\overline{x}) - \overline{w}_h(\overline{x})\,\overline{w}_g'(\overline{x})]\Bigg\}, \\
\qquad\qquad \text{for } g \neq h, \\
\dfrac{1}{4l\overline{\mu}_g}\Bigg\{3\overline{x}\overline{w}_g'(\overline{x})\,\overline{w}_g'(\overline{y}) + \overline{w}_g(\overline{x})\,\overline{w}_g(\overline{y})\left[\overline{x}\overline{\mu}_g - \dfrac{1}{2}\overline{w}_g'(\overline{x})\,\overline{w}_g(\overline{x})\right] \\
\qquad + \dfrac{1}{2}\overline{w}_g(\overline{x})\,\overline{w}_g'(\overline{y})\,[-6 + \overline{w}_g^2(\overline{x})]\Bigg\}, \quad \text{for } g = h.
\end{cases} \tag{2.114}
$$

In (2.114) we have put

$$\bar{x} = \frac{x}{l}, \qquad \bar{y} = \frac{y}{l}, \qquad \bar{\mu}_g = (g\pi)^2, \qquad \bar{w}_g(\bar{x}) = \sqrt{2}\sin(g\pi\bar{x}),$$

$$\bar{\mu}_{gh} = \bar{\mu}_g - \bar{\mu}_h .$$

For the calculation of $D_{gh}(x, y)$ the following equations are used:

$$\int_0^x w_g(t)\, w_h^2(t)\, w_g(t-x)\, \mathrm{d}t$$

$$= \begin{cases} \dfrac{1}{\sqrt{2\bar{\mu}_g}\, l\bar{\mu}_{gh}} \left\{ \bar{\mu}_{gh}\bar{x}\bar{w}_g'(\bar{x}) + \bar{w}_g(\bar{x}) \left[\bar{\mu}_h - \dfrac{\bar{\mu}_g}{2} \bar{w}_h^2(\bar{x}) \right] \right. \\ \qquad \left. - \dfrac{\bar{\mu}_g}{2\bar{\mu}_h} \bar{w}_h(\bar{x})\, \bar{w}_h'(\bar{x})\, \bar{w}_g'(\bar{x}) \right\}, \qquad \text{for } g \neq h, \\ \dfrac{1}{\sqrt{2\bar{\mu}_g}\, 2l} \left\{ 3\bar{x}\bar{w}_g'(\bar{x}) - 3\bar{w}_g(\bar{x}) + \dfrac{1}{2}\bar{w}_g^3(\bar{x}) \right\}, \quad \text{for } g = h, \end{cases}$$

$$\int_0^l w_g^2(t)\, w_h^2(t)\, \mathrm{d}t = \begin{cases} \dfrac{1}{l}, & \text{for } g \neq h, \\ \dfrac{3}{2l}, & \text{for } g = h, \end{cases}$$

$$\int_0^l (l-t)\, w_g^2(t)\, w_h(t)\, w_h'(t)\, \mathrm{d}t = \begin{cases} \dfrac{\bar{\mu}_g}{2l\bar{\mu}_{gh}}, & \text{for } g \neq h, \\ \dfrac{3}{8l}, & \text{for } g = h, \end{cases}$$

$$\int_0^l (l-t)^2\, w_g(t)\, w_h(t)\, w_g'(t)\, w_h'(t)\, \mathrm{d}t = \begin{cases} \dfrac{\bar{\mu}_g\bar{\mu}_h}{l\bar{\mu}_{gh}^2}, & \text{for } g \neq h, \\ \dfrac{1}{2l}\left(\dfrac{1}{3}\bar{\mu}_g - \dfrac{1}{8}\right), & \text{for } g = h, \end{cases}$$

as well as

$$\int_0^x (l-t)\, w_g(t)\, w_h(t)\, w_h'(t)\, w_g(t-x)\, \mathrm{d}t$$

$$= \begin{cases} \dfrac{1}{4l\bar{\mu}_{gh}} \sqrt{\dfrac{1}{2}\bar{\mu}_g} \left\{ 2[\bar{x} + (1-\bar{x})\, \bar{w}_h^2]\, \bar{w}_g' - \dfrac{4\bar{\mu}_h}{\bar{\mu}_{gh}} \left[1 - \dfrac{1}{2}\bar{w}_h^2 \right] \bar{w}_g \right. \\ \qquad \left. + \left[\dfrac{2}{\bar{\mu}_{gh}} - \dfrac{1}{\bar{\mu}_h} \right] \bar{w}_h \bar{w}_g' \bar{w}_h' - 2(1-\bar{x})\, \bar{w}_h \bar{w}_h' \bar{w}_g \right\}, \\ \qquad\qquad \text{for } g \neq h, \\ \dfrac{1}{8l\sqrt{2\bar{\mu}_g}} \left\{ [2\bar{\mu}_g\bar{x}(\bar{x}-2) - 3]\, \bar{w}_g + \dfrac{3}{2}\bar{w}_g^3 + 3\bar{x}\bar{w}_g' + 2(1-\bar{x})\, \bar{w}_g'\bar{w}_g^2 \right\}, \\ \qquad\qquad \text{for } g = h . \end{cases}$$

By the aid of these results it follows that

$$D_{gh}(x, y)$$

$$= \begin{cases} \dfrac{l^2}{8c^2\bar{\mu}_g\bar{\mu}_h}\Bigg\{-\bar{x}\bar{y}\bar{w}_g'(\bar{x})\,\bar{w}_h'(\bar{y}) + \dfrac{3\bar{\mu}_g^2 - 4\bar{\mu}_g\bar{\mu}_h - 3\bar{\mu}_h^2}{4\bar{\mu}_{gh}^2}\bar{w}_g(\bar{x})\,\bar{w}_h(\bar{x}) \\ \qquad + \dfrac{\bar{x}}{2}\bar{w}_g'(\bar{x})\,\bar{w}_h(\bar{y}) - \bar{y}\left(\dfrac{1}{2} + \dfrac{2\bar{\mu}_h}{\bar{\mu}_{gh}}\right)\bar{w}_h'(\bar{y})\,\bar{w}_g(\bar{x}) \\ \qquad + \dfrac{\bar{\mu}_g}{\bar{\mu}_{gh}}\left[\left(-\dfrac{1}{\bar{\mu}_h} + \dfrac{1}{\bar{\mu}_{gh}}\right)\bar{w}_h(\bar{y})\,\bar{w}_g'(\bar{x}) + \dfrac{1}{\bar{\mu}_h}\bar{y}\bar{w}_h'(\bar{y})\,\bar{w}_g'(\bar{x}) \right. \\ \qquad\qquad \left. + (\bar{x} - 1)\,\bar{w}_h(\bar{y})\,\bar{w}_g(\bar{x})\right]\bar{w}_h'(\bar{x})\,\bar{w}_h(\bar{x}) \\ \qquad + \dfrac{\bar{\mu}_g}{\bar{\mu}_{gh}}\left[\bar{y}\bar{w}_h'(\bar{y})\,\bar{w}_g(\bar{x}) + (1 - \bar{x})\,\bar{w}_h(\bar{y})\,\bar{w}_g'(\bar{x}) \right. \\ \qquad\qquad \left. + \left(\dfrac{\bar{\mu}_h}{\bar{\mu}_{gh}} - \dfrac{1}{2}\right)\bar{w}_h(\bar{y})\,\bar{w}_g(\bar{x})\right]\bar{w}_h^2(\bar{x})\Bigg\}, \\ \qquad\qquad\qquad\qquad \text{for} \quad g \neq h, \\ \dfrac{l^2}{16c^2\bar{\mu}_g^2}\Bigg[-3\bar{x}\bar{y}\bar{w}_g'(\bar{x})\,\bar{w}_g'(\bar{y}) + \Bigg\{\bar{\mu}_g\left(\bar{x}(\bar{x} - 2) + \dfrac{1}{3}\right) \\ \qquad\qquad - \dfrac{25}{8} + \dfrac{5}{4}\bar{w}_g^2(\bar{x})\Bigg\}\bar{w}_g(\bar{x})\,\bar{w}_g(\bar{y}) \\ \qquad + \left\{\dfrac{3}{2}\bar{x} + (1 - \bar{x})\,\bar{w}_g^2(\bar{x})\right\}\bar{w}_g'(\bar{x})\,\bar{w}_g(\bar{y}) \\ \qquad + \bar{y}\left\{\dfrac{9}{2} - \bar{w}_g^2(\bar{x})\right\}\bar{w}_g(\bar{x})\,\bar{w}_g'(\bar{y})\Bigg], \quad \text{for} \quad g = h. \end{cases} \tag{2.115}$$

Furthermore, we have for $S_{gh}(x)$:

$$S_{gh}(x) = \begin{cases} \dfrac{\sqrt{l}}{4c\bar{\mu}_{gh}}\left[\left(1 + \dfrac{\bar{\mu}_h}{2\bar{\mu}_g}\right)\bar{w}_g - \dfrac{1}{2}\bar{w}_g\bar{w}_h^2 - \dfrac{1}{2\bar{\mu}_h}\bar{w}_h\bar{w}_h'\bar{w}_g'\right], \\ \qquad\qquad\qquad \text{for} \quad g \neq h, \\ \dfrac{\sqrt{l}}{32c\bar{\mu}_g}(-3 + 2\bar{w}_g^2)\,\bar{w}_g, \quad \text{for} \quad g = h. \end{cases} \tag{2.116}$$

The definition

$$F_{gh}(x, y) \doteq \int_0^x w_g(t)\,w_h(t)\,w_g(t - x)\,w_h(t - y)\,\mathrm{d}t$$

yields

$$\int_0^y w_g(t)\,w_h(t)\,w_g(t - x)\,w_h(t - y)\,\mathrm{d}t = F_{hg}(y, x)\,.$$

By means of (2.114), (2.115), and (2.116) the correlation function $K_{gh}(x, y)$ can be calculated from (2.113). The correlation function of the limit process of

$1/\sqrt{\varepsilon}\ \big(u_g(x, \omega) - w_g(x)\big)$ leads to

$$K_{gg}(x, y) = \frac{1}{8}\left(\frac{l}{c\bar{\mu}_g}\right)^2 \Bigg[\bar{\mu}_g\left(\frac{\bar{x}^2}{2} + \frac{\bar{y}^2}{2} - \bar{y} + \frac{1}{3}\right)\bar{w}_g(\bar{x})\,\bar{w}_g(\bar{y})$$
$$+ 3\bar{x}(1 - \bar{y})\,\bar{w}_g'(\bar{x})\,\bar{w}_g'(\bar{y})$$
$$+ \big(\bar{x}\bar{w}_g'(\bar{x})\,\bar{w}_g(\bar{y}) - (1 - \bar{y})\,\bar{w}_g(\bar{x})\,\bar{w}_g'(\bar{y})\big)$$
$$\times\left(3 - \frac{1}{2}\left(\bar{w}_g^2(\bar{x}) + \bar{w}_g^2(\bar{y})\right)\right)$$
$$+ \frac{5}{8}\,\bar{w}_g(\bar{x})\,\bar{w}_g(\bar{y})\,\{-5 + \bar{w}_g^2(\bar{x}) + \bar{w}_g^2(\bar{y})\}\Bigg] b_{00}^{gg}$$
$$+ \frac{1}{16c\bar{\mu}_g}\left(-3 + \bar{w}_g^2(\bar{x}) + \bar{w}_g^2(\bar{y})\right)\bar{w}_g(\bar{x})\,\bar{w}_g(\bar{y})\,b_{01}^{gg}$$
$$+ \frac{3}{8l^2}\,\bar{w}_g(\bar{x})\,\bar{w}_g(\bar{y})\,b_{11}^{gg}\,, \tag{2.117}$$

for $x \leqq y$, and we have

$$K_{gg}(x, y) = K_{gg}(\min\{x, y\}, \max\{x, y\})\,.$$

The variance of the limit process of $\big(u_g(x, \omega) - w_g(x)\big)/\sqrt{\varepsilon}$ can be computed as

$$K_{gg}(x, x) = \frac{1}{4}\left(\frac{l}{c\bar{\mu}_g}\right)^2 \Bigg\{3\bar{\mu}_g\bar{x}(1 - \bar{x}) + \bar{\mu}_g\left(\frac{1}{6} + 2\bar{x}(\bar{x} - 1)\right)\bar{w}_g^2(\bar{x})$$
$$+ \left(\bar{x} - \frac{1}{2}\right)\left(3 - \bar{w}_g^2(\bar{x})\right)\bar{w}_g'(\bar{x})\,\bar{w}_g(\bar{x})$$
$$+ \frac{5}{8}\left(\bar{w}_g^2(\bar{x}) - \frac{5}{2}\right)\bar{w}_g^2(\bar{x})\Bigg\} b_{00}^{gg}$$
$$+ \frac{1}{16c\bar{\mu}_g}\left(-3 + 2\bar{w}_g^2(\bar{x})\right)\bar{w}_g^2(\bar{x})\,b_{01}^{gg} + \frac{3}{8l^2}\,\bar{w}_g^2(\bar{x})\,b_{11}^{gg}\,. \tag{2.118}$$

In these equations we have to put

$$b_{00}^{gh} = a_{11} - \frac{c}{l^2}(\bar{\mu}_g + \bar{\mu}_h)\,a_{12} + \frac{c^2}{l^4}\bar{\mu}_g\bar{\mu}_h a_{22}\,,$$

$$b_{01}^{gh} = a_{12} - \frac{c}{l^2}\bar{\mu}_g a_{22}\,,$$

$$b_{11}^{gh} = a_{22}\,.$$

In contrast to the eigenvalues, the variance of the g-th eigenfunction is in l as l^2 for $g_{01} \equiv 0$, i.e. the variance increases quadratically as a function of the length l. The variable c often denotes the bending stiffness in stability problems or vibration problems. The correlation function, as a function of c is of the form

$$K_{gg}(x, y) = \frac{1}{c^2}A_{g1}(x, y) + \frac{1}{c}A_{g2}(x, y) + A_{g3}(x, y)\,.$$

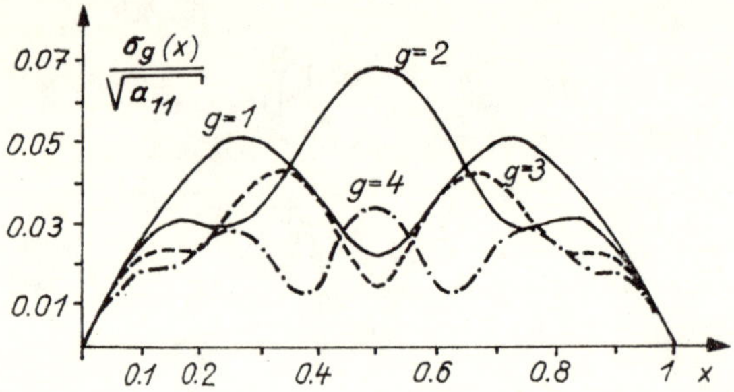

Fig. 2.16a. The dispersion of the eigenfunctions for $g_{01} \equiv 0$

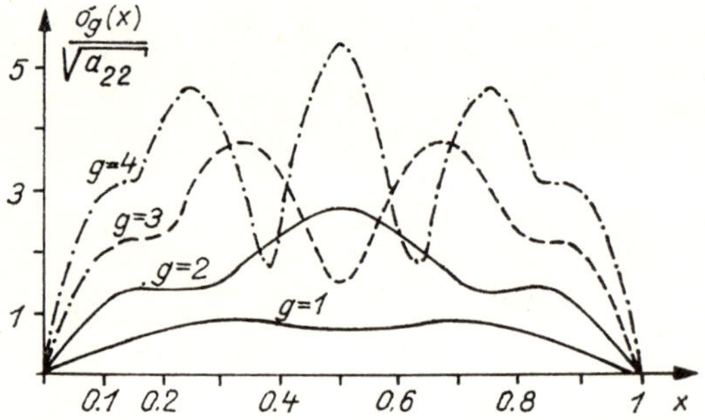

Fig. 2.16b. The dispersion of the eigenfunctions for $f_{01} \equiv 0$

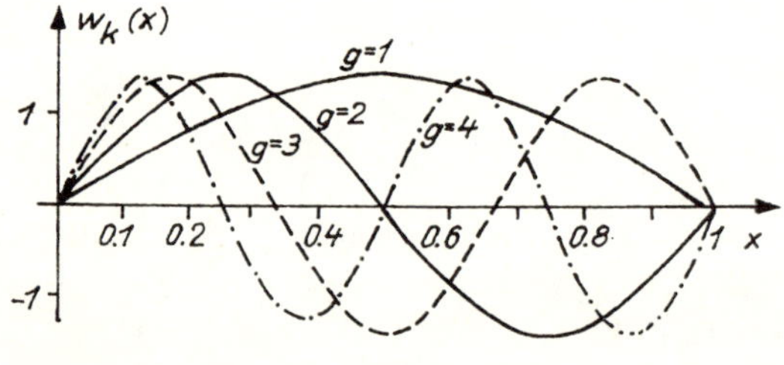

Fig. 2.16c. The eigenfunctions of the averaged eigenvalue problem

$K_{gg}(x, y)$ converges to $A_{g3}(x, y)$ as $c \to \infty$. For small values of c the term determining the function $K_{gg}(x, y)$ is $A_{g1}(x, y)/c^2$. In Fig. 2.16a the dispersion

$$\sigma_g(x) \doteq \sqrt{K_{gg}(x, x)}$$

is plotted for $g = 1, 2, 3, 4$ in the case where $g_{10}(x, \omega) \equiv 0$; and in Fig. 2.16b in the case where $f_{10}(x, \omega) \equiv 0$. We have put $c = l = 1$. The dispersion $\sigma_g(x)$ converges to zero as $g \to \infty$ in the first case; and for $f_{10}(x, \omega) \equiv 0$ the dispersion converges to infinity as $g \to \infty$. For comparision the eigenfunctions of the averaged problem are shown in Fig. 2.16c.

We see that the maximum of $w_1(x)$ for $x = 1/2$ shows a relative stability, i.e. the dispersion $\sigma_1(x)$ has a minimum for $x = 1/2$. Furthermore, the zero of

$w_2(x)$ where $x = 1/2$ indicates a relative instability, since $\sigma_2(x)$ possesses a maximum where $x = 1/2$. In general, the maxima and minima of the eigenfunctions $w_k(x)$ show a relative stability and the zeros a relative instability.

We state that the eigenfunctions for $g_{01} \equiv 0$ are relatively stable in comparison with the case $f_{01} \equiv 0$ since the dispersions are essentially different.

Let $u_g(x, \omega)$ denote the g-th eigenfunction. Then $u_g(x, \omega)$ is normally distributed for each $x \in [0, l]$, and this eigenfunction has the moments

$$\langle u_g(x) \rangle \approx w_g(x) ,$$

$$\langle (u_g(x) - w_g(x))^2 \rangle \approx \varepsilon \sigma_g^2(x) .$$

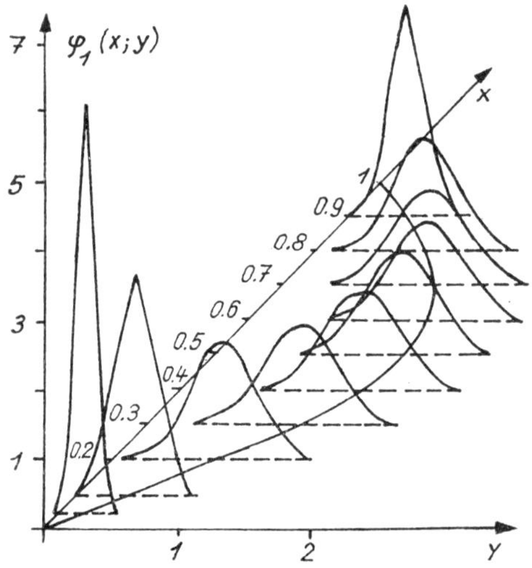

Fig. 2.17a. The one-dimensional density function of $u_1(x, \omega)$

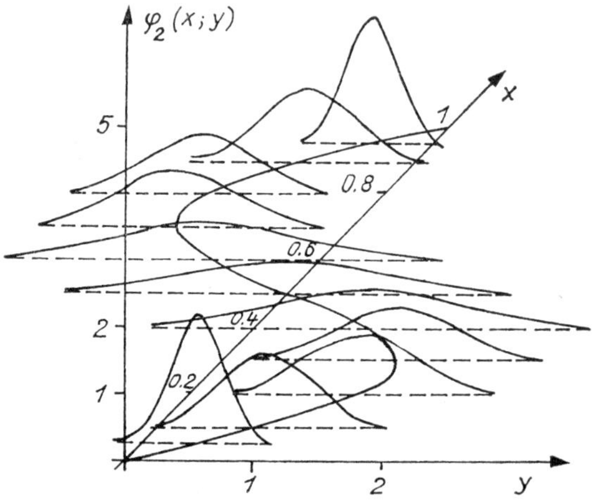

Fig. 2.17b. The one-dimensional density function of $u_2(x, \omega)$

In the case where $f_{01} \equiv 0$, the one-dimensional density function $\varphi_g(x; y)$ of the eigenfunction $u_g(x, \omega)$

$$\varphi_g(x; y) = \frac{1}{\sqrt{2\pi\varepsilon}\,\sigma_g(x)} \exp\left(-\frac{(y - w_g(x))^2}{2\sigma_g^2(x)\,\varepsilon}\right)$$

is plotted in Fig. 2.17a for $g = 1$, and in Fig. 2.17b for $g = 2$, where $\varepsilon a_{22} = 0.1$. The instability discussed above of the zero of $w_2(x)$ for $x = 1/2$ can be seen in Fig. 2.17b. It appears that the zero $x_0(\omega)$ of $u_2(x, \omega)$ for $l = 1$ is located between 0.3 and 0.7, $0.3 < x_0(\omega) < 0.7$, with a high confidence probability. More exact information can obviously be computed.

The correlation coefficient $\varrho_g(x, y)$ resulting from the correlation function (2.117) is shown in Fig. 2.18 for $g = 1, 2$; where we have again put $c = l = 1$. For these figures we can calculate

$$\begin{aligned}
\varrho_g(0, y) &= \lim_{x\to 0} \varrho_g(x, y) \\
&= \frac{\frac{1}{4}\sqrt{\frac{2}{\bar{\mu}_g}}}{\sqrt{K_{gg}(y, y)}\sqrt{\frac{1}{\bar{\mu}_g}\left(\frac{l}{c}\right)^2\left[-\frac{1}{8} + \frac{1}{3}\bar{\mu}_g\right] b_{00}^{gg} - \frac{3}{2c} b_{01}^{gg} + \frac{3}{l^2}\bar{\mu}_g b_{11}^{gg}}} \\
&\quad \times \left[\frac{1}{\mu_g}\left(\frac{l}{c}\right)^2 \left\{\left(\bar{\mu}_g\left(\frac{1}{2}\bar{y}^2 - \bar{y} + \frac{1}{3}\right) - \frac{1}{8}\right)\overline{w}_g(\bar{y})\right.\right. \\
&\qquad \left. + \frac{1}{2}\left(\frac{1}{4}\overline{w}_g(\bar{y}) + (1 - \bar{y})\,\overline{w}_g'(\bar{y})\,\overline{w}_g^2(\bar{y})\right\} b_{00}^{gg} \\
&\qquad \left. + \frac{1}{2c}\overline{w}_g(\bar{y})\,(\overline{w}_g^2(\bar{y}) - 3)\, b_{01}^{gg} + \frac{3}{l^2}\bar{\mu}_g\overline{w}_g(\bar{y})\, b_{11}^{gg}\right]
\end{aligned}$$

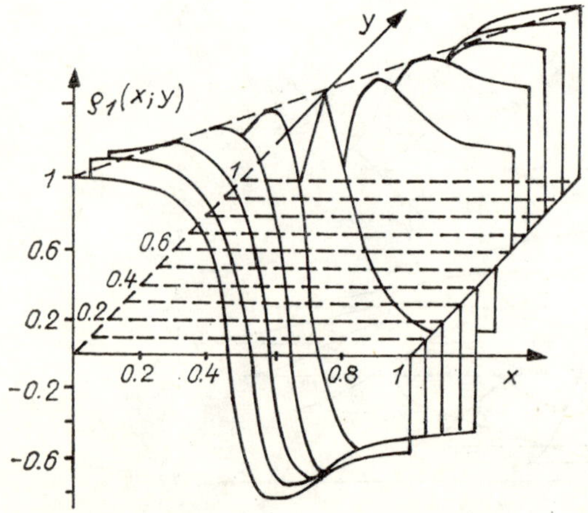

Fig. 2.18a. The correlation coefficient $\varrho_1(x, y)$ for $g_{01} \equiv 0$

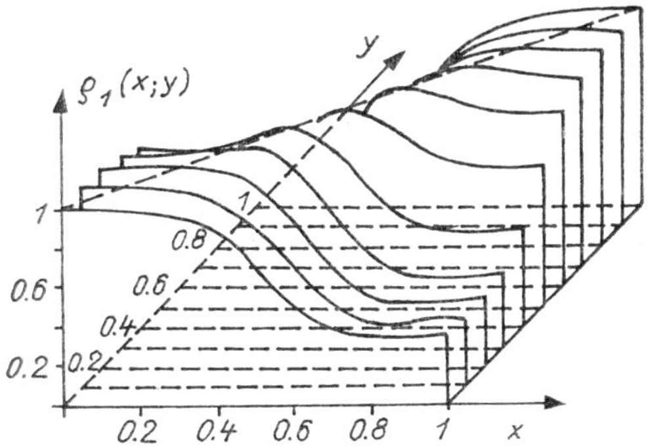

Fig. 2.18b. The correlation coefficient $\varrho_1(x, y)$ for $f_{01} \equiv 0$

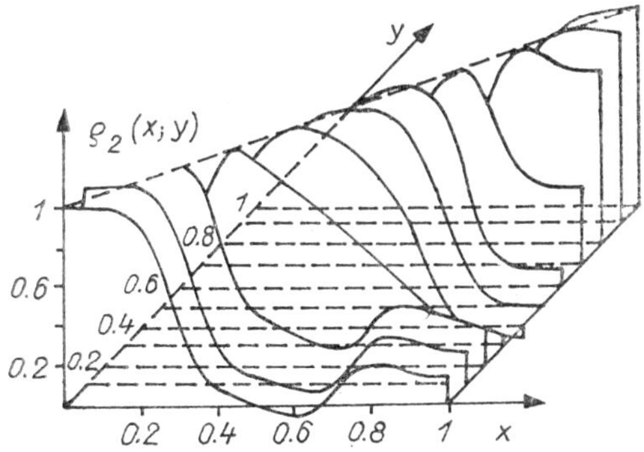

Fig. 2.18c. The correlation coefficient $\varrho_2(x, y)$ for $f_{01} \equiv 0$

and

$$\lim_{y \to 0} \varrho_g(0, y) = 1 \,,$$

$$\lim_{y \to 1} \varrho_g(0, y) = (-1)^{g+1} \frac{105 - 4\bar{\mu}_g}{105 + 8\bar{\mu}_g} \,.$$

The equation

$$\varrho_g(1 - x, 1 - y) = \varrho_g(y, x)$$

follows from

$$K_{gg}(1 - x, 1 - y) = K_{gg}(y, x) \,.$$

For a large g we can obtain the equation

$$\varrho_g(x, y) = \frac{r_g(x, y) + O\left(\frac{1}{g}\right)}{\sqrt{\left(r_g(x, x) + O\left(\frac{1}{g}\right)\right)\left(r_g(y, y) + O\left(\frac{1}{g}\right)\right)}}$$

for $g_{01} \equiv 0$ and the equation

$$\varrho_g(x, y) = \frac{\bar{\mu}_g r_g(x, y) + O(g)}{\sqrt{(\bar{\mu}_g r_g(x, x) + O(g))\,(\bar{\mu}_g r_g(y, y) + O(g))}}$$

for $f_{01} \equiv 0$ where

$$r_g(x, y) = (\tfrac{1}{2}\bar{x}^2 + \tfrac{1}{2}\bar{y}^2 - \bar{y} + \tfrac{1}{3}) \sin(g\pi\bar{x}) \sin(g\pi\bar{y}) + 3\bar{x}(1 - \bar{y}) \cos(g\pi\bar{x}) \cos(g\pi\bar{y}) ,$$

for $x \leqq y$. Hence, in both cases we have for $x \leqq y$

$$\varrho_g(x, y) \approx \bar{\varrho}_g(x, y) = \frac{r_g(x, y)}{\sqrt{r_g(x, x)\, r_g(y, y)}}$$

for g large. For $g = 2$, the form of the correlation functions in the cases $f_{01} \equiv 0$ and $g_{01} \equiv 0$ are very similar.

(c) *for the eigenvalues and eigenfunctions*

In this case, from (2.110) for $r = 0$ and $x_t = x$, we have

$$K_{gh}(x) \doteq \langle \xi_{g0} \xi_h(x) \rangle = -\int_0^l G^h(x, y)\, w_h(y)\, w_g^2(y)\, b_{00}^{gh}(y)\, dy - \tfrac{1}{2} \int_0^l w_g^2(y)\, w_h^2(y)\, b_{01}^{gh}(y)\, dy w_h(x) .$$

Let $(f_{01}(x, \omega), g_{01}(x, \omega))$ again denote a wide-sense stationary vector process and hence the intensities b_{ij}^{gh} are determined as constants. The correlation function between the eigenvalue λ_g and the eigenfunction $u_h(x, \omega)$ can be calculated from the above equation. On the other hand, $K_{gh}(x)$ can be obtained from

$$K_{gh}(x) = \lim_{\varepsilon \downarrow 0} \frac{1}{\varepsilon} \langle \lambda_{g1} u_{h1}(x) \rangle ,$$

where

$$\lambda_{g1} = \int_0^l h_0^g(t, \omega)\, w_g^2(t)\, dt ,$$

and $u_{h1}(x, \omega)$ is given by (2.112). It then follows that

$$K_{gh}(x) = \Bigg[-\sqrt{\frac{l}{2c\mu_h}} \int_0^x w_h(t)\, w_h(t - x)\, w_g^2(t)\, dt + \frac{1}{2\mu_h} \left(x w_g'(x) - \frac{1}{2} w_h(x) \right) \int_0^l w_h^2(t)\, w_g^2(t)\, dt - \frac{1}{2\mu_h} w_h(x) \int_0^l (l - t)\, w_h(t)\, w_h'(t)\, w_g^2(t)\, dt \Bigg] b_{00}^{gh} - \frac{1}{2} w_h(x) \int_0^l w_h^2(t)\, w_g^2(t)\, dt\, b_{01}^{gh} .$$

Using the formulas given in Section (b), the correlation function $K_{gh}(x)$ is determined by

$$K_{gh}(x) = \begin{cases} \dfrac{l}{4c\bar{\mu}_{hg}}\left[\bar{w}_h(\bar{x})\left(\bar{w}_h^2(\bar{x}) - 1\right) + \dfrac{1}{\bar{\mu}_g}\bar{w}_g(\bar{x})\,\bar{w}_g'(\bar{x})\,\bar{w}_h'(\bar{x})\right] b_{00}^{gh} \\ -\dfrac{1}{2\sqrt{l^3}}\bar{w}_h(\bar{x})\,b_{01}^{gh}\,, & \text{for} \quad g \neq h\,, \\ \dfrac{l}{8c\bar{\mu}_h}\left(\dfrac{3}{2} - \bar{w}_h^2(\bar{x})\right)\bar{w}_g(\bar{x})\,b_{00}^{gg} - \dfrac{3}{4\sqrt{l^3}}\bar{w}_g(\bar{x})\,b_{01}^{gg}\,, & \text{for} \quad g = h\,. \end{cases}$$

An approximate correlation function between λ_g and $u_h(x, \omega)$ is given by $\varepsilon K_{gh}(x)$. The correlation coefficient $\varrho_{gh}(x)$ between λ_g and $u_h(x, \omega)$ can be written as

$$\varrho_{gh}(x) = \frac{K_{gh}(x)}{\sqrt{\langle\xi_{g0}^2\rangle\, K_{hh}(x, x)}}\,,$$

where $K_{hh}(x, x)$ follows from (2.118), and

$$\langle\xi_{g0}^2\rangle = \frac{3}{2l}\, b_{00}^{gg}\,.$$

The function $\varrho_{gg}(x)$ is shown in Fig. 2.19 for $g = 1, 2$ with $l = c = 1$. We note the great difference between the correlation values for the cases $f_{01} \equiv 0$ and $g_{01} \equiv 0$. For $g = 2$ the eigenvalue λ_g and $u_g(1/2, \omega)$ are independent, and for $g = 1$ we have a strong dependence between the eigenvalue λ_g and $u_g(1/2, \omega)$.

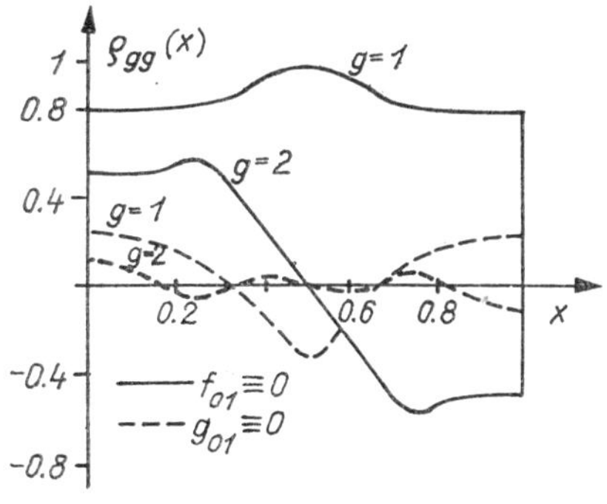

Fig. 2.19. The correlation coefficient between λ_g and $u_g(x, \omega)$ for $g = 1, 2$

Symbol Index

$\doteq$	equal per definitionem	
a.s.	almost sure	
◀	symbol for the end of a proof	
$\{\Omega, \mathfrak{A}, \mathsf{P}\}$	probability space	
ω	element of Ω	
$x(\omega)$, $a(\omega)$, ...	random variables on $\{\Omega, \mathfrak{A}, \mathsf{P}\}$	
$f(x, \omega)$, $a(x, \omega)$, $\xi(x, \omega)$, ...	stochastic processes or fields on $\{\Omega, \mathfrak{A}, \mathsf{P}\}$ with the variable parameter x	
$f_\varepsilon(x, \omega)$, $a_\varepsilon(x, \omega)$, ...	weakly correlated processes or fields of correlation length ε	
$\langle x(\omega)\rangle$	expectation value of $x(\omega)$	
var $x(\omega)$	variance of $x(\omega)$	
σ^2	variance	
$R_a(x, y)$	correlation function of the process or field $a(x, \omega)$	
$R_a(u)$	correlation function in the case of a stationary process $a(x, \omega)$	
$a(y)$	intensity of a weakly correlated field	
$\lambda_g(\omega)$	eigenvalues of a random eigenvalue problem	
$\Lambda_g(\omega)$	eigenvalues of a random matrix eigenvalue problem	
$u_g(x, \omega)$	eigenfunctions of a random eigenvalue problem	
$U_g(\omega)$, ${}^g x(\omega)$	eigenvectors of a random matrix eigenvalue problem	
μ_g	eigenvalues	of the averaged eigenvalue problem corresponding to a given random eigenvalue problem
$w_g(x)$	eigenfunctions	
Λ_{g0}	eigenvalues	of the averaged eigenvalue problem corresponding to a given random matrix eigenvalue problem
U_{g0}	eigenvectors	
${}^n\Lambda_g$	eigenvalues	of a random matrix eigenvalue problem of type n by n (Ritz approximations)
${}^nU_g$	eigenvectors	
$\lambda_{gk}(\omega)$	homogeneous terms of k-th order in the perturbations of the eigenvalue $\lambda_g(\omega)$	
$u_{gk}(x, \omega)$	homogeneous terms of k-th order in the perturbations of the eigenfunction $u_g(x, \omega)$	
$\mathbb{N}$	set of the natural numbers	
$\mathbb{R}$	set of the real numbers	
$\mathbb{C}$	set of the complex numbers	
$\mathbb{R}^n$	n-dimensional Euclidean space	

$\mathscr{D}, \mathscr{B}, \ldots$	bounded domains in $\mathbb{R}^n$
$\mathscr{D}^k$	equal $\mathscr{D} \times \mathscr{D} \times \ldots \times \mathscr{D}$ if $\mathscr{D}$ is a domain in $\mathbb{R}^n$
$\partial\mathscr{B}, \partial\mathscr{D}$	boundary of $\mathscr{B}, \mathscr{D}$
$x_1, x_2, \ldots$	points in $\mathbb{R}^n$
$\lvert x\rvert$	absolute value of $x \in \mathbb{R}$
$\lvert x\rvert$	norm in $\mathbb{R}^n$
$\mathscr{X}, \mathscr{H}, \ldots$	abstract Banach or Hilbert spaces
$\mathbf{C}^n[a, b]$	space of n-times continuously differentiable functions on $[a, b]$
$\mathbf{C}[a, b]$	space of continous functions on $[a, b]$
$\mathbf{L}_2(a, b)$, $\mathbf{L}_2(\mathscr{D})$	space of squared integrable functions on (a, b) or $\mathscr{D} \subset \mathbb{R}^n$
$\mathring{\mathbf{W}}_2^m(a, b)$	Sobolev space of all functions on (a, b) with support $\subset (a, b)$ the generalized m-th derivatives of which are from $\mathbf{L}_2\,(a, b)$
$\lVert f\rVert$	norm in function spaces
$\rightharpoonup$	weak convergence in a Banach space
$\to$	strong convergence in a Banach space
$(\!(u, v)\!)$	scalar product in $\mathbf{L}_2(a, b)$
$(\!(u, v)\!)_r$	wighted scalar product: $\int_a^b u(x)\, v(x)\, r(x)\, \mathrm{d}x$
$M, N, \ldots$	differential operators
$\mathrm{D}(M)$	definition domain of the operator M
δ_{gh}	Kronecker's symbol: $\begin{cases} 1 & \text{for} \quad g = h \\ 0 & \text{for} \quad g \neq h \end{cases}$
$\delta(t)$	Dirac's δ-distribution
O_k	terms of order $\geqq k$ in the perturbations
I	unity matrix
A^{T}	transpose of the matrix A
$U_i[u] = \ldots$	boundary conditions
$G(x, y)$	Green's function
$1_{\mathscr{B}}(y)$	characteristic function of $\mathscr{B} \subset \mathbb{R}^n$
V_1	volume of the sphere with radius one
$[a]$	the greatest integer not greater than a, $a \in \mathbb{R}$

Bibliography

ARNOLD, L.

[1] Zur asymptotischen Verteilung der Eigenwerte zufälliger Matrizen. Habilitationsschrift, Stuttgart 1969.

BALDERESCHI, A., and MASCHKE, K.

[1] Band structure of semiconductor alloys beyond the virtual crystal approximation. Effect of compositional disorder on the energy gaps in GaP_xAs_{1-x}. Solid State Communications **16** (1975), 99–102.

BENDERSKIJ, M. M., and PASTUR, L. A. (Бендерский, M. M., Пастур, Л. A.)

[1] О спектре одномерного уравнения Шредингера со случайным потенциалом (On the spectrum of the one-dimensional Schrödinger equation with random potential). Математический сборник **82** (**124**) (1970) 2, 273–284.

BHARUCHA-REID, A. T.

[1] On the theory of random equations. Prob. Symp. Appl. Math. 16th, 1963, Amer. Math. Soc., Providence, R. I. 1964, 40–69.

[2] Random algebraic equations. In: BHARUCHA-REID, A. T. (ed.), Probabilistic Methods in Applied Mathematics, Vol. 2. Academic Press, New York 1970, 1–52.

[3] Random Integral Equations. Academic Press, New York 1972.

BHARUCHA-REID, A. T., and SAMBANDHAM, M.

[1] Random Polynomials. Academic Press, New York (to appear).

BOYCE, W. E.

[1] Random vibrations of elastic strings and bars. Proc. U.S. Nat. Congr. Appl. Mech., 4th Berkeley, 1962, 77–85.

[2] Stochastic nonhomogeneous Sturm-Liouville problem. J. Franklin Inst. **282** (1966), 206–215.

[3] A "dishonest" approach to certain stochastic eigenvalue problem. SIAM J. Appl. Math. **15** (1967), 143–152.

[4] Random eigenvalue problems. In: BHARUCHA-REID, A. T. (ed.), Probabilistic Methods in Applied Mathematics, Vol. 1. Academic Press. New York 1968, 1–73.

[5] On a conjecture concerning the means of the eigenvalues of random Sturm-Liouville boundary value problems. Quart. Appl. Math. (to appear).

BOYCE, W. E., and GOODWIN, B. E.

[1] Random transverse vibrations of elastic beams. SIAM J. **12** (1964), 613–629.

BOYCE, W. E., and NING-MAO XIA

[1] The approach to normality of the solutions of random boundary and eigenvalue problems with weakly correlated coefficients. Quart. Appl. Math. (to appear).

BUNKE, H.

[1] Gewöhnliche Differentialgleichungen mit zufälligen Parametern. Akademie-Verlag, Berlin 1972.

COLLATZ, L.

[1] Eigenwertaufgaben mit technischen Anwendungen. Akad. Verlagsgesellschaft Geest & Portig, Leipzig 1963.

CRAMER, H., and LEADBETTER, M. R.

[1] Stationary and Related Stochastic Processes. John Wiley, New York 1967.

CHRISTENSEN, M. J., and BHARUCHA-REID, A. T.

[1] Stability of the roots of random algebraic polynomials. Commun. Statist.-Simula. Computa, **B 9** (1980) 2, 179–192,

DAY, W. B.

[1] Asymptotic expansions of eigenvalues and eigenfunctions of random boundary-value problems. Quart. Appl. Math. **38** (1980), 169–177.

ERMAKOV, S. M. (Ермаков, С. М.)

[1] Метод Монте-Карло и смежные вопросы. Наука, Москва 1971 (German translation: Die Monte-Carlo-Methode und verwandte Fragen. Deutscher Verlag d. Wiss., Berlin 1975).

FRÉCHET, M.

[1] Les éléments aléatoires de nature quelconque dans un espace distancié. Ann. Inst. H. Poincaré **10** (1947), 215–310.

[2] Abstrakte Zufallselemente. In: GNEDENKO, B. V. (ed.), Bericht über die Tagung Wahrscheinlichkeitsrechnung und Mathematische Statistik in Berlin, Deutscher Verlag d. Wiss., Berlin 1956, 23–28.

FRISCH, U.

[1] Wave propagation in random media. In: BHARUCHA-REID, A. T. (ed.), Probabilistic methods in Applied Mathematics, Vol. 1. Academic Press, New York 1968, 75–198.

GIHMAN, I. I., and SKOROHOD, A. V. (Гихман, И. И., Скороход, А. В.)

[1] Случайные Процессы, Том 1. Наука, Москва 1971 (English translation: The theory of Stochastic Processes 1. Springer-Verlag, Berlin/Heidelberg/New York 1974).

GIRKO, V. L. (Гирко, В. Л.)

[1] Случайные матрицы (Random matrices). Вища школа, Киев 1975.

GOODWIN, B. E., and BOYCE, W. E.

[1] Vibrations of random elastic strings: method of integral equations. Quart. Appl. Math **22** (1964), 261–266.

HAINES, C. W.

[1] An analysis of stochastic eigenvalue problems. Ph. D. Thesis, Rensselaer Polytechnic Inst., Troy, N.Y. 1965.

[2] Hierarchy methods for random vibrations of elastic strings and beams. J. Eng. Math. **1** (1967), 293–305.

HAMBLEN, J. W.

[1] Distribution of roots of quadratic equations with random coefficients. Ann. Math. Statist. **13** (1942), 235–238.

HAMMERSLEY, J.

[1] The zeros of a random polynomial. Proc. 3rd Berkeley Sympos. Math. Statist. and Prob. Vol. 2. Univ. of California Press, Berkeley 1956, 89–111.

HANŠ, O.

[1] Generalized random variables. Trans. 1st Prague Conf. on Information Theory, Statist. Decision Functions and Random Processes. Czech. Acad. Sci. Prague 1957, 61–103.

[2] Random fixed point theorems. Trans. 1st Prague Conf. on Information Theory, Statist. Decision Functions and Random Processes. Czech. Acad. Sci. Prague 1957, 105–125.

[3] Inverse and adjoint transforms of linear bounded random transforms. Trans. 1st Prague Conf. on Information Theory, Statist. Decision Functions and Random Processes. Czech. Acad. Sci. Prague 1957, 127–133.

[4] Random operator equations. Proc. 4th Berkeley Sympos. Math. Statist. and Prob, Vol. 2, Univ. of California Press, Berkeley 1961, 185—202.

IBRAGIMOV, I. A., and ROZANOV, J. A. (Ибрагимов, И. А., Розанов, Ю. А.)

[1] Гауссовские случайные процессы. Наука, Москва 1970 (English translation: Gaussian random processes. Springer-Verlag, New York/Heidelberg/Berlin 1978).

KELLER, J. B.

[1] Wave propagation in random media. Proc. Symp. Appl. Math., 13th, 1960, Amer. Math. Soc., Providence, R. I. 1962, 217—246.

[2] Stochastic equations and wave propagation in random media. Proc. Symp. Appl. Math., 16th, 1963, Amer. Math. Soc., Providence, R. I. 1964, 145—170.

KORAL, F. C., JR., and KELLER, J. B.

[1] Elastic, electromagnetic, and other waves in a random medium. J. Math. Phys. **5** (1964), 537—547.

KRAICHNAN, R. H.

[1] The closure problem in turbulence theory. Proc. Symp. Appl. Math., 13th, 1960, Amer. Math. Soc., Providence, R I. 1962, 199—225.

KRETSCHMAR, B., and VOM SCHEIDT, J.

[1] Über die Verteilung von Eigenwerten bei stochastischen Eigenwertproblemen. Wiss. Beitr. IH Zwickau **3** (1977) 1, 51—61.

VAN LEAR, G. A., Jr., and UHLENBECK, G. E.

[1] The Brownian motion of strings and elastic rods. Phys. Rev. **38** (1931), 1583 to 1598.

LIESE, F., and VOM SCHEIDT, J.

[1] A limit theorem for sequences of weakly dependent stochastic processes. Serdica Bulgaricae mathematicae publications (to appear).

VAN DER LINDE, R. H.

[1] Eigenfunctions of random eigenvalue problems and their statistical properties. SIAM J. Appl. Math. **6** (1969) 17, 1298—1304.

METHA, M. L.

[1] Random Matrices and the Statistical Theory of Energy Levels. Academic Press, New York 1967.

MEUSEL, B.

[1] Grenzverteilungsaussagen für die Lösungen stochastischer Rand- und Rand-Anfangswertprobleme partieller Differentialgleichungen mit schwach korrelierten Feldern. Diss., Leipzig 1982.

MEUSEL, B., and VOM SCHEIDT, J.

[1] Ein Vergleich von Entwicklungs- und Hierarchiemethoden am Beispiel eines stochastischen Eigenwertproblems. Wiss. Beitr. IH Zwickau **3** (1977) 1, 36—50.

[2] Transversalschwingungen eines dünnen Balkens unter Zufallsbelastung. Wiss. Beitr. IH Zwickau **6** (1980) 2, 64—67.

MIHLIN, S. G. (Михлин, С. Г.)

[1] Вариационные методы математической физики. Наука, Москва 1957 (German translation: Variationsmethoden der Mathematischen Physik. Akademie-Verlag, Berlin 1962).

[2] Численная реализация вариационных методов. Наука, Москва 1966 (German translation: Numerische Realisierung von Variationsmethoden. Akademie-Verlag, Berlin 1969; English translation: The numerical performance of variational methods. Wolters-Noordhoff Publishing, Groningen 1971).

MOURIER, E.

[1] Eléments aléatoires dans un espace de Banach. Ann. Inst. H. Poincaré **13** (1953), 161—244.

NAKE, F.

[1] Über die Anzahl der reellen Lösungen zufälliger Gleichungssysteme. Diss., Stuttgart 1967.

PASTUR, L. A. (Пастур, Л. А.)

[1] Самоусредняемость числа состоянии уравнения Шредингера со случайным потенциалом (The self-averaging property of the number of states of the Schrödinger equation with random potential). Матем. Физика и функц. анализ (сб. трудов ФТИНТ АН СССР) **2** (1971), 111–116.

[2] О спектре случайных матриц (On the spectrum of random matrices). Теорет. и матем. физика **10** (1972) 1, 102–112.

[3] Спектры случайных самосопряженных операторов. Успехи матем. наук **28** (1973) 1, 3–64 (English translation: Spectra of random self-adjoint operators. Russ. Math. Surveys **28** (1973) 1, 1–67).

[4] О распределении собственных значений уравнения Шредингера со случайным потенциалом. Функц. анализ и прилож. **6** (1972) 2, 93–94 (English translation: On the distribution of the eigenvalues of the Schrödinger equation with a random potential. Functional Analysis Appl. **6** (1972), 163–165).

PURKERT, W., and VOM SCHEIDT, J.

[1] Über die Meßbarkeit der Eigenwerte vollstetiger stochastischer Operatoren. Wiss. Beitr. IH Zwickau **2** (1976) 1, 63–70.

[2] Eine Klasse stochastischer Eigenwertprobleme. Wiss. Beitr. IH Zwickau **2** (1976) 3, 74–88.

[3] Zum Mittelungsproblem bei stochastischen Eigenwertaufgaben. In: KLÖTZLER, R., TUTSCHKE, W., WIENER, K. (editors), Beiträge zur Analysis, Vol. 10. Deutscher Verlag d. Wiss., Berlin 1977, 47–61.

[4] Eine Störungsrechnung für die Eigenwerte und Eigenvektoren zufälliger Matrizen. In: KLÖTZLER, R., TUTSCHKE, W., WIENER, K. (editors), Beiträge zur Analysis, Vol. 11. Deutscher Verlag d. Wiss., Berlin 1977, 113–135.

[5] Zur approximativen Lösung des Mittelungsproblems für die Eigenwerte stochastischer Differentialoperatoren. Z. Angew. Math. und Mech. **57** (1977), 515–526.

[6] Randwertprobleme mit schwach korrelierten Prozessen als Koeffizienten. Trans. 8th Prague Conf. on Information Theory, Statist. Decision Functions and Random Processes. Czech. Acad. Sci. Prague 1978, 107–118.

[7] Stochastische Eigenwertprobleme. Dissertation B, Leipzig 1978.

[8] Stochastic eigenvalue problems for differential equations. Reports on Math. Physics **15** (1979) 2, 205–227.

[9] Ein Grenzverteilungssatz für stochastische Eigenwertprobleme. Z. Angew. Math. und Mech. **59** (1979), 611–623.

[10] Limit theorems for solutions of stochastic differential equation problems. Internat. J. Math. & Math. Sci. **1** (1980) 3, 113–149.

[11] Schwach korrelierte Prozesse und ihre Anwendungen. Sitzungsber. d. Akad. d. Wiss. der DDR, Mathem.-Naturw.-Technik, Nr. 23 N, Akademie-Verlag, Berlin 1980.

[12] Ein Eigenwertproblem mit weißem Rauschen als Koeffizienten. Wiss. Beitr. IH Zwickau **6** (1980) 4, 74–82.

[13] Eigenwerte zufälliger normalverteilter Matritzen. In: KLÖTZLER, R., TUTSCHKE, W., WIENER, K. (editors), Beiträge zur Analysis, Vol. 17. Deutscher Verlag d. Wiss., Berlin 1981, 135–143.

[14] Eigenwerte und Eigenvektoren zufällig gestörter Matrizen mit mehrfachen Eigenwerten der gemittelten Matrix. In: KUHNERT, F., SCHMIDT, J. W. (editors), Beiträge zur Numerischen Mathematik, Vol. 10. Deutscher Verlag d. Wiss., Berlin 1981, 139–151.

RELLICH, F.
[1] Störungsrechnung der Spektralzerlegung. I. Mitteilung. Math. Ann. **113** (1937), 600–619.
[2] Störungsrechnung der Spektralzerlegung. II. Mitteilung. Math. Ann. **113** (1937), 685–698.
[3] Störungsrechnung der Spektralzerlegung. III. Mitteilung. Math. Ann. **116** (1939), 555–570.
[4] Störungsrechnung der Spektralzerlegung. IV. Mitteilung. Math. Ann. **117** (1940), 356–382.

RICE, S. O.
[1] Mathematical analysis of random noice. Bell. System. Techn. J. **23** (1944), 282–332; **24** (1945), 46–156. Reprinted in: Selected papers on Noise and Stochastic Processes, Dover, New York 1954.

RICHARDSON, J. H.
[1] The application of truncated hierarchy techniques in the solution of a stochastic linear differential equation. Proc. Symp. Appl. Math., 16th, 1963, Amer. Math. Soc., Providence, R. I. 1964, 290–302.

RYLL-NARDZEWSKI, C.
[1] An analogue of Fubini's theorem and its application to random linear equations. Bull. Acad. Sci. Ser. Sci. Math. Astronom. Phys. **8** (1960), 511–513.

SAMUELS, J. C., and ERINGEN, A. C.
[1] On stochastic linear systems. J. Math . Physics **38** (1959), 83–103.

VOM SCHEIDT, J.
[1] Lineare Schwingungsgleichungen mit schwach korreliertem inhomogenem Term. Wiss. Beitr. IH Zwickau **4** (1978) 2, 63–77.
[2] Grenzwertsätze bei Differentialgleichungen mit stochastischen Koeffizienten. Bericht der 7. Tagung über Probleme und Methoden der Mathematischen Physik, Teil II, TH Karl-Marx-Stadt, 1979, 121–126.
[3] Stochastische Stabilität technischer Systeme. Wiss. Beitr. IH Zwickau **7** (1981) 1, 56–67.
[4] A limit theorem of solutions of stochastic boundary-initial-value problems. In: ARATÓ, M., VERMES, D., BALAKRISHNAN, A. V. (editors), Stochastic Differential Systems. Proc. 3rd. IFIP-WG 7/1 Working Conference Visegrad 1980. Springer-Verlag, Berlin/Heidelberg/New York 1981, 189–201.

SCHULZE, K. R., and UNGER, K.
[1] Electron states and lattice structure of semiconductor alloys. Proc. Conf. Mixed Crystals, Reinhardtsbrunn, 1975, 41–51.

SKOROHOD, A. V. (Скороход, А. В.)
[1] Случайные линейные операторы (Random linear operators). Наукова думка, Киев 1978.

SPAČEK, A.
[1] Zufällige Gleichungen. Czech. Math. J. **5** (80) (1955), 462–466.

UHLENBECK, G. E., and ORNSTEIN, L. S.
[1] On the theory of the Brownian motion. Physical Review **36** (1930), 823–841.

WIGNER, E. P.
[1] On the distribution of the roots of certain symmetric matrices. Ann. of Math. **67** (1958), 325–327.

ZEIDLER, E.
[1] Vorlesungen über nichtlineare Funktionalanalysis I. Fixpunktsätze. Teubner-Verlagsgesellschaft, Leipzig 1976.
[2] Vorlesungen über nichtlineare Funktionalanalysis II. Monotone Operatoren. Teubner-Verlagsgesellschaft, Leipzig 1977.
[3] Vorlesungen über nichtlineare Funktionalanalysis III. Variationsmethoden und Optimierung. Teubner-Verlagsgesellschaft, Leipzig 1977.

Author Index

Subject Index